STUDY GUIDE AND WORKBOOK:
AN INTERACTIVE APPROACH

for Starr's

BIOLOGY
Concepts and Applications
SECOND EDITION

STUDY GUIDE AND WORKBOOK:
AN INTERACTIVE APPROACH

for Starr's

BIOLOGY

Concepts and Applications

SECOND EDITION

JOHN D. JACKSON
North Hennepin Community College

JANE B. TAYLOR
Northern Virginia Community College

Wadsworth Publishing Company
Belmont, California
A Division of Wadsworth, Inc.

Biology Editor: Jack Carey
Editorial Assistant: Kristin Milotich
Production Editor: Carol Carreon Lombardi
Designers: Lisa Berman, Ann Butler
Print Buyer: Diana Spence
Art Editor: Roberta Broyer
Permissions Editor: Peggy Meehan
Copy Editor: Jan McDearmon
Technical Illustrator: Alexander Teshin Associates
Compositor: Joan Olson, Wadsworth Digital Productions
Printer: Courier Company
Cover Photograph: ©Jim Brandenburg—Minden Pictures

International Thomson Publishing
The trademark ITP is used under license.

This book is printed on acid-free paper that meets Environmental Protection Agency standards for recycled paper.

ISBN 0-534-17617-8

Printed in the United States of America

1 2 3 4 5 6 7 8 9 10—98 97 96 95 94

CONTENTS

PREFACE

Tell me and I will forget, show me and I might remember,
involve me and I will understand.
Chinese Proverb

The proverb outlines three levels of learning, each successively more effective than the method preceding it. The writer of the proverb understood that humans learn most efficiently when they involve themselves in the material to be learned. This study guide is like a tutor: when properly used, it increases the efficiency of your study periods.

The Interactive Exercises actively involve you in the most important terms and central ideas of your text. Specific tasks ask you to recall key concepts and terms and apply them to life; they test your understanding of the facts and indicate items to reexamine or clarify. Because they cover the same specific material presented in the text, they can help you estimate (and improve) your next test score. Most important, though, this study guide and text together help you make informed decisions about matters that affect your own well-being and the well-being of your environment. In the years to come, our survival will require administrative and managerial decisions based on an informed biological background.

HOW TO USE THIS STUDY GUIDE

Following this preface, you will find an outline that will show you how the study guide is organized to help you use it efficiently. Each chapter begins with a title and a list of the 1- and 2-level headings in that chapter. The *Interactive Exercises* follow, wherein each main (1-level) heading is labeled 1-I, 1-II, and so on. Interactive exercise sections each begin with a list of page-referenced chapter terms under each main (1-level) heading; space is provided next to each term for you to write a definition in your own words. These exercises include completion/short answer, true-false, fill-in-the-

blanks, matching, choice, dichotomous choice, label-match, crossword puzzles, problems, labeling, sequencing, multiple choice, and completion of tables. A *Self-Quiz* immediately follows the interactive exercises. This quiz is composed primarily of multiple-choice questions, although sometimes we may present another examination device. Any wrong answers in the Self-Quiz indicate portions of the text you need to reexamine.

A series of *Chapter Objectives/Review Questions* follows each Self-Quiz section. These are tasks that you should be able to accomplish if you have understood the assigned reading in the text. Some objectives require you to compose a short answer or long essay; others may require you to draw a sketch or supply correct words. The section titled *Integrating and Applying Key Concepts* invites you to try your hand at applying major concepts to situations in which there is not necessarily a single pat answer and so none is provided in the chapter answer section (except for a problem in Chapter 9). Your text generally will provide enough clues to get you started on an answer, but these sections are intended to stimulate your thought and provoke group discussions. Next will appear a *Critical Thinking Exercise*, which presents you with problem situations that concentrate on the critical and higher-level thinking skills used by scientists. Solving these problems requires application of chapter information to new perspectives, analysis of data, drawing conclusions, making predictions, and identification of basic assumptions. *Answers*—to Interactive Exercises, Self-Quiz, and Critical Thinking—follow. Here the answers to all interactive exercises may be conveniently and quickly checked at the end of each chapter.

A person's mind, once stretched by a new idea, can never
return to its original dimension.
Oliver Wendell Holmes

STRUCTURE OF THE STUDY GUIDE

The outline below indicates how each chapter in this study guide is organized.

Chapter Number ⟶   **8**

Chapter Title ⟶   **MEIOSIS**

Chapter Outline ⟶

ON SEXUAL AND ASEXUAL REPRODUCTION

OVERVIEW OF MEIOSIS
  Think "Homologues"
  Overview of the Two Divisions

STAGES OF MEIOSIS
  Prophase I Activities
  Separating the Homologues During Meiosis I
  Separating the Sister Chromatids During Meiosis II

GAMETE FORMATION
  Gamete Formation in Animals
  Gamete Formation in Plants

MORE GENE SHUFFLINGS AT FERTILIZATION

MEIOSIS AND MITOSIS COMPARED

Interactive Exercises ⟶ Divided into sections by main (1-level) headings; includes the page-referenced important terms from the chapter, followed by groups of interactive exercises that vary in type and require constant interaction with the important chapter information

Self-Quiz ⟶ Usually a set of multiple-choice questions that sample important blocks of text information

Chapter Objectives/Review Questions ⟶ Tasks consisting of combinations of relative objectives and questions to be answered

Integrating and Applying Key Concepts ⟶ Applications of text materials to questions for which there may be more than one correct answer

Critical Thinking Exercises ⟶ Problem situations that concentrate on thinking skills used by scientists

Answers to Interactive Exercises, Critical ⟶ Answers for all Interactive Exercises found under Thinking Analysis, and Self-Quiz   main (1-level) headings in each chapter, answers to Self-Quiz, and Critical Thinking Analysis

STUDY GUIDE AND WORKBOOK:
AN INTERACTIVE APPROACH

for Starr's

BIOLOGY

Concepts and Applications

SECOND EDITION

1

METHODS AND CONCEPTS IN BIOLOGY

SHARED CHARACTERISTICS OF LIFE
    Energy, DNA, and Biological Organization
    Metabolism: Life-Sustaining Energy Transfers
    Interdependency Among Organisms
    Sensing and Responding to Change
    Reproduction
    Mutation: Source of Variations in Heritable
      Traits

LIFE'S DIVERSITY
    Five Kingdoms, Millions of Species
    An Evolutionary View of Diversity

THE NATURE OF BIOLOGICAL INQUIRY
    On Scientific Methods
    About the Word "Theory"
    Focus on Science: Darwin's Theory and Doing
      Science
    The Limits of Science

Interactive Exercises

Note: In the answer sections of this book, a specific molecule is most often indicated by its abbreviation. For example, adenosine triphosphate is ATP.

SHARED CHARACTERISTICS OF LIFE (1-I, pp. 3–7)

Terms

The following page-referenced terms are important; they were in boldface type in the chapter. Compose a *written* definition for each term *in your own words* without looking at the text. Next, compare your definition with that given in the chapter or in the text glossary. If your definition seems to lack accuracy, allow some time to pass and repeat this procedure until you can define each term rather quickly (rapidity of answering is a gauge of the effectiveness of your learning).

energy (3):_____

DNA (3): _____

cell (4):_____

multicelled organism (4):_____

population (4):_____

community (4):_____

ecosystem (4):_____

biosphere (4):_____

metabolism (4): _____

photosynthesis (4): _____

ATP (4):_____

aerobic respiration (5):_____

producers (5):_____

consumers (5): _____

decomposers (5):_____

receptors (6): _____

homeostasis (6):_____

reproduction (6): _____

inheritance (7):_____

mutations (7):_____

adaptive trait (7):_____

Matching

Choose the most appropriate answer to match with each term.

1. ___ organ system
2. ___ cell
3. ___ community
4. ___ ecosystem
5. ___ molecule
6. ___ DNA
7. ___ organelle
8. ___ population
9. ___ subatomic particle
10. ___ tissue
11. ___ biosphere
12. ___ energy
13. ___ multicellular organism
14. ___ organ
15. ___ atom

A. One or more tissues interacting as a unit
B. A proton, neutron, or electron
C. A well-defined structure within a cell, performing a particular function
D. All of the regions of Earth where organisms can live
E. A capacity to make things happen, to do work
F. The smallest unit of life
G. Two or more organs whose separate functions are integrated to perform a specific task
H. Two or more atoms bonded together
I. All of the populations interacting in a given area
J. The smallest unit of a pure substance that has the properties of that substance
K. A special molecule; sets living things apart from the nonliving world
L. A community interacting with its nonliving environment
M. An individual composed of cells arranged in tissues, organs, and often organ systems
N. A group of individuals of the same species in a particular place at a particular time
O. A group of cells that work together to carry out a particular function

Fill-in-the-Blanks

(16) _____ refers to the cell's capacity to extract and transform energy from its environment and use energy to maintain itself, grow, and reproduce. (17) _____ is the chemical process whereby some organisms are able to trap sunlight energy and convert it to chemical energy. During photosynthesis, molecules of (18) _____ are constructed to act as energy carriers and transfer energy to other molecules. In most organisms, stored energy is released and transferred to ATP by way of a metabolic process known as aerobic (19) _____. Plants and some other photosynthetic organisms are food (20) _____ for the world of life. Animals feed directly or indirectly on energy stored in plant parts; these animals are known as (21) _____. (22) _____ are bacteria and fungi that feed on tissues or remains of other organisms. Thus, there is an (23) _____ among organisms based on a one-way flow of energy through them and a cycling of materials among them. Molecules and structures called (24) _____ permit organisms to detect specific information about the environment. (25) _____ is the capacity to maintain internal conditions within some tolerable range, even when external conditions vary. The production of offspring by one or more parents is known as (26) _____. DNA is the molecule of (27) _____, which is defined as the transmission of traits through generations. (28) _____ are changes that occur in the structure or number of DNA molecules. An (29) _____ trait is one that improves the survival and reproduction of an organism in a certain environment.

LIFE'S DIVERSITY (1-II, pp. 8–10)

Terms

species (8):_____

genus (8):_____

Monera (8):_____

Protista (8):_____

Fungi (8):_____

Plantae (8):_____

Animalia (8):_____

evolution (10):_____

natural selection (10): _____

Fill-in-the-Blanks

Different "kinds" of organisms are referred to as (1) _____. A (2) _____ is the first part of an organism's two-part name and encompasses all the species having perceived similarities to one another. The pronghorn antelope is known by the two-part name *Antilocapra americana*. *Antilocapra* is the (3) _____ name and *americana* is the (4) _____ name.

Complete the Table

5. Fill in the table below by entering the correct name of each kingdom of life described.

Kingdom	Description
a.	multicelled consumers
b.	complex single cells; producers or consumers
c.	mostly multicelled decomposers
d.	internally simple, single cells; either producers or decomposers
e.	mostly multicelled producers

Sequence

Arrange in correct hierarchical order with the largest, most inclusive category first and the smallest, most exclusive category last. This exercise classifies a plant with the common name of "false Solomon's seal." Refer to p. 8 in the text and Appendix II, pp. A3–A5.

6. ___   A. Class: Monocotyledonae

7. ___   B. Family: Liliaceae

8. ___   C. Genus: *Smilacina*

9. ___   D. Kingdom: Plantae

10. ___   E. Order: Liliales

11. ___   F. Division: Anthophyta

12. ___   G. Species: *racemosa*

Short Answer

13. Compose a short definition for evolution (p. 10): _____

_____

True/False

If the statement is true, place a T in the blank. If the statement is false, make it correct by changing the underlined word(s) and writing the correct word(s) in the answer blank.

_____ 14. The most inclusive (largest) taxonomic category is the <u>phylum</u>.

_____ 15. There is a <u>larger</u> number of different species in a class than in an order.

_____ 16. If some organisms in a population inherit traits that lend them a survival advantage, they will be <u>less</u> likely to produce offspring.

_____ 17. Darwin described the selection occurring in pigeons as <u>natural</u> selection.

_____ 18. <u>Evolution</u> occurs when forms of traits in populations increase, decrease, or disappear.

_____ 19. <u>Natural</u> selection is the selection of adaptive traits in nature.

THE NATURE OF BIOLOGICAL INQUIRY (1-III, pp. 10–13)

Terms

hypotheses (10):_____

prediction (10):_____

theory (11): _____

experiment (12):_____

control group (12):_____

variable (12):_____

Complete the Table

1. Complete the following table of concepts important to understanding the scientific method of problem solving. Choose from scientific experiment, variable, prediction, control group, hypothesis, and theory.

Concept	Definition
a.	an educated guess about what the answer (or solution) to a scientific problem might be
b.	a statement of what one should be able to observe in nature if one looks; the "if-then" process
c.	a related set of hypotheses that, taken together, form a broad explanation of a fundamental aspect of the natural world
d.	a carefully designed test that manipulates nature into revealing one of its secrets
e.	used in scientific experiments to evaluate possible side effects of a test being performed on an experimental group
f.	The control group is identical to the experimental group except for the key factor under study

Sequence

Arrange the following steps of the scientific method in correct chronological sequence. Write the letter of the first step next to 2, the letter of the second step next to 3, and so on.

2. ___

3. ___

4. ___

5. ___

6. ___

7. ___

8. ___

A. Develop one or more hypotheses about what the solution or answer to a problem might be.

B. Devise ways to test the accuracy of predictions drawn from the hypothesis (use of observations, models, and experiments).

C. Repeat or devise new tests (different tests might support the same hypothesis).

D. Make a prediction using the hypothesis as a guide.

E. If the tests do not provide the expected results, check to see what might have gone wrong.

F. Objectively report the results from tests and the conclusions drawn.

G. Identify a problem or ask a question about nature.

Labeling

Assume that you have to identify objects hidden inside a sealed, opaque box. Your only tools to test the contents are a bar magnet and a triple-beam balance. Label each of the following with an O (for observation) or a C (for conclusion).

9. ___ The object has two flat surfaces.

10. ___ The object is composed of nonmagnetic metal.

11. ___ The object is not a quarter, a half-dollar, or a silver dollar.

12. ___ The object weighs x grams.

13. ___ The object is a penny.

Completion

14. Questions that are _____ in nature do not readily lend themselves to scientific analysis.

15. Scientists often stir up controversy when they explain a part of the world that was considered beyond natural explanation—that is, belonging to the "_____."

Self-Quiz

___ 1. About 12 to 24 hours after a meal, a person's blood-sugar level normally varies from about 60 to 90 mg per 100 ml of blood, although it may attain 130 mg/100 ml after meals high in carbohydrates. That the blood-sugar level is maintained within a fairly narrow range despite uneven intake of sugar is due to the body's ability to carry out _____.
   a. predictions
   b. inheritance
   c. metabolism
   d. homeostasis

___ 2. Different species of Galapagos Island finches have different beak types to obtain different kinds of food. One species removes tree bark with a sharp beak to forage for insect larvae and pupae; another species has a large, powerful beak capable of crushing large, heavy coated seeds. These statements illustrate _____.
   a. adaptation
   b. metabolism
   c. puberty
   d. homeostasis

___ 3. A boy is color-blind just as his grandfather was, even though his mother had normal vision. This situation is the result of _____.
a. adaptation
b. inheritance
c. metabolism
d. homeostasis

___ 4. The digestion of food, the production of ATP by respiration, the construction of the body's proteins, cellular reproduction by cell division, and the contraction of a muscle are all part of _____.
a. adaptation
b. inheritance
c. metabolism
d. homeostasis

___ 5. Which of the following does *not* involve using energy to do work?
a. atoms bonding together to form molecules
b. the division of one cell into two cells
c. the digestion of food
d. none of these

___ 6. The experimental group and control group are identical except for _____.
a. the number of variables studied
b. the variable under study
c. two variables under study
d. the number of experiments performed on each group

___ 7. A hypothesis should *not* be accepted as valid if _____.
a. the sample studied is determined to be representative of the entire group
b. a variety of different tools and experimental designs yield similar observations and results

c. other investigators can obtain similar results when they conduct the experiment under similar conditions
d. several different experiments, each without a control group, systematically eliminate each of the variables except one

___ 8. The principal point of evolution by natural selection is that _____.
a. it measures the difference in survival and reproduction that has occurred among individuals who differ from one another in one or more traits
b. even bad mutations can improve survival and reproduction of organisms in a population
c. evolution does not occur when some forms of traits increase in frequency and others decrease or disappear with time
d. individuals lacking adaptive traits make up more of the reproductive base for each new generation

___ 9. Which match is incorrect?
a. Kingdom Animalia—multicelled consumers
b. Kingdom Plantae—mostly multicelled producers
c. Kingdom Monera—relatively simple, multicelled organisms
d. Kingdom Fungi—mostly multicelled decomposers
e. Kingdom Protista—complex single cells

___ 10. The least inclusive of the taxonomic categories listed is _____.
a. family
b. phylum
c. class
d. order
e. genus

Chapter Objectives/Review Questions

This section lists general and detailed chapter objectives that can be used as review questions. You can make maximum use of these items by writing answers on a separate sheet of paper. Fill in answers where blanks are provided. To check for accuracy, compare your answers with information given in the chapter or glossary.

Page   *Objectives/Questions*

(3)   1. _____ interactions among molecules bind the parts of all structures together—they hold a rock together and they hold a frog together

(4)   2. A _____ is the basic living unit.

(4)   3. Distinguish between single-celled organisms and multicelled organisms.

(4)   4. Arrange in order, from smallest to largest, the levels of organization that occur in nature. Define each as you list it.

(4)   5. _____ means "energy transfers" within the cell.

(4)   6. Organisms use a molecule known as _____ to transfer chemical energy from one molecule to another.

(5)   7. Explain how the actions of producers, consumers, and decomposers create an interdependency among organisms.

(5)   8. Describe the general pattern of energy flow through Earth's life forms and explain how Earth's resources are used again and again (cycled).

(7)   9. _____ means transmitting the DNA instructions for producing new organisms to the next generation.

(7)   10. Explain the origin of trait variations that function in inheritance.

(8)   11. Explain the use of genus and species names by considering your Latin name, *Homo sapiens*.

(8)   12. Arrange in order, from greater to fewer organisms included, the following categories of classification: class, family, genus, kingdom, order, phylum, species.

(10)   13. As organisms move through time in successive generations, the character of populations changes; this is called _____.

(10)   14. Darwin used _____ selection as a model for natural selection.

(10)   15. Define natural selection and briefly describe what is occurring when a population is said to evolve.

(8)   16. Explain what is meant by the term *diversity* and speculate about what caused the great diversity of life forms on Earth.

(11)   17. Define what is meant by a theory; cite an actual example.

(12)   18. Tests performed to reveal nature's secrets are called _____.

(12)   19. Generally, members of a control group should be identical to those of the experimental group except for the key factor under study, the _____.

(12)   20. Explain the advantages of the "uncertainty" related to scientific endeavors.

(12)   21. Explain how the methods of science differ from answering questions by using subjective thinking and systems of belief.

Interpreting and Applying Key Concepts

1. Humans have the ability to maintain body temperature very close to 37°C.
   a. What conditions would tend to make the body temperature drop?
   b. What measures do you think your body takes to raise body temperature when it drops?
   c. What conditions would cause body temperature to rise?
   d. What measures do you think your body takes to lower body temperature when it rises?
2. Do you think that all humans on Earth today should be grouped in the same species?
3. What topics are usually regarded by scientists as untestable by the methods that scientists generally use?

Critical Thinking Exercise

1. One day you are watching a wasp drag a grasshopper down a hole in the ground. A friend sees you and asks what you are doing. You reply, "I am watching that wasp store grasshoppers in her nest to feed her offspring." Which of the following is the best word to describe your statement?

a. observation   b. hypothesis   c. theory   d. assumption   e. prediction

Answers

Answers to Interactive Exercises

SHARED CHARACTERISTICS OF LIFE (1–I)
1. G; 2. F; 3. I; 4. L; 5. H; 6. K; 7. C; 8. N; 9. B; 10. O; 11. D; 12. E; 13. M; 14. A; 15. J; 16. Metabolism; 17. Photosynthesis; 18. ATP; 19. respiration; 20. producers; 21. consumers; 22. Decomposers; 23. interdependency; 24. receptors; 25. Homeostasis; 26. reproduction; 27. inheritance; 28. Mutations; 29. adaptive.

LIFE'S DIVERSITY (1-II)
1. species; 2. genus; 3. genus; 4. species; 5. a. Animalia; b. Protista; c. Fungi; d. Monera; e. Plantae; 6. D; 7. F; 8. A; 9. E; 10. B; 11. C; 12. G; 13. Evolution occurs when the features that characterize populations of organisms change through successive generations. 14. kingdom; 15. T; 16. more; 17. artificial; 18. T; 19. T.

THE NATURE OF BIOLOGICAL INQUIRY (1-III)
1. a. Hypothesis; b. Prediction; c. Theory; d. Scientific experiment; e. Control group; f. Variable; 2. G; 3. A; 4. D; 5. B; 6. E; 7. C; 8. F; 9. O; 10. O; 11. C; 12. O; 13. C; 14. subjective; 15. supernatural.

Answers to Self-Quiz

1. d; 2. a; 3. b; 4. c; 5. d; 6. b; 7. d; 8. a; 9. c; 10. e.

Critical Thinking Analysis

a. Observations are the only basis for evidence in science. They are perceptions of actual properties of the external world. They are the things we can see, hear, smell, taste, or feel, sometimes aided by instruments. Your statement contains an observation—that you are watching a wasp—but the most important part of your statement is an explanation of what the wasp is doing. This part of the statement is from inside your own brain, not from the external world; hence, it is not an observation. A statement of your observation of the wasp would be limited to only what you could see her do.

b. Hypotheses are possible explanations of observations or tentative answers to questions. They go beyond the available facts. Your words "store grasshoppers in her nest to feed her offspring" explain your observation. You have observed no wasp offspring, you have no evidence that the hole is the wasp's nest, and "store" implies a future time that you have not observed. Most of your statement is a possible, tentative answer to the question "Why is the wasp doing that?"

c. A theory is a collection of related hypotheses that together explain a broad range of observations. Your statement explains only this one behavior in this one species, or perhaps this one individual organism. A theory of animal behavior would have to include hypotheses that explain all the things that all animals do.

d. An assumption is a statement that is accepted without evidence in order to proceed in science. Assumptions are always added to observations in order to make hypotheses, make predictions, and interpret experiments. They do not explain observations, but they are necessary in order to create explanations, because we never have *all* the facts. For example, in order to make your stated hypothesis, you had to assume that the behavior of the wasp was useful to the wasp. Without that assumption, you would also have to accept the hypothesis that the grasshopper had provided a stimulus that made the wasp carry the grasshopper into a sheltered location.

e. A prediction is a statement of expected observations given that some hypothesis is true. It always has the "if-then" form and must contain a verb in the future tense. Your statement has none of these characteristics.

2

CHEMICAL FOUNDATIONS FOR CELLS

ORGANIZATION OF MATTER
    The Structure of Atoms
    Isotopes: Variant Forms of Atoms
    Focus on Science: Dating Fossils, Tracking
        Chemicals, and Saving Lives

BONDS BETWEEN ATOMS
    The Nature of Chemical Bonds
    Ionic Bonding
    Covalent Bonding
    Hydrogen Bonding
    Properties of Water

ACIDS, BASES, AND SALTS
    Acids and Bases

The pH Scale
Dissolved Salts
Buffers
CARBON COMPOUNDS
    Functional Groups
    Families of Small Organic Compounds
    Condensation and Hydrolysis
    Carbohydrates
    Lipids
    Focus on Health: Cholesterol Invasions of Your
        Arteries
    Proteins
    Nucleotides and Nucleic Acids

Interactive Exercises

ORGANIZATION OF MATTER (2-I, pp. 17–19)

Terms

The page-referenced terms are important; they were in boldface type in the chapter. Refer to the instructions given in Chapter 1, p. 1 of this workbook.

atom (17): *the smallest unit of an element that can exist alone or w/ other elements*

molecule (17): *smallest particle of a compound that still retains props. of a substance.*

compound (17): *occurs when atoms are held together w/ a chemical bond*

proton (17): *subatomic particle w/ smallest pos. charge, usually found in nucleus*

electron (17): *" " w/ neg. charge & found in orbital; lost or found when ions are made*

isotopes (18): *atoms of an element w/ identical chem. props but w/ diff. masses; has diff. # of neutrons*

Matching

Choose the one most appropriate answer for each. Not every letter may be matched with a number.

1. _A_ atom
2. _B_ atomic number
3. ___ electron
4. ___ element
5. ___ isotope
6. _G_ mass number
7. ___ molecule
8. _C_ neutron
9. ___ compound
10. ___ proton
11. ___ radioisotopes

A. Smallest unit of matter that is unique to a particular element
B. The number of protons in the nucleus of one atom of an element
C. Subatomic particle having no charge
D. Two or more like or unlike atoms linked together by one or more chemical bonds into an electrically neutral substance
E. A positively charged subatomic particle
F. A term applied to unstable isotopes; they tend to decay
G. The sum of protons and neutrons in an atom
H. A substance in which the relative percentages of two or more elements never vary
I. A negatively charged subatomic particle
J. Fundamental substance of which ninety-two different types occur in nature
K. A form of an element, the atoms of which contain a different number of neutrons than other forms of the same element

BONDS BETWEEN ATOMS (2-II, pp. 20–24)

Terms

ion (21):_____

ionic bond (21):_____

covalent bond (22):_____

hydrogen bond (22):_____

hydrophilic (22): _____

hydrophobic (23):_____

solutes (24): _____

Complete the Table

1. Complete the following table (refer to Table 2.1, p. 18, and Table 2.2, p. 21, in the text) by entering the name of the element and its symbol in the appropriate spaces.

Element	Symbol	Atomic Number	Mass Number	Electron Distribution		
				First Shell	Second Shell	Third Shell
a.		1	1	1		
b.		6	12	2	4	
c.		7	14	2	5	
d.		8	16	2	6	
e.		15	31	2	8	5
f.		16	32	2	8	6

Identification

2. Following the model below (number of protons and neutrons shown in the nucleus), identify the atoms of the elements illustrated below by entering appropriate electrons in the form (2e⁻).

    MODEL:

    He

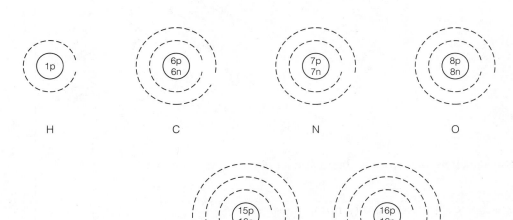

    H   C   N   O

               P   S

Identification

3. Following the model below, complete the sketch, identifying the transfer of electron(s) (by arrows) to show how positive magnesium and negative chlorine ions form ionic bonds to create a molecule of $MgCl_2$ (magnesium chloride).

MODEL:

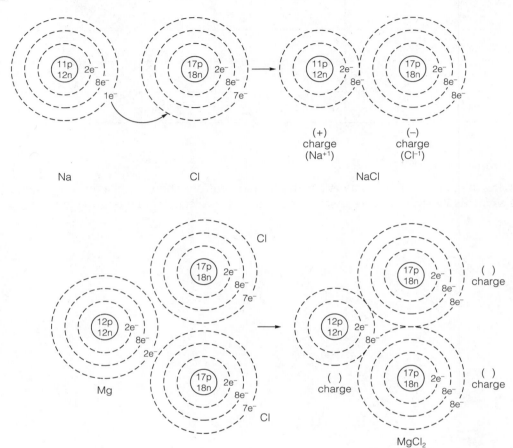

Short Answer

4. Distinguish between a nonpolar covalent bond and a polar covalent bond (p. 22).

_____

_____

_____

Identification

5. Following the model of hydrogen gas below, complete the sketch by placing electrons (as dots) in the outer shells to identify the nonpolar covalent bonding that forms oxygen gas; similarly identify polar covalent bonds by completing electron structures to form a water molecule.

MODEL:

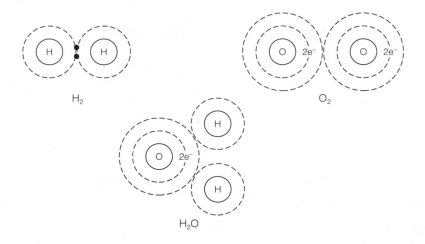

Short Answer

6. Describe one example of a large molecule within which hydrogen bonds exist (p. 22).

_____

_____

_____

Fill-in-the-Blanks

The (7) _____ of water molecules allows them to hydrogen-bond with each other. Water molecules hydrogen-bond with polar substances that are (8) hydrophilic (water-loving). Polarity causes water to repel oil and other nonpolar substances that are (9) hydrophobic (water-dreading). Water changes its temperature more slowly than air because of the great amount of heat required to break the (10) _____ bonds between water molecules. The escape of water molecules from a fluid surface to the surrounding air is known as (11) _____. Below 0°C, water molecules become locked in the less dense bonding pattern of (12) _____. (13) _____ is the property of water that explains how insects walk on water and how long, narrow water columns rise to the tops of tall trees. Water is an excellent (14) _____ in which ions and polar molecules readily dissolve. Dissolved substances in water are known as (15) _____. A substance is (16) _____ in water when spheres of (17) _____ form around its individual ions or molecules.

ACIDS, BASES, AND SALTS (2-III, pp. 24–25)

Terms

hydrogen ions (24):_____

acid (24):_____

base (24):_____

pH scale (24):_____

salt (25): _____

buffer (25): _____

Fill-in-the-Blanks

When acids dissolve in water, they release (1) _____ ions; when bases dissolve in water, they release (2) _____ ions. The (3) _____ scale is used to express the (4) _____ ion concentration of solutions. Most living cells maintain an H^+ concentration close to pH (5) _____. A pH of 8 has an H^+ concentration one hundred times higher than a pH of (6) _____. A (7) _____ is formed when an acid reacts with a base. (8) _____ are molecules that combine with or release hydrogen ions to prevent rapid shifts in pH.

Complete the Table

9. Complete the following table by consulting Figure 2.13 in the text.

Fluid	pH Value	Acid/Base
a. Blood		
b. Saliva		
c. Urine		
d. Stomach acid		

CARBON COMPOUNDS (2-IV, pp. 26–33)

Terms

functional groups (26):_____

enzymes (27):_____

condensation (27):_____

hydrolysis (27): _____

carbohydrate (27):_____

lipids (29):_____

fatty acid (29): _____

proteins (30):_____

amino acid (30):_____

polypeptide chain (30):_____

denaturation (32): _____

nucleotide (33): _____

ATP (33):_____

coenzymes (33):_____

nucleic acids (33):_____

RNA (33):_____

DNA (33):_____

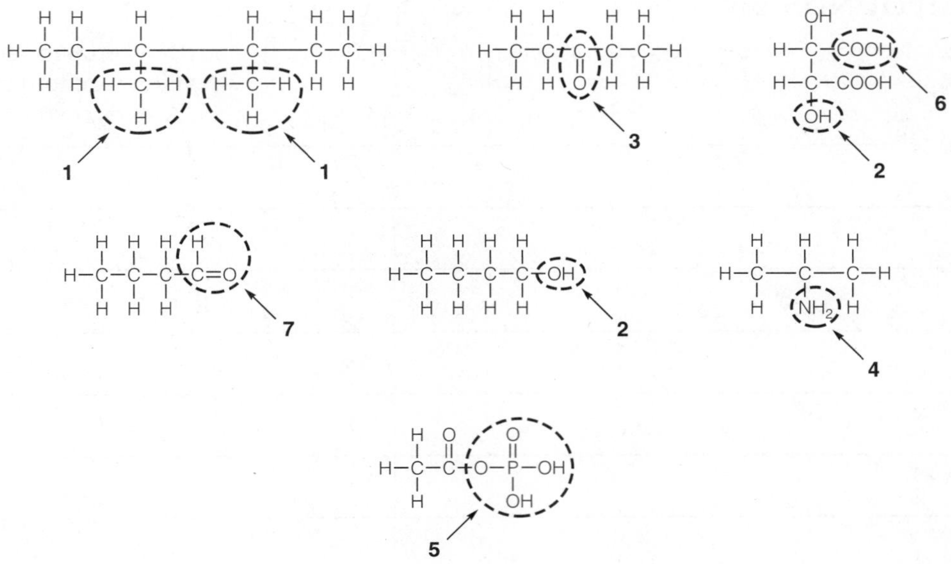

Labeling

Study the above structural formulas of organic compounds; by reference to Figure 2.16 in the text, identify the circled functional groups (sometimes repeated) by entering the correct name in the blanks with matching numbers below the sketches; complete the exercise by circling the compounds (in the parentheses following each blank) in which the functional group might appear.

1. _____ (fats - waxes - oils - sugars - amino acids - proteins - phosphate compounds, e.g. ATP)

2. _____ (fats - waxes - oils - sugars - amino acids - proteins - phosphate compounds, e.g. ATP)

3. _____ (fats - waxes - oils - sugars - amino acids - proteins - phosphate compounds, e.g. ATP)

4. _____ (fats - waxes - oils - sugars - amino acids - proteins - phosphate compounds, e.g. ATP)

5. _____ (fats - waxes - oils - sugars - amino acids - proteins - phosphate compounds, e.g. ATP)

6. _____(fats - waxes - oils - sugars - amino acids - proteins - phosphate compounds, e.g. ATP)

7. _____(fats - waxes - oils - sugars - amino acids - proteins - phosphate compounds, e.g. ATP)

Short Answer

8. State the general role of enzymes as they relate to organic compounds (p. 27).

_____

_____

Identification

9. The structural formulas of two adjacent amino acids are shown below. Identify how enzyme action causes formation of a covalent bond and a water molecule (through a condensation reaction) by circling an H atom from one amino acid and an -OH group from the other amino acid. Also circle the covalent bond that formed the dipeptide.

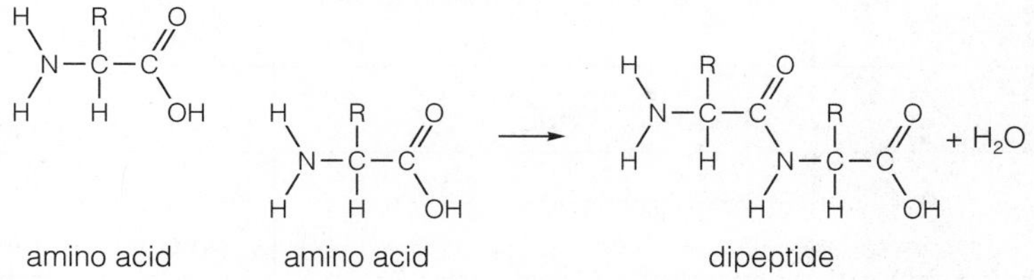

amino acid   amino acid   dipeptide

Short Answer

10. Describe hydrolysis through enzyme action for the molecules in exercise 9 (p. 27).

_____

_____

Identification

11. In the diagram below, identify condensation reaction sites between the two glucose molecules by circling the components of the water removed that allow a covalent bond to form between the glucose molecules. Note that the reverse reaction is hydrolysis and that both condensation and hydrolysis reactions require enzymes in order to proceed.

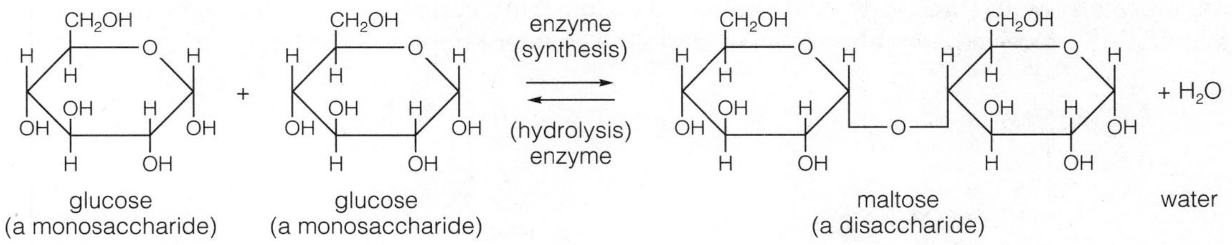

glucose   glucose   maltose   water
(a monosaccharide)   (a monosaccharide)   (a disaccharide)

e Table

e below, enter the name of the carbohydrate described by its carbohydrate class and

	Carbohydrate Class	Function
a.	oligosaccharide (disaccharide)	most plentiful sugar in nature; transport form of carbohydrates in leafy plants
b.	monosaccharide	five-carbon sugar occurring in RNA
c.	monosaccharide	main energy source for most organisms; precursor of many organic compounds
d.	polysaccharide	structural material of plant cell walls
e.	monosaccharide	five-carbon sugar occurring in DNA
f.	oligosaccharide (disaccharide)	sugar present in milk
g.	polysaccharide	main structural material in external skeletons and other hard body parts of some animals and fungi
h.	branched polysaccharide	animal starch
i.	polysaccharide	sugar-storage form in plants

Identification

13. Combine glycerol with three fatty acids below to form a triglyceride by circling the participating atoms that will identify three covalent bonds; also circle the covalent bonds in the triglyceride.

$$
\begin{array}{ccc}
\underset{\text{glycerol}}{\begin{array}{c} H \\ | \\ H-C-OH \\ | \\ H-C-OH \\ | \\ H-C-OH \\ | \\ H \end{array}}
&
+
\underset{\substack{\text{three fatty} \\ \text{acids}}}{\begin{array}{c} O \\ || \\ HO-C-R \\ \\ O \\ || \\ HO-C-R \\ \\ O \\ || \\ HO-C-R \end{array}}
&
\xrightarrow{\text{yields}}
\underset{\substack{\text{triglyceride} \\ \text{(a complete fat} \\ \text{molecule)}}}{\begin{array}{c} H \quad O \\ | \quad || \\ H-C-O-C-R \\ | \quad O \\ | \quad || \\ H-C-O-C-R \;+\; 3H_2O \\ | \quad O \\ | \quad || \\ H-C-O-C-R \\ | \\ H \end{array}}
\end{array}
$$

Labeling

14. In the appropriate blanks, label the molecules shown at the right as saturated or unsaturated.

a. _____   b. _____

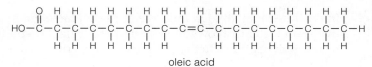

oleic acid

a

Short Answer

15. Define *phospholipid*; describe the structure and biological functions of these molecules (p. 30).

stearic acid

b

_____

_____

Matching

Choose the appropriate answer for each.

16. ___ cholesterol
17. ___ cutin
18. ___ phospholipid
19. ___ saturated fat
20. ___ unsaturated fat

A. Basic fabric for all membranes
B. Butter and bacon
C. Vegetable oil
D. Wax
E. Steroid

Matching

For exercises 21, 22, and 23, match the major parts of *every* amino acid by entering the letter of the part in the blank corresponding to the number on the molecule.

21. ___
22. ___
23. ___

A. R group (a symbol for a characteristic group of atoms that differ in number and arrangement from one amino acid to another)
B. Carboxyl group (ionized)
C. Amino group (ionized)

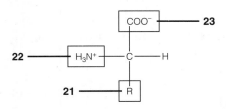

Identification

24. In the illustration of four amino acids in cellular solution (ionized state) below, circle the atoms and ions that form water to allow identification of covalent (peptide) bonds between adjacent amino acids that form a polypeptide. On the completed polypeptide, circle the newly formed peptide bonds.

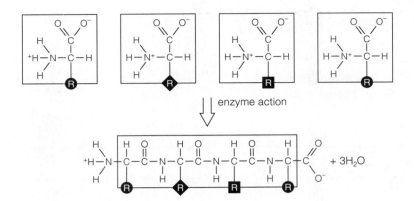

enzyme action

$+ 3H_2O$

Matching

25. ___ primary protein structure

26. ___ secondary protein structure

27. ___ tertiary protein structure

28. ___ quaternary protein structure

A. A coiled or extended pattern, based on regular hydrogen bonding

B. Incorporates two or more polypeptide chains yielding a protein that is globular, fiberlike, or both

C. A unique sequence of amino acids in their polypeptide chains

D. Bending and twisting of the protein chain due to R-group interactions

Short Answer

29. Describe protein denaturation (p. 32). _____

_____

30. Define *nucleotide* (p. 33). _____

_____

Matching

For exercises 31, 32, and 33, match the following answers to the parts of a nucleotide shown in the diagram at right.

31. ___

32. ___

33. ___

A. A five-carbon sugar (ribose or deoxyribose)

B. Phosphate group

C. A nitrogen-containing base that has either a single-ring or a double-ring structure

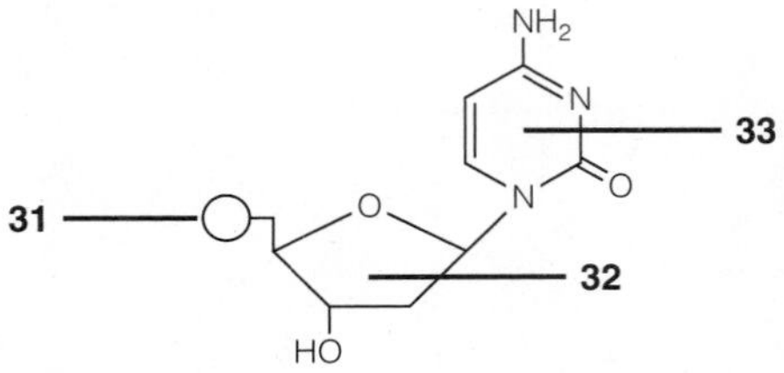

Identification

34. In the diagram of a single-stranded nucleic acid molecule at the right, encircle as many complete nucleotides as is possible. How many complete nucleotides are presented?

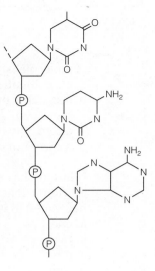

Matching

Choose the appropriate answer for each.

35. ___ adenosine triphosphate

36. ___ RNA

37. ___ DNA

38. ___ NAD$^+$ and FAD

A. Single nucleotide strand; functions in processes by which genetic instructions are used to build proteins
B. ATP, a cellular energy carrier
C. Single nucleotide units; coenzymes; transport hydrogen ions and their associated electrons from one cell reaction site to another
D. Double nucleotide strand; encodes genetic instructions with base sequences

Complete the Table

39. Complete the table below by entering the correct name of the major cellular organic compounds suggested in the "types" column (choose from carbohydrates, proteins, nucleic acids, and lipids).

Cellular Organic Compounds	Types
a.	phospholipids
b.	antibodies
c.	enzymes
d.	genes
e.	glycogen, starch, cellulose, and chitin
f.	glycerides
g.	saturated and unsaturated fats
h.	coenzymes
i.	steroids, oils, and waxes
j.	glucose and sucrose

Self-Quiz

___ 1. A molecule is _____.
  a. a combination of two or more atoms
  b. less stable than its constituent atoms separated
  c. electrically charged
  d. a carrier of one or more extra neutrons

___ 2. If lithium has an atomic number of 3 and an atomic mass of 7, it has _____ neutron(s) in its nucleus.
  a. one
  b. two
  c. three
  d. four
  e. seven

___ 3. A hydrogen bond is _____.
  a. a sharing of a pair of electrons between a hydrogen nucleus and an oxygen nucleus
  b. a sharing of a pair of electrons between a hydrogen nucleus and either an oxygen or a nitrogen nucleus
  c. formed when the weak charge of an atom of a molecule interacts weakly with a neighboring hydrogen atom that is already taking part in a polar covalent bond
  d. none of the above

___ 4. A solution with a pH of 10 is _____ times as basic as one with a pH of 7.
  a. 2
  b. 3
  c. 10
  d. 100
  e. 1,000

___ 5. Any molecule that combines with or releases hydrogen ions, or both, and helps to stabilize pH is known as a(n) _____.
  a. neutral molecule
  b. salt
  c. base
  d. acid
  e. buffer

___ 6. Carbon is part of so many different substances because _____.
  a. carbon generally forms two covalent bonds with a variety of other atoms
  b. carbon generally forms four covalent bonds with a variety of atoms
  c. carbon ionizes easily
  d. carbon is a polar compound

___ 7. Amino, carboxyl, phosphate, and hydroxyl are examples of _____.
  a. enzymes
  b. sugar units
  c. functional groups
  d. coenzymes

___ 8. _____ are compounds used by cells as transportable packets of quick energy, storage forms of energy, and structural materials.
  a. Lipids
  b. Nucleic acids
  c. Carbohydrates
  d. Proteins

___ 9. Hydrolysis could be correctly described as the _____.
  a. heating of a compound in order to drive off its excess water and concentrate its volume
  b. breaking of a long-chain compound into its subunits by adding water molecules to its structure between the subunits
  c. linking of two or more molecules by the removal of one or more water molecules
  d. constant removal of hydrogen atoms from the surface of a carbohydrate

___ 10. Genetic instructions are encoded in the bases of _____; molecules of _____ function in processes using genetic instructions to construct proteins.
  a. DNA; DNA
  b. DNA; RNA
  c. RNA; DNA
  d. RNA; RNA

Chapter Objectives/Review Questions

Page   *Objectives/Questions*

(17)   1. All forms of matter are composed of one or more _____; an _____ is the smallest unit of matter peculiar to a particular element; a _____ is two or more joined-together atoms of the same or different element.

(20)   2. Describe the distribution of electrons in the space around the nucleus of an atom; the union between the electron structures of atoms is known as the chemical _____.

(21)   3. An atom tends to react with other atoms when its outermost shell is only partly filled with _____.

(22–23)   4. Explain what is meant by the polarity of the water molecule; describe how the polarity of water molecules allows them to interact with one another.

(23)   5. Explain why ice floats on water in terms of the bonding between adjacent water molecules.

(24–25)   6. The pH of hair remover is 13; it is a(n) [choose one] ( ) base, ( ) acid. The pH of vinegar is 3; it is a(n) ( ) base, ( ) acid; define *buffer;* cite an example; explain how a salt is formed.

(26)   7. Name the three most abundant elements in living things.

(26)   8. Each carbon atom can form as many as _____ covalent bonds with other carbon atoms as well as with atoms of other elements.

(27)   9. List the four main families of small organic molecules.

(27)   10. _____ reactions result in covalent bonding between small molecules and, often, the formation of water; describe what occurs during hydrolysis reactions.

(27)   11. Define *carbohydrates;* list their general functions.

(27–28) 12. The simplest carbohydrates are _____; be able to give examples and their functions.

(28)   13. An _____ is a short chain of two or more sugar monomers; be able to give examples of well-known disaccharides and their functions.

(28)   14. A _____ is a straight or branched chain of hundreds or thousands of sugar monomers, of the same or different kinds; be able to give common examples and their functions.

(29)   15. Define *lipids;* list their general functions.

(29)   16. Describe a "fatty acid"; a _____ molecule has one to three fatty acid tails attached to a backbone of glycerol; distinguish a saturated fat from an unsaturated fat.

(29)   17. A _____ has two fatty acid tails and a hydrophilic head attached to a glycerol backbone; cite the importance of these molecules; _____ have long-chain fatty acids linked to long-chain alcohols or to carbon rings; list functions of these molecules.

(30)   18. Describe general functions of lipids without fatty acids.

(30)   19. Define *steroids* and describe their chemical structure; cite the importance of the steroids known as cholesterol and hormones.

(30)   20. Describe proteins and cite their general functions; be able to sketch the three parts of every amino acid.

(32)   21. _____ refers to the loss of a molecule's three-dimensional shape through disruption of the weak bonds responsible for it.

(33)   22. Describe the three parts of every nucleotide; give the general functions of DNA and RNA molecules.

Integrating and Applying Key Concepts

1. Explain what would happen if water were a nonpolar molecule instead of a polar molecule. Would water be a good solvent for the same kinds of substances? Would the nonpolar molecule's specific heat likely be higher or lower than that of water? Would surface tension be affected? cohesive nature? ability to form hydrogen bonds? Is it likely that the nonpolar molecules could form unbroken columns of liquid? What implications would that hold for trees?
2. Humans can obtain energy from many different food sources. Do you think this ability is an advantage or a disadvantage in terms of long-term survival? Why?
3. If the ways that atoms bond affect molecular shapes, do the ways that molecules behave toward one another influence the shapes of organelles? Do the ways that organelles behave toward one another influence the structure and function of the cells?

Critical Thinking Exercise

1. A protein chemist studied a protein that had no regions of helically coiled chain. Because protein molecules made only of the amino acid glutamate (Figure 2.22) are totally helical, the chemist concluded that the protein contained no glutamate. Which of the following would be the best criticism of this conclusion?

   a. Other amino acids besides glutamate can form helically coiled chains.
   b. Helically coiled chains are stabilized by hydrogen bonds along the backbone of the string of amino acids.
   c. No natural protein contains only glutamate.
   d. Glutamate can participate in other types of secondary structure, depending on the kinds of amino acids next to it.
   e. Peptide bonds are always identical.

Answers

Answers to Interactive Exercises

ORGANIZATION OF MATTER (2-I)
1. A; 2. B; 3. I; 4. J; 5. K; 6. G; 7. D; 8. C; 9. H; 10. E; 11. F.

BONDS BETWEEN ATOMS (2-II)
1. a. Hydrogen; b. Carbon; c. Nitrogen; d. Oxygen; e. Phosphorus; f. Sulfur.

2.

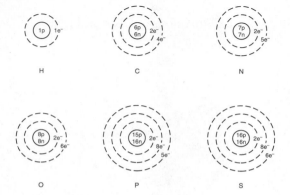

3.

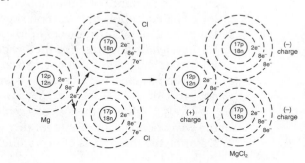

4. In a nonpolar covalent bond, atoms attract shared electrons equally. In a polar covalent bond, atoms do not share electrons equally and the bond is positive at one end, negative at the other (for example, the water molecule).

5.

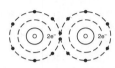

6. In the DNA molecule, the two nucleotide chains are held together by hydrogen bonds. 7. polarity; 8. hydrophilic; 9. hydrophobic; 10. hydrogen; 11. evaporation; 12. ice; 13. Cohesion; 14. solvent; 15. solutes; 16. dissolved; 17. hydration.

ACIDS, BASES, AND SALTS (2-III)
1. hydrogen (H^+); 2. hydroxyl (OH^-); 3. pH; 4. hydrogen; 5. 7; 6. 10; 7. salt; 8. Buffers; 9. a. blood: 7.3–7.5, slightly basic; b. saliva: 6.2–7.4, slightly acid or slightly basic; c. urine: 5.0–7.0, slightly acid or neutral; d. stomach acid: 1.0–3.0, acid.

CARBON COMPOUNDS (2-IV)
1. methyl (fats - waxes - oils); 2. hydroxyl (sugars); 3. ketone (sugars); 4. amino (amino acids, proteins); 5. phosphate (phosphate compounds); 6. carboxyl (fats, sugars, amino acids); 7. aldehyde (sugars).
8. Enzymes are proteins that speed up the chemical reactions occurring among the organic molecules involved in life's chemistry. These reactions usually occur at the functional groups of the organic compounds.
9.

amino acid   amino acid   dipeptide   + H₂O

10. Hydrolysis reactions reverse the chemistry of condensation reactions; in the presence of water, large molecules are split to their component smaller molecules. Both condensation and hydrolysis require the presence of enzymes specific to the particular molecules involved.
11.

glucose   glucose   maltose   water
(a monosaccharide) (a monosaccharide)   (a disaccharide)

12. a. sucrose; b. ribose; c. glucose; d. cellulose; e. deoxyribose; f. lactose; g. chitin; h. glycogen; i. starch.
13.

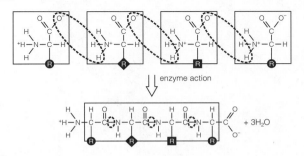

glycerol   three fatty   triglyceride
            acids   (a complete fat
                            molecule)

14. a. unsaturated; b. saturated.
15. Phospholipids have two fatty acid tails attached to a glycerol backbone; they have hydrophilic heads that dissolve in water. Phospholipids are the main structural materials of cell membranes. 16. E(B); 17. D; 18. A; 19. B; 20. C; 21. A; 22. C; 23. B.
24.

enzyme action

+ 3H₂O

25. C; 26. A; 27. D; 28. B; 29. Protein denaturation refers to the loss of a molecule's three-dimensional shape through disruption of the weak bonds responsible for it; high temperatures and solutions of strong pH can denature proteins. 30. Nucleotides are essential to life; each nucleotide has three parts: a five-carbon sugar (ribose or deoxyribose), a nitrogen-containing base that has either single-ring or double-ring structure, and a phosphate group; 31. B; 32. A; 33. C. 34. Three as shown:

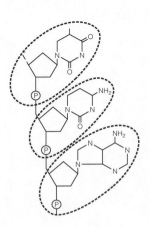

35. B; 36. A; 37. D; 38. C; 39. a. lipids; b. proteins; c. proteins; d. nucleic acids; e. carbohydrates; f. lipids; g. lipids; h. nucleic acids; i. lipids; j. carbohydrates.

Answers to Self Quiz

1. a; 2. d; 3. c; 4. e; 5. e; 6. b; 7. c; 8. c; 9. b; 10. b.

Critical Thinking Analysis

a. This statement is true, but it does not criticize the conclusion. If the conclusion was that a protein that did contain helical coil must contain glutamate, this statement would be a criticism.

b. This is also true, but it does not depend on the types of amino acids present. Any segment of helical coil, with or without glutamate, would have backbone hydrogen bonds.

c. This is true, too, but it is not a criticism of the conclusion. Polyglutamate can be made and does form only the helically coiled structure. The question is whether glutamate can be present in a mixed polymer when helical coil is not present.

d. This is true and it is a criticism. The protein might contain glutamate in an amino acid sequence that forms other types of secondary structure.

e. This is true in some senses, but peptide bonds can twist in various ways and lead to various secondary structures, depending on which amino acids are participating. If all peptide bonds were completely identical, all proteins would have the same secondary structure.

3

CELL STRUCTURE AND FUNCTION

Interactive Exercises

THE NATURE OF CELLS (3-I, pp. 37–39)

Terms

The page-referenced terms are important; they were in boldface type in the chapter. Refer to the instructions given in Chapter 1, p. 1 of this workbook.

plasma membrane (37): _____

nucleus (37): _____

cytoplasm (37): _____

cell theory (38): _____

Label-Match

Although cells vary in many specific ways, they are all alike in a few basic respects. Identify each part of the illustration below. Complete the exercise by matching and entering the letter of the proper description in the parentheses following each label.

1. _____ _____ ( )

2. _____ ( )

3. _____ ( )

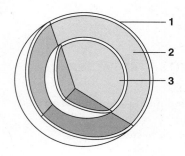

A. Everything in the cell that is enclosed by the plasma membrane, except the nucleus (in all but bacterial cells, specific compartments exist where different metabolic reactions occur)

B. A membrane-bound compartment that contains the hereditary instructions (DNA)

C. Outermost membrane of the cell; separates the internal events from the external environment

Short Answer

4. Bacterial cells are an exception to the concept illustrated above. Describe why this is so (p. 37).

_____

5. Most cells, chloroplasts, mitochondria, and bacteria are measured in micrometers (μm). Which type(s) of microscope would you use to view these structures and which would allow you to obtain the most information about these structures (pp. 38–39)?

_____

6. Which metric unit is used to measure structures only visible through the electron microscope (p. 39)?

_____

7. Why cannot larger organisms such as humans and elephants be composed of just one huge cell (p. 37)?

_____

Complete the Table

8. Summarize important contributions to the emergence of the cell theory by completing the following table.

Contributor	Years	Contribution
a.	(1564–1642)	first to record a microscopic biological observation
b.	(1635–1703)	looked at cork cells; originated the word "cell"
c.	(1632–1723)	skilled in lens construction; observed a bacterium
d.	(1804–1881) (1810–1882)	concluded that all plant and animal tissues are composed of cells
e.	(1821–1902)	established that all cells arise from preexisting cells

CELL MEMBRANES (3-II, pp. 40–44)

Terms

lipid bilayer (40):_____

fluid mosaic model (40): _____

concentration gradients (41):_____

diffusion (41):_____

osmosis (43):_____

passive transport (43):_____

active transport (43):_____

vesicle (44):_____

exocytosis (44):_____

endocytosis (44): _____

Short Answer

1. What are the main components of cell membranes (p. 40)? _____

2. What is responsible for the "fluid" aspect of the fluid mosaic model of membrane structure? The "mosaic" quality of the fluid mosaic model (p. 40)?_____

_____

3. Study the illustration below and complete the table by entering the name of the membrane structure involved in the transport mechanism described. Choose from lipid bilayer, channel protein, carrier protein, transport system of carrier proteins, receptor protein, and recognition protein.

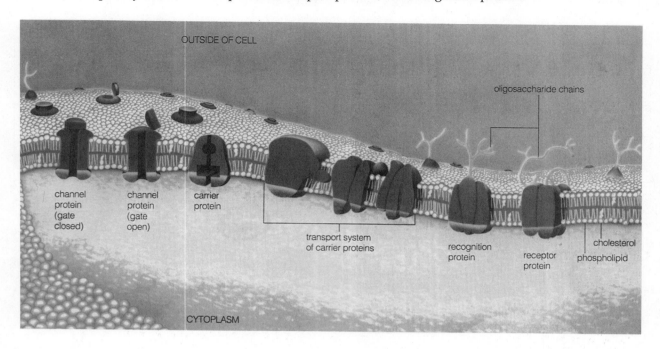

Membrane Structures(s) Involved	Function
a.	pumps solutes across the membrane against a gradient (energized by ATP)
b.	solute molecules pass through by facilitated diffusion
c.	solute molecules pass through by mediated facilitated diffusion
d.	electron transfer over several adjacent membrane proteins
e.	bind with hormones; cause metabolic or behavioral changes
f.	allows cells to recognize one another
g.	simple diffusion

Fill-in-the-Blanks

(4) _____ refers to the number of molecules (or ions) of a substance in a specified volume of fluid. A (5) _____ means that one region of the fluid contains more molecules than a neighboring region. The net movement of like molecules down their concentration gradient is called (6) _____. (7) _____ is the movement of water across membranes in response to concentration gradients, a pressure gradient, or both.

True/False

If the statement is true, write a T in the blank. If the statement is false, make it correct by changing the underlined word(s) and writing the correct word(s) in the answer blank.

_____ 8. Osmosis occurs in response to a concentration gradient that involves <u>unequal</u> concentrations of water molecules.

_____ 9. An animal cell placed in a <u>hypertonic</u> solution would swell and perhaps burst.

_____ 10. Physiological saline is 0.9% NaCl; red blood cells placed in such a solution will not gain or lose water; therefore, one could state that the fluid in red blood cells is <u>hypertonic</u>.

_____ 11. A solution of 80% solvent, 20% solute is <u>more</u> concentrated than a solution of 70% solvent, 30% solute.

_____ 12. Red blood cells shrivel and shrink when placed in a <u>hypotonic</u> solution.

_____ 13. Plant cells placed in a <u>hypotonic</u> solution will swell.

Labeling

In the blank following each ion, molecule, or structure, enter the name(s) of the correct membrane transport mechanism. Choose from

diffusion   osmosis   facilitated diffusion   active transport   exocytosis   endocytosis

14. H_2O _____

15. CO_2 _____

16. Na^+ _____

17. protein particles _____

18. glucose _____

19. O_2 _____

20. K^+ _____

21. amino acids _____

22. small prey _____

23. ingestion of an alien cell by a white blood cell _____

24. secretion of mucus _____

PROKARYOTIC CELLS—THE BACTERIA (3-III, p. 45)

Terms

cell wall (45):_____

ribosome (45):_____

Fill-in-the-Blanks

(1) _____ are the smallest and most structurally simple cells. Such prokaryotic cells have a wall that surrounds the (2) _____ membrane; this membrane controls movement of substances into and out of the (3) _____. In all cells, (4) _____ are structures composed of two molecular subunits; proteins are synthesized there. The DNA in bacterial cells is located not in a nucleus but in an irregularly shaped region of (5) _____. The word *prokaryotic* means "before the (6) _____," which implies that bacteria evolved before cells possessing nuclei existed.

EUKARYOTIC CELLS (3-IV, pp. 46–58)

Terms

organelles (46): _____

nucleus (50): _____

nuclear envelope (50):_____

nucleolus (50):_____

chromosome (50):_____

cytomembrane system (51):_____

endoplasmic reticulum (52):_____

Golgi body (52): _____

lysosomes (52):_____

mitochondria (53):_____

chloroplast (54):_____

central vacuole (54):_____

cytoskeleton (55): _____

microtubules (55):_____

intermediate filaments (55):_____

microfilaments (55):_____

flagella (56):_____

cilia (56):_____

centrioles (56):_____

cell wall (57):_____

Complete the Table

1.   Complete this table about the eukaryotic nucleus. Enter the name of each nuclear component described.

Nuclear Component	Description
a.	sites where the ribosome subunits are assembled
b.	DNA molecules and associated proteins
c.	two lipid bilayer membranes thick; proteins and protein pores span the bilayers
d.	total fluid portion of the interior of the nucleus

Matching

Study the illustration below and match each component of the cytomembrane system with the most correct description. Some components may be used more than once.

a.   smooth ER (SER)
b.   nucleus
c.   Golgi body
d.   vesicles from Golgi
e.   vesicles from rough ER (RER)

f.   endocytosis
g.   exocytosis
h.   rough ER (RER)
i.   ribosomes
j.   vesicles budding from ER or Golgi
k.   lysosomes

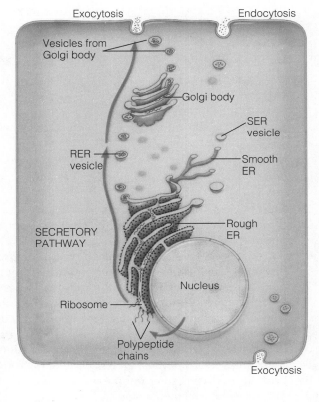

___   2. Assembly of polypeptide chains

___   3. Lipid assembly

___   4. DNA instructions for building polypeptide chains

___   5. Transport substances into the cytoplasm

___   6.   Initiate protein modification

___   7. Proteins and lipids take on final form

___   8. Sort and package lipids and proteins for transport

___   9. Vesicles formed at plasma membrane transport substances into cytoplasm

___ 10. Special vesicle budding from Golgi body; major digestion organelle

___ 11. Transport unfinished proteins and lipids to a Golgi body

___ 12. Transport finished Golgi products to the plasma membrane

___ 13. Release Golgi products at the plasma membrane

___ 14. May transport proteins and lipids to other organelles or to the plasma membrane

True/False

If the statement is true, write a T in the blank. If the statement is false, make it correct by changing the underlined word(s) and writing the correct word(s) in the answer blank.

_____ 15. The bacterium-sized mitochondrion is an organelle in eukaryotic cells that specializes in liberating energy stored in sugars and using it to form many <u>DNA</u> molecules.

_____ 16. Mitochondria use <u>CO_2</u> to extract far more energy than can be done by any other means.

_____ 17. The different enzymes in <u>lysosomes</u> break down virtually all of life's major molecules.

_____ 18. In the <u>stroma</u> of the chloroplast, enzymes speed carbohydrate assembly.

_____ 19. Central vacuoles of living <u>plant</u> cells store amino acids, sugars, ions, and toxic wastes in addition to water.

_____ 20. Grana in <u>mitochondria</u> are involved with capture of sunlight energy.

_____ 21. During growth, plant cell walls enlarge under the force of water pressure that builds up inside the <u>chloroplast</u>.

_____ 22. <u>Chloroplasts</u> have an outer membrane and an inner membrane of deep folds called the cristae.

_____ 23. Energy-demanding muscle cells generally have many more <u>mitochondria</u> than less active cells.

_____ 24. <u>Mitochondria</u> fuse with other vesicles containing cell substances or foreign particles brought into the cell by endocytosis.

Dichotomous Choice

Circle one of two possible answers given between parentheses in each statement.

25. The cytoskeleton gives (prokaryotic/eukaryotic) cells their shape and internal organization.
26. (Protein/Carbohydrate) subunits form the basic components of microtubules, microfilaments, and intermediate filaments.
27. (Flagella/Cilia) are long, tail-like motile structures on cells.
28. (Flagella/Cilia) are short but very numerous motile structures on cells.
29. The cross-sectional pattern of 9 + 2 microtubules is found in (cilia/centrioles).
30. The human respiratory tract is lined with (flagella/cilia).
31. (Cilia/Centrioles) are cylinders of triplet microtubules.
32. In most animal cells, an MTOC also includes a pair of (flagella/centrioles).
33. In (unicellular/multicellular) organisms, location of centrioles influences the plane of cell division.
34. (Centrioles/Basal bodies) give rise to the microtubules of cilia and flagella.
35. Prior to producing microtubules of flagella and cilia, the centriole becomes positioned near the (plasma/nuclear) membrane.
36. The structure remaining at the bases of cilia and flagella is called a (centriole/basal body).
37. The (primary/secondary) wall of plants is pliable and porous.
38. The (primary/secondary) wall of plants is rigid and helps maintain cell shape.
39. Adjacent plant cell walls are cemented together by (pectins/waxes).
40. Numerous channels cross adjacent walls and connect the cytoplasm of neighboring (plant/animal) cells.
41. (Plant/Animal) cells are meshed together in an "extracellular matrix."
42. The extracellular matrix may be described as being like (bone/jelly).
43. Nutrients, hormones, and other substances move through the ground substance by (diffusion/osmosis).
44. (Tight/Adhering) junctions link cells of epithelial tissues.
45. Open channels of the (adhering/gap) junctions allow rapid flow of signals and substances.
46. Like spot welds, the (tight/adhering) junctions hold cells in tissues of skin, heart, and other organs together.

Label-Match

Identify each indicated part of the accompanying illustrations. Complete the exercise by matching and entering the letter of the proper function description in the parentheses following each label. Some letter choices must be used more than once.

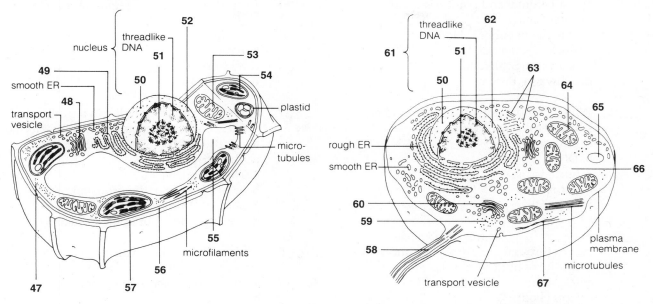

47. _____ ( )

48. _____ _____ ( )

49. _____ _____ _____ ( )

50. _____ _____ ( )

51. _____ ( )

52. _____ _____ ( )

53. _____ ( )

54. _____ ( )

55. _____ _____ ( )

56. _____ _____ ( )

57. _____ _____ ( )

58. _____ ( )

59. _____ ( )

60. _____ _____ ( )

61. _____ ( )

62. _____ _____ ( )

63. _____ ( )

64. _____ ( )

65. _____ ( )

66. _____ ( )

67. _____ ( )

A. Two-membrane structure; outermost part of the nucleus
B. Protection and structural support
C. Increasing cell surface area and storage
D. Everything enclosed by the plasma membrane, except the nucleus
E. Formed as buds from Golgi membranes; contain enzymes for intracellular digestion
F. Small cylinders composed of triplet microtubules; determine cell division plane, produce microtubules of cilia and flagella; act as basal bodies
G. Microtubular structures for propelling eukaryotic cells; longer than cilia but with similar microtubule structure
H. Site of protein synthesis
I. A membrane-bound compartment that houses DNA in eukaryotic cells
J. Further modification, sorting, and shipping of proteins and lipids for secretion or for use in the cell
K. Sites where the protein and RNA subunits of ribosomes are assembled
L. A major component of the cytoskeleton
M. Photosynthesis and some starch storage
N. Control of material exchanges; mediates cell–environment interactions
O. Site of aerobic respiration
P. Initial modification of protein structure after formation in ribosomes
Q. Protein openings that span both bilayers of the nuclear envelope

Self-Quiz

___ 1. _____ and _____ concluded that all animal and plant tissues are composed of cells.
 a. van Leeuwenhoek; Hooke
 b. Brown; Schleiden
 c. Schwann; Schleiden
 d. Schwann; Brown

___ 2. Which of the following is *not* found as a part of prokaryotic cells?
 a. ribosomes
 b. DNA
 c. nucleus
 d. cytoplasm
 e. cell wall

___ 3. The _____ is free of ribosomes, curves through the cytoplasm, and is the main site of lipid synthesis.
 a. lysosome
 b. Golgi body
 c. smooth ER
 d. rough ER

___ 4. Which of the following is *not* present in all cells?
 a. cell wall
 b. plasma membrane
 c. ribosomes
 d. DNA molecules

___ 5. Mitochondria convert energy stored in _____ to forms that the cell can use, principally ATP.
 a. water
 b. carbon compounds
 c. NADPH$_2$
 d. carbon dioxide

___ 6. In a lipid bilayer, tails point inward and form a(n) _____ region that excludes water.
 a. acidic
 b. basic
 c. hydrophilic
 d. hydrophobic

___ 7. A protistan adapted to life in a freshwater pond is collected in a bottle and transferred to a saltwater bay. Which of the following is likely to happen?
 a. The cell bursts.
 b. Salts flow out of the protistan cell.
 c. The cell shrinks.
 d. Enzymes flow out of the protistan cell.

___ 8. Which of the following is *not* a form of active transport?
 a. sodium-potassium ion pump
 b. endocytosis
 c. exocytosis
 d. osmosis

___ 9. O_2 and CO_2 and other small, electrically neutral molecules move across the cell membrane by _____.
 a. facilitated diffusion
 b. endocytosis
 c. simple diffusion
 d. active transport

___ 10. Ions such as H^+, Na^+, K^+, and Ca^{++} move across cell membranes by _____.
 a. facilitated diffusion
 b. endocytosis
 c. simple diffusion
 d. active transport

Chapter Objectives/Review Questions

Page	Objectives/Questions
(38–39)	1. Briefly describe the operation of light microscopes, phase-contrast microscopes, transmission electron microscopes, and scanning electron microscopes.
(39)	2. A _____ is one-millionth of a meter long.
(37)	3. Be able to explain the reason there is a limit to the size that a cell can attain by growth.
(39)	4. A _____ is one-billionth of a meter long.
(40)	5. Materials are exchanged between cytoplasm and external cell environment across the _____.

(40)   6. Describe the general structure of a phospholipid molecule.
(40)   7. Explain how behavior of many phospholipid molecules in water provides the basis for the structure of membranes.
(40)   8. Membrane functions are carried out by membrane _____ embedded in the bilayer or positioned at its surfaces.
(41)   9. Molecules moving to regions where they are less concentrated are moving down their _____ gradient.
(41)   10. _____ is the net movement of like molecules or ions down their concentration gradient.
(42)   11. When salt is dissolved in water, which is the solute and which is the solvent?
(43)   12. _____ is the movement of water across a selectively permeable _____ in response to concentration gradients, pressure gradients, or both.
(43)   13. Movement of solute molecules directly across the lipid bilayer down their concentration gradient is known as simple _____; facilitated diffusion through insides of channel proteins and facilitated diffusion mediated by carrier proteins are types of _____ transport; pumping solutes through carrier proteins with ATP boosts against the concentration gradient is known as _____ transport.
(44)   14. During _____, a vesicle moves to the plasma membrane and fuses with it, releasing its contents to the outside; _____ cells such as amoebas "eat" other cells by means of _____.
(46–58)   15. Give the functions and cellular location of the following basic eukaryotic organelles and structures: nucleus, endoplasmic reticulum, Golgi bodies, lysosomes, transport vesicles, mitochondria, ribosomes, cytoskeleton, chloroplasts, central vacuoles, and cell walls.
(45)   16. In all cells, not just bacteria, _____ are the sites of protein synthesis.
(50)   17. Between cell divisions, eukaryotic _____ is threadlike, with many enzymes and proteins attached to it like beads on a string; prior to cell division, DNA molecules are _____; before the _____ molecules are sorted into two sets, they fold and twist into condensed structures.
(55)   18. Microtubules, microfilaments, and intermediate filaments are all main components of the _____.
(56)   19. _____ and _____ propel eukaryotic cells through their environment.
(56)   20. _____ are cytoplasmic masses that organize microtubules and sometimes include _____ in animal cells.
(57)   21. Indicate whether the following occur in plants or in animals: cell walls, extracellular matrix, cell junctions.

Integrating and Applying Key Concepts

1. If there were no such thing as active transport, how would the lives of organisms be affected?
2. How did the existence of a nucleus, compartments, and extensive internal membranes confer selective advantages on cells that developed these features?

Critical Thinking Exercise

1. A certain enzyme (E) can catalyze the formation of a product (P) from the amino acid (A). A solution of E is placed at position 2 on a strip of gel, as shown in the diagram. A solution of A is placed at position 7. At which position is P most likely to appear first? Assume that both A and E can diffuse through the water in the gel.

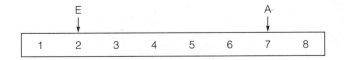

Answers

Answers to Interactive Exercises

THE NATURE OF CELLS (3-I)
1. plasma membrane (C); 2. cytoplasm (A); 3. nucleus (B). 4. Bacterial cells lack an organized nucleus; bacterial DNA is in an organized part of the cytoplasm called the nucleoid. 5. These structures are within the resolution limits of the light microscope, but the magnifying power of the electron microscope would provide far more structural information. 6. The nanometer (one billionth of one meter). 7. As a cell expands in diameter, its volume increases more rapidly than its surface area does (a matter of supply and demand). 8. a. Galileo Galilei; b. Robert Hooke; c. Antony van Leeuwenhoek; d. Matthias Schleiden (a botanist) and Theodor Schwann (a zoologist); e. Rudolf Virchow.

CELL MEMBRANES (3-II)
1. Phospholipids and proteins. 2. The bilayer represents the "fluid" aspect of the model because lipid molecules in water show much movement. The membrane is a "mosaic" because it is a composite of lipids and proteins. 3. a. carrier protein; b. channel protein; c. carrier protein; d. transport system of carrier proteins; e. receptor proteins; f. recognition protein; g. lipid bilayer; 4. Concentration; 5. gradient; 6. diffusion; 7. Osmosis; 8. T; 9. hypotonic; 10. isotonic; 11. less; 12. hypertonic; 13. T; 14. osmosis; 15. diffusion; 16. active transport; 17. endocytosis (phagocytosis); 18. facilitated diffusion; 19. diffusion; 20. active transport; 21. facilitated diffusion; 22. endocytosis (phagocytosis); 23. endocytosis (phagocytosis); 24. exocytosis.

PROKARYOTIC CELLS—THE BACTERIA (3-III)
1. Bacteria; 2. plasma; 3. cytoplasm; 4. ribosomes; 5. cytoplasm; 6. nucleus.

EUKARYOTIC CELLS (3-IV)
1. a. nucleolus; b. chromosome; c. nuclear envelope; d. nucleoplasm; 2. i; 3. a; 4. b; 5. d, e, f; 6. h; 7. c; 8. c; 9. f; 10. k; 11. a, e, h, j; 12. d; 13. d, g; 14. j; 15. ATP; 16. O_2; 17. T; 18. T; 19. T; 20. chloroplasts; 21. central vacuole; 22. Mitochondria; 23. T; 24. Lysosomes; 25. eukaryotic; 26. Protein; 27. Flagella; 28. Cilia; 29. cilia; 30. cilia; 31. Centrioles; 32. centrioles; 33. multicellular; 34. Basal bodies; 35. plasma; 36. basal body; 37. primary; 38. secondary; 39. pectins; 40. plant; 41. Animal; 42. jelly; 43. diffusion; 44. Tight; 45. gap; 46. adhering; 47. ribosomes (H); 48. Golgi complex (body) (J); 49. rough endoplasmic reticulum (P); 50. nuclear pore (Q); 51. nucleolus (K); 52. nuclear envelope (A); 53. mitochondrion (O); 54. chloroplast (M); 55. central vacuole (C); 56. cell (plasma) membrane (N); 57. cell wall (B); 58. flagellum (G); 59. ribosomes (H); 60. Golgi complex (body) (J); 61. nucleus (I); 62. nuclear envelope (A); 63. centrioles (F); 64. mitochondrion (O); 65. lysosome (or vacuole) (E); 66. cytoplasm (D); 67. microfilaments (L)

Answers to Self-Quiz

1. c; 2. c; 3. c; 4. a; 5. b; 6. d; 7. c; 8. d; 9. c; 10. d.

Critical Thinking Analysis

P will form first at the point where the diffusing E and A molecules first meet. This will be between 2 and 7. Because E is a protein, a polymer of amino acids, it is larger than A. Larger molecules diffuse more slowly than smaller molecules. Therefore, A will diffuse farther in a given time than E, and the meeting point will be closer to 2 than to 7. Unless you know how much faster A moves than E, you cannot choose between positions 3 and 4.

4

GROUND RULES OF METABOLISM

Interactive Exercises

ENERGY AND LIFE (4-I, pp. 61–62)

Terms

The page-referenced terms are important; most were in boldface type in the chapter. Refer to the instructions given in Chapter 1, p. 1 of this workbook.

metabolism (61) _____

energy (61) _____

first law of thermodynamics (61) _____

second law of thermodynamics (62)_____

entropy (62) _____

Short Answer

1.   The world of life maintains a high degree of organization only because

_____

True/False

If the statement is true, write a T in the blank. If the statement is false, make it correct by changing the underlined word(s) and writing the correct word(s) in the answer blank.

_____ 2. The <u>first</u> law of thermodynamics states that entropy is constantly increasing in the universe.

_____ 3. Your body steadily gives off heat equal to that from a <u>100-watt</u> light bulb.

_____ 4. When you eat a potato, some of the stored chemical energy of the food is converted into <u>mechanical</u> energy that moves your muscles.

_____ 5. The amount of low-quality energy in the universe is <u>decreasing</u>.

Labeling

In the blank preceding each item, indicate whether the first law of thermodynamics (I) or the second law of thermodynamics (II) is best described.

6.____ Cooling of a cup of coffee

7.____ Evaporation of gasoline into the atmosphere

8.____ A hydroelectric plant at a waterfall producing electricity

9.____ The creation of a snowman by children

10.____ The death and decay of an organism

THE NATURE OF METABOLISM (4-II, pp. 62–63)

Terms

metabolic pathway (64)_____

reactants (64) _____

intermediates (64)_____

enzymes (64) _____

cofactors (64)_____

energy carriers (64)_____

end products (64) _____

Short Answer

1. Define exergonic reactions; give an example._____

_____

2. Define endergonic reactions; give an example. _____

_____

3. A (3) _____ _____ is an orderly series of chemical reactions, with each reaction catalyzed by a

   specific (4) _____.

Labeling

Classify each of the following reactions as *endergonic* or *exergonic*.

5._____ The product of a chemical reaction has more energy than the reactants.

6._____ Glucose + oxygen → carbon dioxide + water + energy

7._____ The reactants of a chemical reaction have more energy than the product.

Labeling

Choose the most appropriate answer for each: A or B.

8.___ a degradative reaction

9.___ an endergonic reaction

10.___ a biosynthetic reaction

11.___ an exergonic reaction

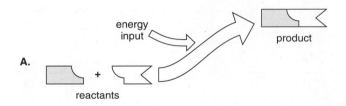

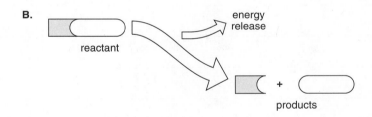

Matching

Study the sequence of reactions below. Identify the components of the reactions by selecting items from the following list and entering the correct letter in the appropriate blank.

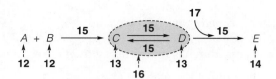

12.___   15.___   A. cofactor   D. end product
13.___   16.___   B. intermediates   E. reversible reaction
                      C. reactants   F. enzymes
14.___   17.___

ENZYMES (4-III, pp. 64–67)

Terms

substrates (64)_____

active site (64)_____

activation energy (65) _____

feedback inhibition (66) _____

hormones (67)_____

coenzymes (67) _____

NAD$^+$ (67)_____

FAD (67)_____

NADP$^+$ (67)_____

Short Answer

1. List four characteristics that enzymes have in common._____

_____

Matching

Match the items on the sketch below with the list of descriptions. Some answers may require more than one letter.

2.___

3.___

4.___

5.___

6.___

7.___

A. Transition state, the time of the most precise fit between enzyme and substrate
B. Complementary active site of the enzyme
C. Enzyme, a protein with catalytic power
D. Product or reactant molecules that an enzyme can specifically recognize
E. Product or reactant molecule
F. Bound enzyme-substrate complex

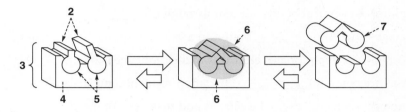

Fill-in-the-Blanks

(8)_____ are highly selective proteins that act as catalysts, which means that they greatly enhance the rate at which specific reactions approach (9)_____. The specific substance on which a particular enzyme acts is called its (10)_____; this substance fits into the enzyme's crevice, which is called its (11)_____ _____. The (12)_____-_____ model describes how a substrate contacts the substrate without a perfect fit. Enzymes increase reaction rates by lowering the required (13)_____ _____. (14)_____ and (15)_____ are two important factors that influence the rates of enzyme activity. Extremely high fevers can destroy the three-dimensional shape of an enzyme, which may adversely affect (16)_____ and cause death. (17)_____ enzymes have control sites where specific substances can bind and alter enzyme activity. The situation in which the end product binds to the first enzyme in a metabolic pathway and prevents product formation is known as (18)_____ _____. Nonprotein substances that aid enzymes in their catalytic task are called (19)_____; they include some large organic molecules that function as (20)_____. (21)_____ and (22)_____ are coenzymes that have roles in the breakdown of glucose and other carbohydrates. (23)_____ is a coenzyme with a central role in photosynthesis; its abbreviation is (24)_____ when it is loaded with protons and electrons. Some metal (25)_____ such as Fe^{++} also serve as cofactors as components of cytochrome molecules serving as carrier proteins in cell membranes.

ELECTRON TRANSFERS IN METABOLIC PATHWAYS (4-IV, pp. 67–68)

Terms

electron transport systems (67)_____

Fill-in-the-Blanks

The release of energy from glucose in cells proceeds in controlled steps, so that (1)_____ molecules

form along the route from glucose to carbon dioxide and water. At each step in a metabolic pathway, a spe-

cific (2)_____ lowers the activation energy for the formation of an intermediate compound. At each

step in the pathway, only some energy is released. In chloroplasts and mitochondria, the liberated electrons

released from the breaking of chemical bonds are sent through (3)_____ _____ systems; these

systems consist of enzymes and (4)_____, bound in a cell membrane, that transfer electrons in a highly

organized sequence. A molecule that donates electrons in the sequence is being (5)_____, while mole-

cules accepting electrons are being (6)_____. Oxidation-reduction means an (7)_____ transfer.

Electron transport systems "intercept" excited electrons and make use of the (8)_____ they release. If

we think of the electron transport system as a staircase, electrons at the top of the staircase have the (9)

[choose one] ( ) most, ( ) least energy. As the electrons are transferred from one electron carrier to another,

some (10)_____ can be harnessed to do biological (11)_____. One type of biological work occurs

when energy released during electron transfers is used to bond a (12)_____ group to ADP.

Matching

Match the lettered statements to the numbered items on the sketch below.

13.____   A. Represent the cytochrome molecules in an electron transport system
            B. Electrons at their highest energy level
14.____   C. Released energy harnessed and used to produce ATP
15.____   D. Electrons at their lowest level
            E. The separation of hydrogen atoms into protons and electrons
16.____

17.____

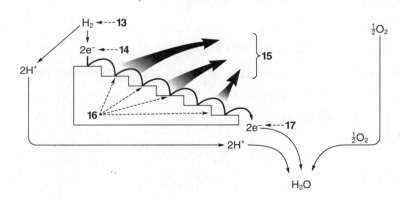

ATP: THE MAIN ENERGY CARRIER (4-V, p. 68)

Terms

ATP (68)_____

ATP/ADP cycle (68) _____

phosphorylation (68)_____

Fill-in-the-Blanks

ATP is constructed of the nitrogen-containing molecule (1)_____, the sugar (2)_____, and three (3)_____ groups. When ATP is hydrolyzed, a molecule of (4)_____ in the presence of an appropriate (5)_____ is used to split ATP into (6)_____, a (7)_____ group, and (most important) usable (8)_____, which is easily transferred to other molecules in the cell. The hydrolysis of ATP provides (9)_____ for biosynthesis, active transport across cell membranes, and molecular displacements such as those required for muscle contraction. ATP directly or indirectly delivers energy to almost all (10)_____ pathways. In the (11)_____/_____ cycle, a phosphate group is linked to adenosine diphosphate, and adenosine triphosphate donates a phosphate group elsewhere and reverts back to adenosine diphosphate. Adding a phosphate to a molecule is called (12)_____. When this occurs, the molecule increases its store of (13)_____ and becomes primed to enter a specific (14)_____. The (15)_____/_____ cycle provides a renewable means of conserving energy and transferring it to specific reactions.

Labeling

Identify the molecule at the right and label its parts.

16._____

17._____ _____

18._____

19. The name of this molecule is _____

_____.

Self-Quiz

____1.   An important principle of the second law of thermodynamics states that _____.

   a. energy can be transformed into matter, and because of this we can get something for nothing

   b. energy can be destroyed only during nuclear reactions, such as those that occur inside the sun

   c. if energy is gained by one region of the universe, another place in the universe must also gain energy in order to maintain the balance of nature

   d. matter tends to become increasingly more disorganized

___2. Essentially, the first law of thermodynamics states that _____.
   a. one form of energy cannot be converted into another
   b. entropy is increasing in the universe
   c. energy cannot be created or destroyed
   d. energy cannot be converted into matter or matter into energy

___3. An enzyme is best described as _____.
   a. an acid
   b. a protein
   c. a catalyst
   d. a fat
   e. both (b) and (c)

___4. Which is not true of enzyme behavior?
   a. Enzyme shape may change during catalysis.
   b. The active site of an enzyme orients its substrate molecules, thereby facilitating interaction of their reactive parts.
   c. All enzymes have an active site where substrates are temporarily bound.
   d. An individual enzyme can catalyze a wide variety of different reactions.

___5. When NAD^+ combines with hydrogen, the NAD^+ is _____.
   a. reduced
   b. oxidized
   c. phosphorylated
   d. denatured

___6. A substance that gains electrons is _____.
   a. oxidized
   b. a catalyst
   c. reduced
   d. a substrate

___7. In _____ pathways, carbohydrates, lipids, and proteins are broken down in stepwise reactions that lead to products of lower energy.
   a. intermediate
   b. biosynthetic
   c. induced
   d. degradative

___8. With regard to major function, NAD^+, FAD, and $NADP^+$ are classified as _____.
   a. enzymes
   b. phosphate carriers
   c. cofactors that function as coenzymes
   d. end products of metabolic pathways

___9. If a phosphate bond is linked to ADP, the bond formed _____.
   a. absorbs a large amount of free energy when the phosphate group is attached during hydrolysis
   b. is oxidized when ATP is hydrolyzed to ADP and one phosphate group
   c. is usually found in each glucose molecule; this is why glucose is chosen as the starting point for glycolysis
   d. later will release a large amount of usable energy when the phosphate group is split off during hydrolysis.

___10. An allosteric enzyme _____.
   a. has an active site where substrate molecules bind and another site that binds with intermediate or end-product molecules
   b. is an important energy-carrying nucleotide
   c. carries out either oxidation reactions or reduction reactions but not both
   d. raises the activation energy of the chemical reaction it catalyzes

Chapter Objectives/Review Questions

This section lists general and detailed chapter objectives that can be used as review questions. You can make maximum use of these items by writing answers on a separate sheet of paper. Fill in answers where blanks are provided. To check for accuracy, compare your answers with information given in the chapter or glossary.

Page   *Objectives/Questions*

(61)   1. _____ is the controlled capacity to acquire and use energy for stockpiling, breaking apart, building, and eliminating substances in ways that contribute to survival and reproduction.

(61)   2. Define energy; be able to state the first and second laws of thermodynamics.

(62)   3. _____ is a measure of the degree of randomness or disorder of systems.

(62)   4. Explain how the world of life maintains a high degree of organization.

(64)   5. Reactions that show a net loss in energy are said to be _____; reactions that show a net gain in energy are said to be _____.

(63–64)   6. What is the function of metabolic pathways in cellular chemistry?

(64)   7. Give the function of each of the following participants in metabolic pathways: reactants, intermediates, enzymes, cofactors, energy carriers, and end products.

(65)   8. Explain the effects of enzymes on activation energy.

(64–65)   9. Describe the induced-fit hypothesis.

(66)   10. Explain what happens to enzymes if temperature and pH continually increase.

(66)   11. What are "allosteric enzymes," and what is their function?

(66)   12. Describe the control mechanism known as feedback inhibition.

(67)   13. Cite examples of coenzymes.

(67)   14. Explain the differences between NAD^+ and NADH; $NADP^+$ and NADPH; FAD and $FADH_2$. [Peek ahead to text pp. 71 and 87 for the answer.]

(67)   15. Describe the components, organization, and functions of an electron transport system.

(68)   16. ATP is composed of _____, a five-carbon sugar, three _____ groups, and _____, a nitrogen-containing compound.

(68)   17. ATP directly or indirectly delivers _____ to almost all metabolic pathways.

(68)   18. Explain the functioning of the ATP/ADP cycle.

(68)   19. Adding a phosphate to a molecule is called _____.

Integrating and Applying Key Concepts

A piece of dry ice left sitting on a table at room temperature vaporizes. As the dry ice vaporizes into CO_2 gas, does its entropy increase or decrease? Tell why you answered as you did.

Critical Thinking Exercise

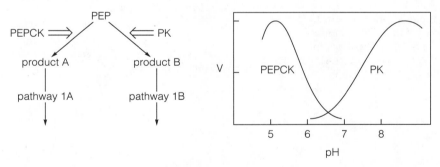

Metabolic pathways branch when two different enzymes have the same substrate but produce different products. For example, the substance phosphoenolpyruvate (PEP) is the substrate for two different enzymes—phosphoenolpyruvate carboxykinase (PEPCK) and pyruvate kinase (PK). Each enzyme catalyzes the formation of a different product, and each product is metabolized by the enzymes of a different pathway. These two enzymes are both dependent on pH but in very different ways. In the graph above, V is a measure of the percentage of enzyme activity.

Suppose an oyster, at high tide, has a tissue pH of about 7.3. When the tide goes out, the oyster closes its shells, acidic wastes can no longer escape to the sea, and the internal pH drops to about 6. What happens to PEP metabolism under these conditions?

Answers

Answers to Interactive Exercises

ENERGY AND LIFE (4-I)

1. The world of life maintains a high degree of organization only because it is being resupplied with energy lost from someplace else. 2. second; 3. T; 4. T; 5. increasing; 6. II; 7. II; 8. I; 9. I; 10. II.

THE NATURE OF METABOLISM (4-II)

1. Exergonic reactions show a net loss of energy (energy out); an example is the breakdown of food molecules in the human body as the reactants become products.
2. Endergonic reactions show a net gain in energy (energy in); an example is the construction of starch and other large molecules from smaller, energy-poor molecules. 3. metabolic pathway; 4. enzyme; 5. endergonic; 6. exergonic; 7. exergonic; 8. B; 9. A; 10. A; 11. B; 12. C; 13. B; 14. D; 15. F; 16. E; 17. A.

ENZYMES (4-III)

1. Enzymes are usually *protein molecules* with *enormous catalytic power* (a few RNA forms have been found to act as enzymes). Enzymes do not cause reactions that would not happen on their own; enzymes are not changed in reactions—they can be used over and over; each enzyme is *highly selective about its substrates*; an enzyme can recognize both the reactants and the products of a given reaction as its substrate. Enzymes work most effectively at *specific temperatures and pH values*. Enzymes *lower the energy of activation* of the reactions they catalyze.
2. D; 3. F; 4. C; 5. B; 6. A; 7. E; 8. Enzymes; 9. equilibrium; 10. substrate; 11. active site; 12. induced-fit; 13. activation energy; 14. Temperature (pH); 15. pH (temperature); 16. metabolism; 17. Allosteric; 18. feedback inhibition; 19. cofactors; 20. coenzymes; 21. NAD$^+$ (FAD); 22. FAD (NAD$^+$); 23. NADP$^+$; 24. NADPH; 25. ions.

ELECTRON TRANSFERS IN METABOLIC PATHWAYS (4-IV)

1. intermediate; 2. enzyme; 3. electron transport; 4. cofactors; 5. oxidized; 6. reduced; 7. electron; 8. energy; 9. most; 10. energy; 11. work; 12. phosphate; 13. E; 14. B; 15. C; 16. A; 17. D.

ATP: THE MAIN ENERGY CARRIER (4-V)

1. adenine; 2. ribose; 3. phosphate; 4. water; 5. enzyme; 6. ADP; 7. phosphate; 8. energy; 9. energy; 10. metabolic; 11. ATP/ADP; 12. phosphorylation; 13. energy; 14. reaction; 15. ATP/ADP; 16. three phosphate groups; 17. ribose sugar; 18. adenine, a nitrogen-containing molecule; 19. adenosine triphosphate.

Answers to Self-Quiz

1. d; 2. c; 3. e; 4. d; 5. a; 6. c; 7. d; 8. c; 9. d; 10. a.

Critical Thinking Analysis

At pH 7.3, PK is functioning at about 50 percent of maximal activity, and PEPCK is almost inactive. Thus, almost all the PEP goes on pathway 1B. When the pH drops to 6, the situation is reversed; PEPCK functions at about 50 percent of maximum, and PK is inactivated. Under these conditions, PEP follows pathway 1A. The change in pH resulting ultimately from the movement of the tide causes a switch in the metabolism of the oyster. Notice that neither enzyme normally encounters its optimum pH in the living animal.

Reference: Hochachka, P. W., and G. N. Somero (1973). *Strategies of Biochemical Adaptation* (Philadelphia: Saunders), p. 49.

5

ENERGY-ACQUIRING PATHWAYS

PHOTOSYNTHESIS: AN OVERVIEW
  Sources of Energy and Materials for the
    Reactions
  Where the Reactions Occur

LIGHT-DEPENDENT REACTIONS
  Absorbing Sunlight Energy
  ATP and NADPH: Loading Up Energy,
    Hydrogen, and Electrons

LIGHT-INDEPENDENT REACTIONS
  Capturing Carbon
  Building the Glucose Subunits
  Focus on the Environment: Pastures of the Seas
  C4 Plants: Squirreling Away Carbon on Hot,
    Dry Days

CHEMOSYNTHESIS

Interactive Exercises

PHOTOSYNTHESIS: AN OVERVIEW (5-I, pp. 71–73)

Terms

The page-referenced terms are important; they were in boldface type in the chapter. Refer to the instructions given in Chapter 1, p. 1 of this workbook.

autotrophs (71)_____

heterotrophs (71)_____

chloroplast (73)_____

stroma (73) _____

thylakoid membrane system (73)_____

Fill-in-the-Blanks

(1)_____ obtain carbon and energy from the physical environment; their carbon source is

(2)_____ _____. (3) _____ autotrophs obtain energy from sunlight. (4)_____ autotrophs

are represented by a few kinds of bacteria; they obtain energy by stripping (5)_____ from sulfur or

other inorganic substances. (6)_____ feed on autotrophs, each other, and organic wastes; representa-

tives include (7)_____, fungi, many protistans, and most bacteria. Although energy stored in organic

compounds such as glucose may be released by several pathways, the pathway known as (8)_____

_____ releases the most energy.

9. In the space below, supply the missing information to complete the summary equation for photosynthesis:

$$12 \underline{\hspace{1cm}} + \underline{\hspace{1cm}} CO_2 \rightarrow O_2 + C_6H_{12}O_6 + 6\underline{\hspace{1cm}}$$

10. Supply the appropriate information to state the equation (above) for photosynthesis in words:

(a)_____ molecules of water plus six molecules of (b)_____ _____ (in the presence of pig-

ments, enzymes, and sunlight) yield six molecules of (c)_____ plus one molecule of (d)_____ plus

(e)_____ molecules of water.

Fill-in-the-Blanks

The two major sets of reactions in photosynthesis are the (11)_____-_____ reactions and the

(12)_____-_____ reactions. (13)_____ _____ and (14)_____ are the reactants of

photosynthesis, and the end product is usually given as (15)_____. The internal membranes and chan-

nels of the chloroplast are the (16)_____ membrane system and are organized into stacks, called

(17)_____. Spaces inside the thylakoid disks and channels form a continuous compartment where

(18)_____ ions accumulate to be used to produce ATP. The semifluid interior area surrounding the

grana is known as the (19)_____ and is the area where the products of photosynthesis are produced.

LIGHT-DEPENDENT REACTIONS (5-II, pp. 74–78)

Terms

light-dependent reactions (74)_____

pigment (74)_____

chlorophyll (74)_____

carotenoid (74)_____

photosystem (76)_____

electron transport systems (76) _____

cyclic pathway of ATP formation (76)_____

noncyclic pathway of ATP formation (76) _____

photolysis (76)_____

Fill-in-the-Blanks

The light-capturing phase of photosynthesis takes place on a system of (1)_____ membranes. A(n)
(2)_____ is a packet of light energy. Thylakoid membranes contain (3)_____, which absorb photons of light. The principal pigments are the (4)_____, which reflect green wavelengths but absorb
(5)_____ and (6)_____ wavelengths. (7)_____ are pigments that absorb violet and blue
wavelengths but reflect yellow, orange, and red. A cluster of 200 to 300 of these pigment proteins is a(n)
(8)_____. When pigments absorb (9)_____ energy, an (10)_____ is transferred from a photosystem to a(n) (11)_____ molecule. (12)_____ refers to the attachment of phosphate to ADP or
other organic molecules. Due to the input of light energy, electrons flow through a transport system that
causes protons (H^+) to be simultaneously pumped into the thylakoid compartments. Electrons then end up
in (13)_____ chlorophyll at the end of this transport chain. The flow of protons from the thylakoid
compartment through (14)_____ _____ drives the enzyme machinery that phosphorylates
(15)_____ , a sequence of events known as the (16)_____ theory of ATP formation.

Complete the Table

17. In the table below, identify and state the role of each item given in the cyclic pathway of the light-
    dependent reactions.

a. Photosystem I	
b. Electrons	
c. P700	
d. Electron acceptor	
e. Electron transport system	
f. ADP	

Labeling

The diagram below illustrates noncyclic photophosphorylation. Identify each numbered part of the illustration.

18. _____

_____

19. _____

_____

20. _____

_____

21. _____

_____

22. _____

23. _____

24. _____

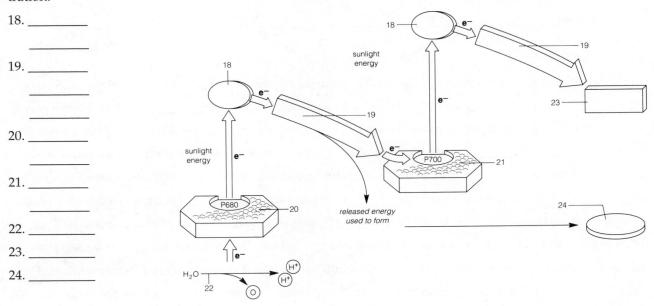

Complete the Table

With a check mark (√) indicate for each phase of the light-dependent reactions all items from the left-hand column that are applicable.

Light-Dependent Reactions:	Cyclic Pathway	Noncyclic Pathway	Photolysis Alone
Uses H_2O as a reactant	(25)	(37)	(49)
Produces H_2O as a product	(26)	(38)	(50)
Photosystem I involved (P700)	(27)	(39)	(51)
Photosystem II involved (P680)	(28)	(40)	(52)
ATP produced	(29)	(41)	(53)
NADPH produced	(30)	(42)	(54)
Uses CO_2 as a reactant	(31)	(43)	(55)
Causes H^+ to be pumped into the thylakoid compartments from the stroma	(32)	(44)	(56)
Produces O_2 as a product	(33)	(45)	(57)
Produces H^+ by breaking apart H_2O	(34)	(46)	(58)
Uses ADP and P_i as reactants	(35)	(47)	(59)
Uses $NADP^+$ as a reactant	(36)	(48)	(60)

LIGHT-INDEPENDENT REACTIONS (5-III, pp. 79–82)

Terms

light-independent reactions (79)_____

RuBP (ribulose bisphosphate) (79)_____

carbon dioxide fixation (79)_____

Calvin-Benson cycle (79)_____

PGA (phosphoglycerate) (79)_____

PGAL (phosphoglyceraldehyde) (79)_____

Label-Match

Identify each part of the illustration at the right. Complete the exercise by matching and entering the letter of the proper function description in the parentheses following each label.

1._____ _____ ( )

2._____ _____ _____ ( )

3._____ ( )

4._____ _____ ( )

5._____ ( )

6._____ ( )

7._____ _____ ( )

8._____-_____ _____ ( )

9._____ _____ ( )

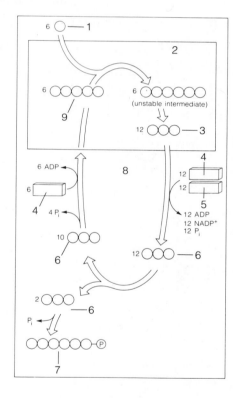

A. A three-carbon sugar, the first sugar produced; goes on to form sugar phosphate and RuBP

B. Typically used at once to form carbohydrate end products of photosynthesis

C. A five-carbon compound produced from PGALs; attaches to incoming CO_2

D. A compound that diffuses into leaves; attached to RuBP by enzymes in photosynthetic cells

E. Includes all the chemical reactions that "fix" carbon into an organic compound

F. Three-carbon compounds formed from the splitting of the six-carbon intermediate compound

G. A molecule that was produced in the noncyclic pathway; furnishes hydrogen atoms to construct sugar molecules

H. A product of the light-dependent reactions; necessary in the light-independent reactions to energize molecules in metabolic pathways

I. Includes all the chemistry that fixes CO_2, converts PGA to PGAL and PGAL to RuBP and sugar phosphates

Fill-in-the-Blanks

The light-independent reactions can proceed without sunlight as long as (10)_____ and (11)_____ are available. The reactions begin when an enzyme links (12)_____ _____ to (13)_____ _____, a five-carbon compound. The resulting six-carbon compound is highly unstable and breaks apart at once into two molecules of a three-carbon compound, (14)_____. This entire reaction sequence is called carbon dioxide (15)_____. ATP gives a phosphate group to each (16)_____. This intermediate compound takes on H^+ and electrons from NADPH to form (17)_____. It takes (18)_____ carbon dioxide molecules and six ribose biophosphate molecules to produce twelve PGAL. Most of the PGAL becomes rearranged into new (19)_____ molecules, which can be used to fix more (20)_____ _____. Two (21) _____ are joined together to form a (22)_____ _____, primed for further reactions. The Calvin-Benson cycle yields enough RuBP to replace those used in carbon dioxide (23)_____. ADP, $NADP^+$, and phosphate leftovers are sent back to the (24)_____-_____ reaction sites, where they are again converted to (25)_____ and (26)_____. (27)_____ _____ formed in the cycle serves as a building block for the plant's main carbohydrates. When RuBP attaches to oxygen instead of carbon dioxide, (28)_____ results; this is typical of (29)_____ plants in hot, dry conditions. If less PGA is available, leaves produce a reduced amount of (30)_____. C4 plants can still construct carbohydrates when the ratio of carbon dioxide to (31)_____ is unfavorable, because of the attachment of carbon dioxide to (32)_____ in certain leaf cells.

Complete the Table

With a check mark (√) indicate for each phase of the light-independent reaction all items from the left-hand column that are applicable.

Light-Independent Reactions:	CO_2 Fixation Alone	Conversion of PGA to PGAL	Regeneration of RuBP	Formation of Glucose and other organic compounds
Requires RuBP as a reactant	(33)	(44)	(55)	(66)
Requires ATP as a reactant	(34)	(45)	(56)	(67)
Produces ADP as a product	(35)	(46)	(57)	(68)
Requires NADPH as a reactant	(36)	(47)	(58)	(69)
Produces NADP$^+$ as a reactant	(37)	(48)	(59)	(70)
Produces PGA	(38)	(49)	(60)	(71)
Produces PGAL	(39)	(50)	(61)	(72)
Requires PGAL as a reactant	(40)	(51)	(62)	(73)
Produces P$_i$ as a product	(41)	(52)	(63)	(74)
Produces H_2O as a product	(42)	(53)	(64)	(75)
Requires CO_2 as a reactant	(43)	(54)	(65)	(76)

CHEMOSYNTHESIS (5-IV, p. 83)

Fill-in-the-Blanks

Organisms that obtain energy from oxidation of inorganic substances such as ammonium ions and iron or sulfur compounds, are known as (1)_____ autotrophs. Such organisms use this energy to build (2)_____ compounds. As an example, some soil bacteria use ammonia molecules as an energy source, stripping them of (3)_____ and (4)_____; this leaves (5)_____ and (6)_____ ions that are readily washed out of the soil, thus lowering its (7) _____.

Self-Quiz

___1. The electrons that are passed to NADPH
        during noncyclic photophosphorylation
        were obtained from _____.
        a. water
        b. CO_2
        c. glucose
        d. sunlight

___2. The cyclic pathway of the light-dependent
        reactions functions mainly to _____.
        a. fix CO_2
        b. make ATP
        c. produce PGAL
        d. regenerate ribulose bisphosphate

___3. Chemosynthetic autotrophs obtain energy
        by oxidizing such inorganic substances as
        _____.
        a. PGA
        b. PGAL
        c. sulfur
        d. water

___4. The ultimate electron and hydrogen accep-
        tor in the noncyclic pathway is _____.
        a. $NADP^+$
        b. ADP
        c. O_2
        d. H_2O

___5. C4 plants have an advantage in hot, dry
        conditions because _____.
        a. their leaves are covered with thicker wax
           layers than those of C3 plants
        b. their stomates open wider than those of
           C3 plants, thus cooling their surfaces
        c. CO_2 is fixed in the mesophyll cells,
           where the C3 pathway occurs, then
           delivered to the bundle sheath cells

        d. they are also capable of capturing CO_2
           by photorespiration

___6. Chlorophyll is _____.
        a. on the outer chloroplast membrane
        b. inside the mitochondria
        c. in the stroma lamellae
        d. part of the thylakoid membrane system

___7. Thylakoid disks are stacked in groups
        called _____.
        a. grana
        b. stroma
        c. lamellae
        d. cristae

___8. Plant cells produce O_2 during photosyn-
        thesis by _____.
        a. breaking apart CO_2 molecules
        b. breaking apart water molecules
        c. degradation of the stroma
        d. breaking apart sugar molecules

___9. Plants need _____ and _____ to
        carry on photosynthesis.
        a. oxygen; water
        b. oxygen; CO_2
        c. CO_2; H_2O
        d. sugar; water

___10. The two products of the light-dependent
        reactions that are required for the light-
        independent chemistry are _____ and
        _____.
        a. CO_2; H_2O
        b. O_2; NADPH; inorganic phosphate
        c. O_2; ATP
        d. ATP; NADPH

Chapter Objectives/Review Questions

This section lists general and detailed chapter objectives that can be used as review questions. You can
make maximum use of these items by writing answers on a separate sheet of paper. Fill in answers where
blanks are provided. To check for accuracy, compare your answers with information given in the chapter
or glossary.

Page   Objectives/Questions

(71)   1. List the major stages of photosynthesis and state what occurs in those sets of reactions.
(71)   2. Study the general equation for photosynthesis as shown on p. 71 of the main text until you
              can remember the reactants and products. Reproduce the equation from memory on another
              piece of paper.

(73)   3. Describe the structural details of the green leaf. Begin with the layers of a leaf cross section and complete your description with the minute structural sites within the chloroplast where the major sets of photosynthetic reactions occur. Explain how each of the reactants needed in various phases of photosynthesis arrive at the place where they are used. Explain what happens to the products of photosynthesis.

(76)   4. Describe how the pigments found on thylakoid membranes are organized into photosystems and how they relate to photon light energy.

(74)   5. Describe the role that chlorophylls and the other chloroplast pigments play in the light-dependent reactions. After consulting Figure 5.3b of the main text, state which colors of the visible spectrum are absorbed by (a) chlorophyll *a*, (b) chlorophyll *b*, and (c) carotenoids.

(75)   6. State what T. Englemann's 1882 experiment with *Cladophora* revealed.

(76)   7. Two energy-carrying molecules produced in the noncyclic pathways are _____ and _____; explain why these molecules are necessary for the light-independent reactions.

(78)   8. Explain how the chemiosmotic theory is related to thylakoid compartments and the production of ATP.

(77)   9. After evolution of the noncyclic pathway, _____ accumulated in the atmosphere and made _____ respiration possible.

(79)   10. Explain why the light-independent reactions are called by that name.

(79)   11. Describe the Calvin-Benson cycle as it is related to the four phases shown in the table in 5-III.

(82)   12. Describe the mechanism by which C4 plants thrive under hot, dry conditions; distinguish this CO_2-capturing mechanism from that of C3 plants.

Integrating and Applying Key Concepts

Suppose that humans acquired all the enzymes needed to carry out photosynthesis. Speculate about the attendant changes in human anatomy, physiology, and behavior that would be necessary for those enzymes to actually carry out photosynthetic reactions.

Critical Thinking Exercise

1. Observation: Plants release oxygen gas. Assumption: Chlorophyll is the pigment that absorbs the light energy for photosynthesis. Hypothesis: Oxygen is a product of photosynthesis. Which of the following is predicted by this hypothesis?
    a. Water is a reactant as well as a product in photosynthesis.
    b. At least some of the oxygen atoms in carbon dioxide appear in the starch formed by photosynthesis.
    c. Oxygen is released at a greater rate when plants are illuminated with red light than when they receive green light.
    d. The leaf surface contains pores through which carbon dioxide and oxygen can move.
    e. The oxygen atoms in water do not appear in the starch formed by photosynthesis.

Answers

Answers to Interactive Exercises

PHOTOSYNTHESIS: AN OVERVIEW (5-I)
1. Autotrophs; 2. carbon dioxide; 3. Photosynthetic; 4. Chemosynthetic; 5. electrons; 6. Heterotrophs; 7. animals; 8. aerobic respiration; 9. $12 H_2O + 6CO_2 \rightarrow 6O_2 + C_6H_{12}O_6 + 6H_2O$; 10. (a) Twelve (b) carbon dioxide (c) oxygen (d) glucose (e) six.; 11. light-dependent (light-independent); 12. light-independent (light-dependent); 13. Carbon dioxide; 14. water; 15. glucose; 16. thylakoid; 17. grana; 18. hydrogen; 19. stroma.

LIGHT-DEPENDENT REACTIONS (5-II)
1. thylakoid; 2. photon; 3. pigments; 4. chlorophylls; 5. red (blue); 6. blue (red); 7. Carotenoids; 8. photosystem; 9. light (photon); 10. electron; 11. acceptor; 12. Phosphorylation; 13. P700; 14. channel proteins; 15. ADP; 16. chemiosmotic; 17. a. Photosystem I: a pigment cluster dominated by P700; b. Electrons: electrons representing energy are ejected from P700 to an electron acceptor but move over the electron transport system, where some of the energy is used to produce ATP; c. P700: a special chlorophyll molecule that absorbs wavelengths of 700 nanometers and then ejects electrons; d. Electron acceptor: a molecule that accepts electrons ejected from chlorophyll P700 and then passes electrons down the electron transport system; e. Electron transport system: electrons flow through this system, which is composed of a series of molecules bound in the thylakoid membrane that drive the phosphorylation of ADP to produce ATP; f. ADP: ADP undergoes phosphorylation in the cyclic pathway to become ATP; 18. electron acceptor; 19. electron transport system; 20. photosystem II; 21. photosystem I; 22. photolysis; 23. NADPH; 24. ATP; (25–60). The following numbers should have a check mark (√): 27, 29, 32, 35, 37, 39, 40, 41, 42, 44, 47, 48, 49, 52, 57, and 58. All others should be blank.

LIGHT-INDEPENDENT REACTIONS (5-III)
1. carbon dioxide (D); 2. carbon dioxide fixation (E); 3. phosphoglycerate (F); 4. adenosine triphosphate (H); 5. NADPH (G); 6. phosphoglyceraldehyde (A); 7. sugar phosphates (B); 8. Calvin-Benson cycle (I); 9. ribulose bisphosphate (C); 10. ATP (NADPH); 11. NADPH (ATP); 12. carbon dioxide; 13. ribulose bisphosphate; 14. PGA; 15. fixation; 16. PGA; 17. PGAL; 18. six; 19. RuBP;

20. carbon dioxide; 21. PGALs; 22. sugar phosphate; 23. fixation; 24. light-dependent; 25. ATP (NADPH); 26. NADPH (ATP); 27. Sugar phosphate; 28. photorespiration; 29. C3 30. food(glucose); 31. oxygen; 32. oxaloacetate; (33–76). The following numbers should have a check mark (√): 33, 38, 43, 45, 46, 47, 48, 50, 52, 56, 57, 62, 63, 64, 73, and 74. All others should be blank.

CHEMOSYNTHESIS (5-IV)
1. chemosynthetic; 2. organic (food); 3. protons (electrons); 4. electrons (protons); 5. nitrate (nitrite); 6. nitrite (nitrate); 7. fertility.

Answers to Self-Quiz

1. a; 2. b; 3. c; 4. a; 5. c; 6. d; 7. a; 8. b; 9. c; 10. d.

Critical Thinking Analysis

a. The hypothesis does not predict the source of the oxygen, only that it will be released when photosynthesis occurs. Furthermore, it does not predict that oxygen atoms will be found in products of photosynthesis. If anything, the hypothesis makes it less likely that oxygen will be released in any form other than oxygen gas.

b. Once again, the hypothesis does not predict the source or other destinations of the oxygen, only that oxygen gas will be released in coordination with the rate of photosynthesis.

c. Chlorophyll absorbs red light but not green light. If we assume that chlorophyll absorbs the light energy that drives photosynthesis, we expect that photosynthesis will go much faster in red light than in green light. The hypothesis then predicts that oxygen evolution will be faster when photosynthesis is faster, that is, in red light.

d. Because plants are observed to release oxygen and absorb carbon dioxide, we expect that there will be a pathway for those gases to leave and enter the plant. This would be true whatever hypothesis was made about how the plant produces the oxygen.

e. The hypothesis can predict only that oxygen atoms from some substrate of photosynthesis will be released as gas when a plant is illuminated. It cannot predict which molecule will be the source or where else oxygen atoms might go.

6

ENERGY-RELEASING PATHWAYS

ATP-PRODUCING PATHWAYS

AEROBIC RESPIRATION
Overview of the Reactions
Glycolysis
Krebs Cycle
Electron Transport Phosphorylation

ANAEROBIC ROUTES
Alcoholic Fermentation
Lactate Fermentation
Anaerobic Electron Transport

ALTERNATIVE ENERGY SOURCES IN THE HUMAN BODY
Commentary: Perspective on Life

Interactive Exercises

ATP-PRODUCING PATHWAYS (6-I, p. 86)

Terms

The page-referenced terms are important; they were in boldface type in the chapter. Refer to the instructions given in Chapter 1, p. 1 in this workbook.

ATP (86) _____

aerobic respiration (86) _____

fermentation (86) _____

anaerobic electron transport (86)_____

glycolysis (86) _____

_____

Short Answer (p. 86)

1. Although various organisms utilize different energy sources, what is the usual form of chemical energy that will drive metabolic reactions?

   _____

2. Describe the function of oxygen in the main degradative pathway, aerobic respiration.

   _____

   _____

3. List the most common anaerobic pathways and describe the conditions in which they function.

   _____

Fill-in-the-Blanks

Virtually all forms of life depend on a molecule known as (4)_____ as their primary energy carrier. Plants produce adenosine triphosphate during (5)_____, but plants and all other organisms can also produce ATP through chemical pathways that degrade (take apart) food molecules. The main degradative pathway requires free oxygen and is called (6) _____ _____. Other degradative pathways are (7)_____, in that something other than oxygen serves as the final electron acceptor in energy-releasing reactions. (8)_____ and anaerobic (9)_____ _____ are the most common anaerobic pathways.

AEROBIC RESPIRATION (6-II, pp. 86–92)

Terms

Krebs cycle (86) _____

_____

electron transport phosphorylation (86) _____

_____

pyruvate (88) _____

PGAL (88) _____

substrate-level phosphorylation (88) _____

_____

NAD$^+$ (88) _____

mitochondrion (90) _____

FAD (91) _____

Completion

1. Complete the equation below, which summarizes the degradative pathway known as aerobic respiration:

$$\underline{\hspace{1cm}} + \underline{\hspace{1cm}} O_2 \rightarrow 6 \underline{\hspace{1cm}} + 6 \underline{\hspace{1cm}}$$

2. Supply the appropriate information to state the equation (above) for aerobic respiration in words:

   One molecule of glucose plus six molecules of _____ (in the presence of appropriate enzymes) yield

   _____ molecules of carbon dioxide plus _____ molecules of water.

Fill-in-the-Blanks

There are three stages of aerobic respiration. In the first stage, (3) _____, glucose is partially degraded to (4) _____. By the end of the second stage, which includes the (5) _____ cycle, glucose has been completely degraded to carbon dioxide and (6) _____. Neither of the first two stages produces much (7) _____. During both stages, protons and (8) _____ are stripped from intermediate compounds and delivered to a (9) _____ system. That system is used in the third stage of reactions, electron transport (10) _____; passage of electrons along the transport system drives the enzymatic "machinery" that phosphorylates ADP to produce a high yield of (11) _____. (12) _____ accepts "spent" electrons from the transport system and keeps the pathway clear for repeated ATP production.

Labeling

In exercises 13–17, identify the structure or location; in exercises 18–21, identify the chemical substance involved. In exercise 22, name the metabolic pathway.

13._____ _____ of mitochondrion   18._____

14._____ _____ of mitochondrion   19._____

15._____ _____ of mitochondrion   20._____

16._____ _____ of mitochondrion   21._____

17._____   22._____ _____ _____

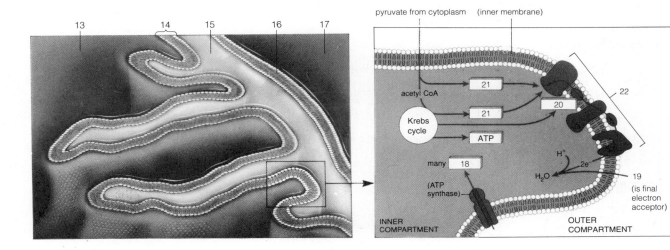

Fill-in-the-Blanks

If sufficient oxygen is present, the end product of glycolysis enters a preparatory step, (23) _____ _____ formation. This step converts pyruvate into acetyl CoA, the molecule that enters the (24) _____ cycle, which is followed by (25) _____ _____ phosphorylation. During these three processes, (26) _____ (number) additional (27) _____ molecules are generated. In the preparatory conversions prior to the Krebs cycle and within the Krebs cycle, the food molecule fragments are further broken down into (28) _____ _____. During these reactions, hydrogen atoms (with their (29) _____) are stripped from the fragments and transferred to the energy carriers (30) _____ and (31) _____. NADH delivers its electrons to the highest possible point of entry into a transport system; from each NADH enough H^+ is pumped to produce (32) _____ (number) ATP molecules. $FADH_2$ delivers its electrons at a lower point of entry into the transport system; fewer H^+ are pumped, and (33) _____ (number) ATPs are produced. The electrons are then sent down highly organized (34) _____ systems located in the inner membrane of the mitochondrion; hydrogen ions are pumped into the outer mitochondrial compartment. According to (35) _____ theory, the hydrogen ions accumulate and then follow a gradient to flow through channel proteins, called ATP (36) _____, that lead into the inner compartment. The energy of the hydrogen ion flow across the membrane is used to phosphorylate ADP to produce (37) _____. Electrons leaving the electron transport system combine with hydrogen ions and (38) _____ to form water. These reactions occur only in (39) _____. From glycolysis (in the cytoplasm) to the final reactions occurring in the mitochondria, the aerobic pathway commonly yields (40) _____ (number) ATP or (41) _____ (number) ATP for every glucose molecule degraded.

ANAEROBIC ROUTES (6-III, pp. 93–94)

Terms

alcoholic fermentation (93) _____

lactate fermentation (94) _____

Fill-in-the-Blanks

(1) _____ organisms can synthesize and stockpile energy-rich carbohydrates and other food molecules from inorganic raw materials. (2) _____ is partially dismantled by the glycolytic pathway, during which process some of its stored energy remains in two (3) _____ molecules. Some of the energy of glucose is released during the breakdown reactions and used in forming the energy carriers (4) _____ and (5) _____. These reactions take place in the cytoplasm. If (6) _____ is not present in sufficient amounts, the end product of glycolysis enters (7) _____ pathways; in some bacteria and muscle cells, pyruvate is converted into such products as (8) _____, or in yeast cells it is converted into (9) _____ and (10) _____ _____.

Anaerobic electron transport is an energy-releasing pathway occurring among the (11) _____.
Sulfate-reducing bacteria produce (12) _____ by stripping electrons from a variety of compounds and
sending them through membrane transport systems. The inorganic compound (13) _____ ($SO_4^=$)
serves as the final electron acceptor and is converted into sulfide (H_2S). Other kinds of bacteria produce
ATP by stripping electrons from nitrate (NO_3^-), leaving (14)_____ (NO_2^-) as the end product. These
bacteria are important in the global cycling of (15)_____.

Complete the Table

Include a check mark (√) in each box that correctly links an occurrence (left-hand column) with a process
(or processes).

	Glycolysis	Lactate Fermentation	Alcoholic Fermentation	Anaerobic Electron Transport
6-C → 3C / 3C	(16)	(29)	(42)	(55)
NADH → NAD^+	(17)	(30)	(43)	(56)
NAD^+ → NADH	(18)	(31)	(44)	(57)
$SO_4^=$ → H_2S	(19)	(32)	(45)	(58)
NO_3^- → NO_2^-	(20)	(33)	(46)	(59)
CO_2 is a waste product	(21)	(34)	(47)	(60)
3-C → 3-C	(22)	(35)	(48)	(61)
3-C → 2-C	(23)	(36)	(49)	(62)
ATP is used as a reactant	(24)	(37)	(50)	(63)
ATP is produced	(25)	(38)	(51)	(64)
pyruvate → ethanol / CO_2	(26)	(39)	(52)	(65)
Occurs in animal cells	(27)	(40)	(53)	(66)
Occurs in yeast cells	(28)	(41)	(54)	(67)

ALTERNATIVE ENERGY SOURCES IN THE HUMAN BODY (6-IV, pp. 94–95)

True/False

If the statement is true, write a T in the blank. If the statement is false, explain why.

_____ 1. Glucose is the only carbon-containing molecule that can be fed into the glycolytic pathway.

_____ 2. Simple sugars, fatty acids, and glycerol that remain after a cell's biosynthetic and storage needs have been met are generally sent to the cell's respiratory pathways for energy extraction.

_____ 3. Carbon dioxide and water, the products of aerobic respiration, generally get into the blood and are carried to gills or lungs, kidneys, and skin, where they are expelled from the animal's body.

_____ 4. Energy is recycled along with materials.

_____ 5. The first forms of life on Earth were most probably photosynthetic eukaryotes.

Labeling

Identify the process or substance indicated in the illustration below.

6._____ _____

7._____

8._____

9._____ _____

10._____ _____

11._____ -_____

12._____

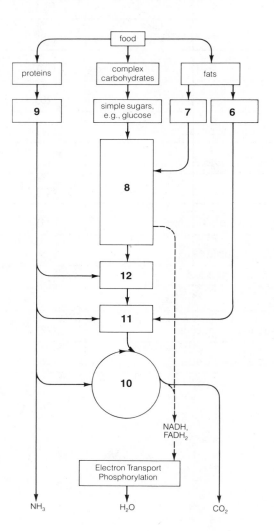

66   Chapter Six

Complete the Table

Across the top of each column are the principal phases of degradative pathways into which food molecules (in various stages of breakdown) enter or in which specific events occur. Put a check mark (√) in each box that indicates the phase into which a specific food molecule is fed, or in which a specific event occurs. For example, if simple sugars can enter the glycolytic pathway, put a check in the top left-hand box; if not, let the box remain blank.

	Glycolysis (includes pyruvate)	Acetyl CoA Formation	Krebs Cycle	Electron Transport Phosphorylation	Fermentation	
					Alcoholic	Lactate
Complex Carbo-hydrates → Simple Sugars, which enter	(13)	(28)	(43)	(58)	(73)	(88)
Fats → Fatty Acids, which enter	(14)	(29)	(44)	(59)	(74)	(89)
Glycerol, which enters	(15)	(30)	(45)	(60)	(75)	(90)
Proteins → Amino Acids, which enter	(16)	(31)	(46)	(61)	(76)	(91)
Intermediate energy carriers are produced (NADH)	(17)	(32)	(47)	(62)	(77)	(92)
(FADH$_2$)	(18)	(33)	(48)	(63)	(78)	(93)
ATPs produced directly as a result of this process alone	(19)	(34)	(49)	(64)	(79)	(94)
NAD$^+$ produced	(20)	(35)	(50)	(65)	(80)	(95)
FAD produced	(21)	(36)	(51)	(66)	(81)	(96)
ADP produced	(22)	(37)	(52)	(67)	(82)	(97)
Unbound phosphate (P$_i$) required	(23)	(38)	(53)	(68)	(83)	(98)
CO$_2$ produced (waste product)	(24)	(39)	(54)	(69)	(84)	(99)
H$_2$O produced (waste product)	(25)	(40)	(55)	(70)	(85)	(100)
Atoms from O$_2$ react here	(26)	(41)	(56)	(71)	(86)	(101)
NADH required to drive this process	(27)	(42)	(57)	(72)	(87)	(102)

Self-Quiz

___1. Glycolysis would quickly halt if the process ran out of _____ , which serves as the hydrogen and electron acceptor.
a. $NADP^+$
b. ADP
c. NAD^+
d. H_2O

___2. The ultimate electron acceptor in aerobic respiration is _____.
a. NADH
b. carbon dioxide (CO_2)
c. oxygen ($\frac{1}{2}O_2$)
d. ATP

___3. When glucose is used as an energy source, the largest amount of ATP is generated by the _____ portion of the entire respiratory process.
a. glycolytic pathway
b. acetyl-CoA formation
c. Krebs cycle
d. electron transport phosphorylation

___4. The process by which about 10 percent of the energy stored in a sugar molecule is released as it is converted into two small organic-acid molecules is _____.
a. photolysis
b. glycolysis
c. fermentation
d. the dark reactions

___5. During which of the following phases of aerobic respiration is ATP produced directly by substrate-level phosphorylation?
a. glucose formation
b. ethyl-alcohol production
c. acetyl-CoA formation
d. glycolysis

___6. What is the name of the process by which reduced NADH transfers electrons along a chain of acceptors to oxygen so as to form water and in which the energy released along the way is used to generate ATP?
a. glycolysis
b. acetyl-CoA formation
c. the Krebs cycle
d. electron transport phosphorylation

___7. Pyruvic acid can be regarded as the end product of _____.
a. glycolysis
b. acetyl-CoA formation
c. fermentation
d. the Krebs cycle

___8. Which of the following is not ordinarily capable of being reduced at any time?
a. NAD
b. FAD
c. oxygen, O_2
d. water

___9. ATP production by chemiosmosis involves _____.
a. H^+ concentration and electric gradients across a membrane
b. ATP synthases
c. formation of ATP in the inner mitochondrial compartment
d. all of the above

___10. During the fermentation pathways, a net yield of two ATP is produced from _____; the NAD^+ necessary for _____ is regenerated during the reactions.
a. the Krebs cycle; glycolysis
b. glycolysis; electron transport phosphorylation
c. the Krebs cycle; electron transport phosphorylation
d. glycolysis; glycolysis

Matching

Match the following components of respiration to the list of words below. Some components may have more than one answer.

11.___ lactic acid, lactate

12.___ $NAD^+ \rightarrow NADH$

13.___ carbon dioxide is a product

14.___ $NADH \rightarrow NAD^+$

15.___ pyruvic acid, pyruvate, used as a reactant

16.___ ATP produced by substrate-level phosphorylation

17.___ glucose

18.___ acetyl-CoA is both a reactant and a product

19.___ oxygen

20.___ water is a product

A. Glycolysis
B. Preparatory conversions prior to the Krebs cycle
C. Fermentation
D. Krebs cycle
E. Electron transport phosphorylation

Chapter Objectives/Review Questions

This section lists general and detailed chapter objectives that can be used as review questions. You can make maximum use of these items by writing answers on a separate sheet of paper. Fill in answers where blanks are provided. To check for accuracy, compare your answers with information given in the chapter or glossary.

Page	Objectives/Questions
(86)	1. No matter what the source of energy might be, organisms must convert it to _____, a form of chemical energy that can drive metabolic reactions.
(86)	2. Give the overall equation for the aerobic respiratory route; indicate where energy occurs in the equation.
(86)	3. In the first of the three stages of aerobic respiration, _____ is partially degraded to pyruvate.
(86)	4. By the end of the second stage of aerobic respiration, which includes the _____ cycle, _____ has been completely degraded to carbon dioxide and water.
(86)	5. Explain, in general terms, the role of oxygen in aerobic respiration.
(86)	6. Glycolysis occurs in the _____ of the cell.
(88)	7. Explain the purpose served by molecules of ATP reacting first with glucose and then with fructose-6-phosphate in the early part of glycolysis (see Figure 6.3 in the text).
(88)	8. Four ATP molecules are produced by _____-_____ phosphorylation for every two used during glycolysis. Consult Figure 6.3 in the text.
(88)	9. Glycolysis produces _____ (number) NADH, _____ (number) ATP (net) and _____ (number) pyruvate molecules for each glucose molecule entering the reactions.
(90)	10. Consult Figures 6.3 and 6.5 in the text. State the events that happen during the preparatory steps and explain how the process of acetyl-CoA formation relates glycolysis to the Krebs cycle.
(90)	11. What happens to the CO_2 produced during acetyl-CoA formation and the Krebs cycle?
(91)	12. Consult Figure 6.5 in the text and predict what will happen to the NADH produced during acetyl-CoA formation and the Krebs cycle.
(92)	13. Explain how chemiosmotic theory operates in the mitochondrion to account for the production of ATP molecules.

(92)   14. Briefly describe the process of electron transport phosphorylation by stating what reactants are needed and what the products are. State how many ATP molecules are produced through operation of the transport system.

(88–92) 15. Be able to account for the total net yield of thirty-six ATP molecules produced through aerobic respiration; that is, state how many ATPs are produced in glycolysis, the Krebs cycle, and electron transport phosphorylation.

(93)   16. List some places where there is very little oxygen present and where anaerobic organisms might be found.

(93–94) 17. Describe what happens to pyruvate in anaerobic organisms. Then explain the necessity for pyruvate to be converted to a fermentative product.

(93–94) 18. State which factors determine whether the pyruvate (pyruvic acid) produced at the end of glycolysis will enter into the alcoholic fermentation pathway, the lactate fermentation pathway, or the acetyl-CoA formation pathway.

(96)   19. After reading "Perspective on Life" in the text, outline the supposed evolutionary sequence of energy-extraction processes.

(96)   20. Closely scrutinize the diagram of the carbon cycle in the *Commentary*; be able to reproduce the cycle from memory.

Integrating and Applying Key Concepts

How is the "oxygen debt" experienced by runners and sprinters related to aerobic and anaerobic respiration in humans?

Critical Thinking Exercise

1. Suppose you inoculate some yeast cells into a nutrient medium with glucose as an energy source. You seal the culture vessel, and at intervals you measure the growth of the cells and express the results as relative cell number. Your data are as shown:

Time (hrs)	Relative Cell Number
0	0.085
1	0.104
2	0.151
3	0.213
4	0.322
6	0.354
8	0.381
10	0.408

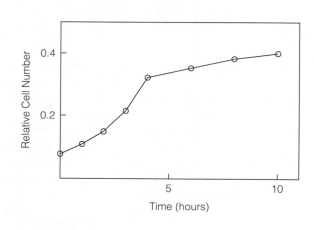

Make a graph of the data. What can you conclude about conditions in the culture vessel?

Answers

Answers to Interactive Exercises

ATP-PRODUCING PATHWAYS (6-I)
1. Adenosine triphosphate (ATP).
2. Oxygen withdraws electrons from the electron transport system and joins with H^+ to form water.
3. Glycolysis followed by some end reactions (called fermentation) and anaerobic electron transport. Some organisms (including humans) use fermentation pathways when oxygen supplies are low; many microbes rely exclusively on anaerobic pathways. 4. ATP;
5. photosynthesis; 6. aerobic respiration; 7. anaerobic;
8. Fermentation; 9. electron transport.

AEROBIC RESPIRATION (6-II)
1. $C_6H_{12}O_6 + 6O_2 \rightarrow 6CO_2 + 6H_2O$
2. One molecule of glucose plus six molecules of **oxygen** (in the presence of appropriate enzymes) yield **six** molecules of carbon dioxide plus **six** molecules of water.
3. glycolysis; 4. pyruvate; 5. Krebs; 6. water; 7. ATP;
8. electrons; 9. transport; 10. phosphorylation; 11. ATP;
12. Oxygen; 13. inner compartment; 14. inner membrane;
15. outer compartment; 16. outer membrane; 17. cytoplasm; 18. ATP; 19. oxygen (O_2); 20. $FADH_2$; 21. NADH;
22. electron transport system; 23. acetyl CoA; 24. Krebs;
25. electron transport; 26. thirty-four; 27. ATP; 28. carbon dioxide; 29. electrons; 30. NAD^+ (FAD); 31. FAD (NAD^+);
32. three; 33. two; 34. transport; 35. chemiosmotic;
36. synthases; 37. ATP; 38. oxygen; 39. mitochondria;
40. thirty-six (thirty-eight); 41. thirty-eight (thirty-six).

ANAEROBIC ROUTES (6-III)
1. Autotrophic; 2. Glucose; 3. pyruvate; 4. ATP (NADH);
5. NADH (ATP); 6. oxygen (O_2); 7. fermentation; 8. lactate; 9. ethanol; 10. carbon dioxide; 11. bacteria; 12. ATP;
13. sulfate; 14. nitrite; 15. nitrogen; (16–67). With a check (√): 16, 18, 24, 25, 27, 28, 30, 35, 40, 43, 47, 49, 52, 54, 56, 58, 59, 64; all others lack a check.

ALTERNATIVE ENERGY SOURCES IN THE HUMAN BODY (6-IV)
1. F. Figure 6.1 shows how any complex carbohydrate or fat can be broken down and at least part of those molecules can be fed into the glycolytic pathway; 2. T; 3. T;
4. F Page 96 (last paragraph) tells us that energy flows through time in one direction—from organized to less organized forms; thus energy cannot be completely recycled; 5. F Page 96 tells us that Earth's first organisms were anaerobic fermenters; 6. fatty acids; 7. glycerol;
8. glycolysis; 9. amino acids; 10. Krebs cycle; 11. acetyl-CoA; 12. pyruvate; (13–102). With a check mark (√):
13, 15, 16, 17, 19, 22, 23, 25, 29, 31, 32, 39, 46, 47, 48, 49, 53, 54, 65, 66, 70, 71, 72, 80, 84, 87, 95, 102; all others lack a check.

Answers to Self-Quiz

1. c; 2. c; 3. d; 4. b; 5. d; 6. d; 7. a; 8. d; 9. d; 10. d; 11. C;
18. B, D; 19. E; 20. A, E.
12. A, B, D; 13. B, (C), D; 14. C, E; 15. B, C; 16. A, D; 17. A;

Critical Thinking Analysis

The graph shows that the yeast grew much faster for the first four hours than from the fourth to the tenth hour. It is possible that some toxic metabolic product accumulated in the flask and inhibited growth, but in that case a steady decrease in growth would be expected, not a sudden change. Another hypothesis is that some essential nutrient was exhausted from the medium, but that would lead to the prediction of no growth at all after the nutrient was finished. Now, remember that yeast can use two ways to metabolize glucose, aerobic and anaerobic, and that aerobic metabolism produces many more ATP molecules per glucose molecule than does anaerobic metabolism. If we assume that growth rate depends on the supply of ATP and that only glucose metabolism produces ATP in this culture, we can make the hypothesis that the culture ran out of oxygen after four hours of aerobic metabolism and rapid growth. After that time, the cells were forced to ferment the glucose and grew more slowly on the limited ATP production.

7

CELL DIVISION AND MITOSIS

Interactive Exercises

DIVIDING CELLS: THE BRIDGE BETWEEN GENERATIONS (7-I, pp. 101–102)

Terms

The page-referenced terms are important; they were in boldface type in the chapter. Refer to the instructions given in Chapter 1, p. 1 in this workbook.

reproduction (101):_____

mitosis (101):_____

meiosis (101):_____

somatic cells (101): Cytoplasmic division of body cell.

germ cells (101): lancage set aside for sexual production

chromosome (102): DNA molecule, along with its attacthed protein

sister chromatids (102): DNA molecule will stay together until late in mitosis

centromere (102): ✔

diploid cell (102): Any cell having two of each type of chromosome

chromosome number (102): tells you how many of each type of chro. present in a cell

Matching

Choose the one most appropriate answer for each.

1. _G_ centromere
2. ___ prokaryotic fission
3. ___ diploid cell
4. ___ chromosome
5. ___ germ cells
6. ___ reproduction
7. ___ somatic cells
8. ___ sister chromatids
9. ___ mitosis and meiosis
10. ___ chromosome number

A. Cell lineage set aside for meiosis
B. Producing a new generation of cells or multicelled individuals
C. Any cell having two of each type of chromosome
D. How many of each type of chromosome is present in a cell
E. Each DNA molecule with attached proteins
F. Sort out and package DNA molecules into daughter cell nuclei
G. Constricted chromosome region with attachment sites for micro-tubules
H. Bacterial cell division
I. The two attached DNA molecules of a duplicated chromosome
J. Body cells that reproduce by mitosis and cytoplasmic division

MITOSIS AND THE CELL CYCLE (7-II, p. 103)

Terms

cell cycle (103):_____

interphase (103):_____

Labeling

Identify the stage in the cell cycle indicated by each number.

1. _____
2. _____
3. _____
4. _____
5. _____
6. _____
7. _____
8. _____
9. _____
10. _____

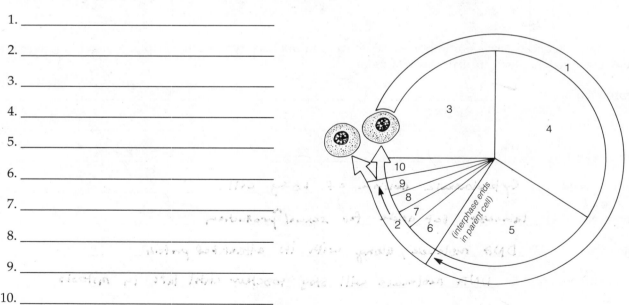

Matching

Link each time span identified below with the most appropriate number in the preceding labeling section.

11. ___ Period after replication of DNA during which the cell prepares for division by further growth and protein synthesis

12. ___ Period of nuclear division

13. ___ DNA replication occurs now

14. ___ Period of cell growth before DNA replication

15. ___ Period when chromosomes are not visible

16. ___ Period of cytoplasmic division

17. ___ Period which includes G_1, S, G_2

STAGES OF MITOSIS (7-III, pp. 104–106)

Terms

prophase (104):_____

metaphase (104):_____

anaphase (104):_____

telophase (104):_____

spindle apparatus (104):_____

Label-Match

Identify each of the mitotic stages shown below by entering the correct stage in the blank beneath the sketch. Select from late prophase, transition to metaphase, interphase—parent cell, metaphase, early prophase, telophase, interphase—daughter cells, and anaphase. Complete the exercise by matching and entering the letter of the correct phase description in the parentheses following each label.

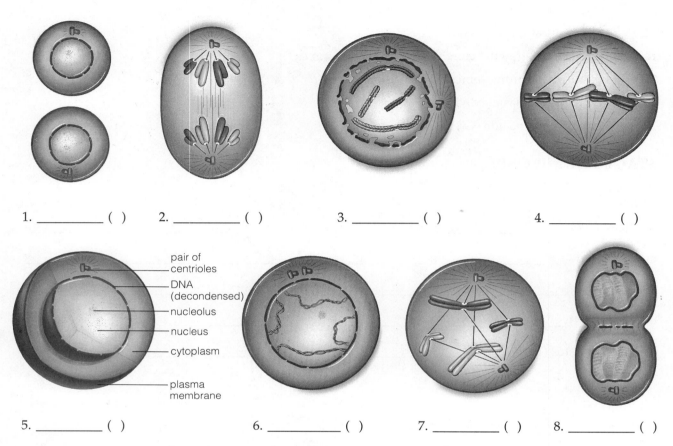

1. _____ ( )   2. _____ ( )   3. _____ ( )   4. _____ ( )

pair of centrioles
DNA (decondensed)
nucleolus
nucleus
cytoplasm
plasma membrane

5. _____ ( )   6. _____ ( )   7. _____ ( )   8. _____ ( )

A. Attachment between two sister chromatids of each chromosome break; the two are now chromosomes in their own right.
B. Microtubules that form the spindle apparatus penetrate the nuclear region; microtubules become attached to the sister chromatids of each chromosome.
C. The DNA and its associated proteins start to condense into the threadlike chromosome form.
D. Chromosomes are now fully condensed and lined up at the equator of the spindle.
E. DNA is duplicated and the cell prepares for division.
F. Two daughter cells have formed, each diploid.
G. Chromosomes continue to condense. New microtubules are assembled, and they move one of two centrioles toward the opposite end of the cell. The nuclear envelope begins to break up.
H. New patches of membrane join to form nuclear envelopes around the decondensing chromosomes. Cytokinesis begins before this stage ends.

DIVISION OF THE CYTOPLASM (7-IV, pp. 106–108)

Terms

cytoplasmic division (106):_____

cleavage furrow (106):_____

cell plate (106):_____

Choice

For questions 1–5, choose from the following.

a. plant cells   b. animal cells

___ 1.   Formation of a cell plate

___ 2.   Microfilaments pull the plasma membrane inward and cut the cell in two

___ 3.   Cellulose deposits form a crosswall between the two daughter cells

___ 4.   Possess rigid walls that cannot be pinched in two

___ 5.   Cleavage furrow

Self-Quiz

___ 1.   The replication of DNA occurs _____.
a. between the growth phases of interphase
b. immediately before prophase of mitosis
c. during prophase of mitosis
d. during prophase of meiosis

___ 2.   In the cell life cycle of a particular cell, _____.
a. mitosis occurs immediately prior to S
b. mitosis occurs immediately prior to G_1
c. G_2 precedes S
d. G_1 precedes S
e. mitosis and S precede G_1

___ 3.   In eukaryotic cells, which of the following can occur during mitosis?
a. two mitotic divisions to maintain the parental chromosome number
b. the replication of DNA
c. a long growth period
d. the disappearance of the nuclear envelope and nucleolus

___ 4.   *Diploid* refers to _____.
a. having two chromosomes of each type in somatic cells
b. twice the parental chromosome number
c. half the parental chromosome number
d. having one chromosome of each type in somatic cells

___ 5.   Somatic cells are _____ cells; germ cells are _____ cells.
a. meiotic; body
b. body; body
c. meiotic; meiotic
d. body; meiotic

___ 6.   If a parent cell has sixteen chromosomes and undergoes mitosis, each resulting cell will have _____ chromosomes.
a. sixty-four
b. thirty-two
c. sixteen
d. eight
e. four

___ 7. The correct order of the stages of mitosis is
_____.
   a. prophase, metaphase, telophase, anaphase
   b. telophase, anaphase, metaphase, prophase
   c. telophase, prophase, metaphase, anaphase
   d. anaphase, prophase, telophase, metaphase
   e. prophase, metaphase, anaphase, telophase

___ 8. "The nuclear envelope breaks up completely into vesicles. Microtubules are now free to interact with the chromosomes." These sentences describe the _____ of mitosis.
   a. prophase
   b. metaphase
   c. transition to metaphase
   d. anaphase
   e. telophase

___ 9. During _____, sister chromatids of each chromosome are separated from each other, and those former partners, now chromosomes, are moved toward opposite poles.
   a. prophase
   b. metaphase
   c. anaphase
   d. telophase

___ 10. In the process of cytokinesis, cleavage furrows are associated with _____ cell division, and cell plate formation is associated with _____ cell division.
   a. animal, animal
   b. plant, animal
   c. plant, plant
   d. animal, plant

Chapter Objectives/Review Questions

This section lists general and detailed chapter objectives that can be used as review questions. You can make maximum use of these items by writing answers on a separate sheet of paper. Fill in answers where blanks are provided. To check for accuracy, compare your answers with information given in the chapter or glossary.

Page   *Objectives/Questions*

(101)   1. Name the substance that contains the instructions for making proteins.
(101)   2. Mitosis and meiosis refer to the division of the cell's _____.
(101)   3. Distinguish between somatic cells and germ cells as to their location and function.
(101)   4. Name the prokaryotic cell division mechanism.
(102)   5. The eukaryotic chromosome is composed of _____ and _____.
(102)   6. The two attached threads of a duplicated chromosome are known as sister _____.
(102)   7. Describe the function of the portion of a chromosome known as a centromere.
(103)   8. Be able to list, in order, the various activities occurring in the eukaryotic cell life cycle.
(103)   9. Interphase of the cell cycle consists of G_1, _____, and G_2.
(103)   10. S is the time in the cell cycle when _____ replication occurs.
(104–106)   11. Be able to describe the cellular events occurring in the prophase, metaphase, anaphase, and telophase of mitosis.
(105)   12. The "_____" is the time when the nuclear envelope breaks up into vesicles prior to metaphase.
(106)   13. Compare and contrast cytokinesis as it occurs in plant mitosis and animal mitosis; use the following terms: cleavage furrow and cell plate formation.

Integrating and Applying Key Concepts

Runaway cell division is characteristic of cancer. Imagine the various points of the mitotic process that might be sabotaged in cancerous cells in order to halt their multiplication. Then try to imagine how one might discriminate between cancerous and normal cells in order to guide those methods of sabotage most effective in combating cancer.

Critical Thinking Exercise

1. DNA content was estimated in several individual cells from a culture of diploid cells.

Cell	DNA Content	Cell	DNA Content
a	7.4	e	10.9
b	5.5	f	8.7
c	9.2	g	5.4
d	11.1	h	5.7

In which part of interphase would you say each cell was measured? How much DNA would you expect to find in a cell in anaphase of mitosis?

Answers

Answers to Interactive Exercises

DIVIDING CELLS: THE BRIDGE BETWEEN GENERATIONS (7-I)
1. G; 2. H; 3. C; 4. E; 5. A; 6. B; 7. J; 8. I; 9. F; 10. D.

MITOSIS AND THE CELL CYCLE (7-II)
1. interphase; 2. mitosis; 3. G_1; 4. S; 5. G_2; 6. prophase; 7. metaphase; 8. anaphase; 9. telophase; 10. cytokinesis; 11. 5; 12. 2; 13. 4; 14. 3; 15. 1; 16. 10; 17. 1.

STAGES OF MITOSIS (7-III)
1. interphase—daughter cells (F); 2. anaphase (A); 3. late prophase (G); 4. metaphase (D); 5. interphase—parent cell (E); 6. early prophase (C); 7. transition to metaphase (B); 8. telophase (H).

DIVISION OF THE CYTOPLASM (7-IV)
1. a; 2. b; 3. a; 4. a; 5. b.

Answers to Self-Quiz

1. a; 2. d; 3. d; 4. a; 5. d; 6. c; 7. e; 8. c; 9. c; 10. d.

Critical Thinking Analysis

These cultured cells are growing by mitosis, which distributes half the parent cell's DNA to each daughter cell. Each daughter then replicates its DNA and divides again. Cells in G_2 have completed DNA replication and are about to enter mitosis. After mitosis, they pass through G_1 with half as much DNA as they had in G_2. Cells in S phase have intermediate amounts of DNA. Cells d and e have approximately twice as much DNA as cells b and g, so we can conclude that d and e are in G_2 and b and g are in G_1. Cells a, c, and f have intermediate values; they are probably in S phase and have replicated only part of their DNA. Cell h has a little more DNA than cells b and g. It might have begun replication and should be assigned to S phase. On the other hand, the accuracy of all measuring methods is limited. The difference between 5.7 units and 5.5 units might be less than the measuring method can accurately resolve, in which case we should regard the two cells as having the same amount of DNA and place both in G_1.

A cell in anaphase should have about 11 units of DNA. It has completed DNA replication, separated the sister chromatids, and distributed them into two groups, but all the chromatin is still in a single cell.

8

MEIOSIS

Interactive Exercises

ON SEXUAL AND ASEXUAL REPRODUCTION (8-I, p. 111)

Terms

The page-referenced terms are important; they were in boldface type in the chapter. Refer to the instructions given in Chapter 1, p. 1 of this workbook.

asexual reproduction (111): _____

genes (111): _____

sexual reproduction (111): _____

allele (111): _____

Choice

For questions 1–10, choose from the following.

a. asexual reproduction   b. sexual reproduction

___ 1. Brings together new combinations of alleles in offspring

___ 2. Involves only one parent

___ 3. Both parents pass on one of each gene to their offspring

___ 4. Commonly involves two parents

___ 5. The production of "clones"

___ 6. Involves meiosis and fertilization

___ 7. Offspring are genetically identical copies of the parent

___ 8. Produces the variation in traits that forms the basis of evolutionary change

___ 9. The instructions in every pair of genes is identical in all individuals of a species

___ 10. New combinations of alleles lead to variations in physical and behavioral traits

OVERVIEW OF MEIOSIS (8-II, p. 112)

Terms

meiosis (112):_____

gametes (112): _____

diploid (112): _____

homologous chromosomes (112):_____

haploid (112):_____

True/False

If the statement is true, write a T in the blank. If the statement is false, make it correct by changing the underlined word(s) and writing the correct word(s) in the answer blank.

_____ 1. Mitosis is the first step leading to gamete formation.

_____ 2. Sperms and eggs are sex cells known as gametes.

_____ 3. Germ cells are haploid as are the rest of the body's cells.

_____ 4. Diploid cells possess pairs of homologous chromosomes.

_____ 5. Meiosis reduces the diploid chromosome number by one-half because gametes are

formed that have one member of each pair of homologous chromosomes possessed by

the species.

Complete the Table

6. To consider the major aspects of meiosis, complete the following table by sketching the chromosome conditions described (left column) in the center column and indicating in the right column whether the cells shown are haploid (*n*) or diploid (*2n*). The cell initially has only one pair of homologous chromosomes. The use of shaded and open chromosomes indicates that one member of the homologous pair came from a maternal source, the other from a paternal source. Thus, this model germ cell has a diploid number of 2. Assume that crossing over does not occur.

Meiosis—Overview Descriptions	Stages	Cell(s) 2n or n
a. One pair of homologous chromosomes prior to S of interphase in a diploid germ cell.		
b. While the germ cell is in S of interphase, chromosomes are duplicated through DNA replication; the two sister chromatids are attached at the centromere.		
c. During meiosis I, each duplicated chromosome lines up with its partner, homologue to homologue.		
d. Also during meiosis I, the chromosome partners separate from each other in anaphase I; cytokinesis occurs, and each chromosome goes to a different cell.		
e. During meiosis II (in two cells), the sister chromatids of each chromosome are separated from each other; four haploid nuclei form; cytokinesis results in four cells (potential gametes).		

STAGES OF MEIOSIS (8-III, pp. 113–116)

Terms

crossing over (113):_____

genetic recombination (113):_____

Label-Match

Identify each of the meiotic stages shown below by entering the correct stage of either meiosis I or meiosis II in the blank beneath the sketch. Choose from prophase I, metaphase I, anaphase I, telophase I, prophase II, metaphase II, anaphase II, and telophase II. Complete the exercise by matching and entering the letter of the correct phase description in the parentheses following each label.

  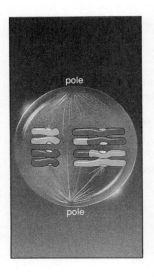 

1. _____ ( ) 2. _____ ( ) 3. _____ ( ) 4. _____ ( )

5. _____ ( ) 6. _____ ( ) 7. _____ ( ) 8. _____ ( )

A. All chromosomes are positioned midway between the spindle poles of one cell.

B. Centrioles have moved to form the poles in each of two cells, a spindle forms, and microtubules attach the duplicated chromosomes to the spindle and begin moving them toward the equator of each cell.

C. Four daughter nuclei form; when the cytoplasm divides, each new cell has a haploid number of chromosomes, all in the unduplicated state. One or all cells may develop into gametes.

D. In one cell, each duplicated chromosome is pulled away from its homologue.

E. Each chromosome is drawn up close to its homologue; crossing over and genetic recombination occur.

F. All of the chromosomes are now aligned at the spindle's equator; this is occurring in two haploid cells.

G. Two haploid cells form, but chromosomes are still in the duplicated state.

H. Chromatids of each chromosome separate; former "sister chromatids" are now chromosomes in their own right and are moved to opposite poles.

GAMETE FORMATION (8-IV, pp. 116–117)

Terms

sperm (116):_____

egg (116): _____

spores (116):_____

Choice

For questions 1–10, choose from the following.

  a. animal life cycle   b. plant life cycle   c. both animal and plant life cycles

____ 1. Meiosis results in the production of haploid spores

____ 2. A zygote divides by mitosis

____ 3. Meiosis results in the production of haploid gametes

____ 4. Haploid gametes fuse in fertilization to form a diploid zygote

____ 5. A zygote divides by mitosis to form a diploid sporophyte

____ 6. A spore divides by mitosis to produce a haploid gametophyte

____ 7. A haploid gametophyte divides by mitosis to produce haploid gametes

____ 8. A haploid spore divides by mitosis to produce a gametophyte

____ 9. A diploid body forms from mitosis of a zygote

____ 10. A gamete-producing body and a spore-producing body develops during the life cycle

Sequence

Arrange the following entities in correct order of development, entering a 1 by the stage that appears first and a 5 by the stage that completes the process of spermatogenesis. Refer to Figures 8.8 and 8.9 in the text.

11. ____ primary spermatocyte

12. ____ sperm

13. ____ spermatid

14. ____ spermatogonium

15. ____ secondary spermatocyte

Matching

Choose the most appropriate answer to match with each oogenesis concept. Refer to Figure 8.9 in the text.

16. ___ primary oocyte

17. ___ oogonium

18. ___ secondary oocyte

19. ___ ovum and three polar bodies

20. ___ first polar body

A. The cell in which synapsis, crossing over, and recombination occur

B. A cell that is equivalent to a diploid germ cell

C. A haploid cell formed after division of the primary oocyte that does not form an ovum at second division

D. Haploid cells, but only one of which functions as an egg

E. A haploid cell formed after division of the primary oocyte, the division of which forms a functional ovum

MORE GENE SHUFFLINGS AT FERTILIZATION (8-V, p. 117)

Terms

fertilization (117):_____

Short Answer

1. List the various mechanisms that contribute to the huge number of new gene combinations that may result from fertilization._____

_____

_____

MEIOSIS AND MITOSIS COMPARED (8-VI, pp. 118–119)

Complete the Table

1. Complete the table below by entering the word "mitosis" or "meiosis" in the blank adjacent to the statement referring to one of these processes.

a. Involves one division cycle	
b. Functions in growth and repair	
c. Daughter cells are haploid	
d. Initiated in germ cells	
e. Involves two division cycles	
f. Daughter cells have one chromosome from each homologous pair	
g. Produces spores in plant life cycles	
h. Daughter cells have the diploid chromosome number	
i. Completed when four daughter cells are formed	

Matching

The cell model used in this exercise has two pairs of homologous chromosomes, one long pair and one short pair. Match the descriptions to the number of chromosomes shown in the sketches below.

2. __ beginning of meiosis II following interkinesis

3. __ a daughter cell at the end of meiosis II

4. __ metaphase I of meiosis

5. __ metaphase of mitosis

6. __ G₁ in a daughter cell following mitosis

7. __ prophase of mitosis

A   B   C   D   E   F

The following questions refer to the sketches above; enter answers in the blanks following each question.

8. How many chromosomes are present in cell E? _____

9. How many chromatids are present in cell E? _____

10. How many chromatids are present in cell C? _____

11. How many chromatids are present in cell D? _____

12. How many chromosomes are present in cell F? _____

Self-Quiz

___ 1. Which of the following does *not* occur in prophase I of meiosis?
a. a cytoplasmic division
b. a cluster of four chromatids
c. homologues pairing tightly
d. crossing over

___ 2. Crossing over is one of the most important events in meiosis because _____.
a. it produces new combinations of alleles on chromosomes
b. homologous chromosomes must be separated into different daughter cells
c. the number of chromosomes allotted to each daughter cell must be halved
d. homologous chromatids must be separated into different daughter cells

___ 3. Crossing over _____.
a. generally results in pairing of homologues and binary fission
b. is accompanied by gene-copying events
c. involves breakages and exchanges being made between sister chromatids
d. alters the composition of chromosomes and results in new combinations of alleles being channeled into the daughter cells

___ 4. The appearance of chromosome ends lapped over each other in meiotic prophase I provides evidence of _____.
a. meiosis
b. crossing over
c. chromosomal aberration
d. fertilization
e. spindle fiber formation

___ 5. Which of the following does *not* increase genetic variation?
   a. crossing over
   b. random fertilization
   c. prophase of mitosis
   d. random homologue alignments at metaphase I

___ 6. Which of the following is the most correct sequence of events in animal life cycles?
   a. meiosis → fertilization → gametes → diploid organism
   b. diploid organism → meiosis → gametes → fertilization
   c. fertilization → gametes → diploid organism → meiosis
   d. diploid organism → fertilization → meiosis → gametes

___ 7. In sexually reproducing organisms, the zygote is _____.
   a. an exact genetic copy of the female parent
   b. an exact genetic copy of the male parent
   c. unlike either parent genetically
   d. a genetic mixture of male parent and female parent

___ 8. Which of the following is the most correct sequence of events in plant life cycles?
   a. fertilization → zygote → sporophyte → meiosis → spores → gametophytes → gametes
   b. fertilization → sporophyte → zygote → meiosis → spores → gametophytes → gametes
   c. fertilization → zygote → sporophyte → meiosis → gametes → gametophyte → spores
   d. fertilization → zygote → gametophyte → meiosis → gametes → sporophyte → spores

___ 9. The cell in the diagram is a diploid that has three pairs of chromosomes. From the number and pattern of chromosomes, the cell _____.

   a. could be in the first division of meiosis
   b. could be in the second division of meiosis
   c. could be in mitosis
   d. could be in neither mitosis nor meiosis, because this stage is not possible in a cell with three pairs of chromosomes

___10. You are looking at a cell from the same organism as in the previous question. Now the cell _____.

   a. could be in the first division of meiosis
   b. could be in the second division of meiosis
   c. could be in mitosis
   d. could be in neither mitosis nor meiosis, because this stage is not possible in this organism

Chapter Objectives/Review Questions

This section lists general and detailed chapter objectives that can be used as review questions. You can make maximum use of these items by writing answers on a separate sheet of paper. Fill in answers where blanks are provided. To check for accuracy, compare your answers with information given in the chapter or glossary.

Page	Objectives/Questions
(111)	1. "One parent always passes on a duplicate of all its genes to offspring" describes _____ reproduction.
(111)	2. Sexual reproduction puts together new combinations of _____ in offspring.
(112)	3. Describe the relationship between the following terms: homologous chromosomes, diploid, and haploid.
(112)	4. While germ cells are in interphase, each chromosome is duplicated by a process called DNA _____; each chromosome is then composed of two sister _____.
(112)	5. What are the two different divisions of meiosis called?

(113)   6. In meiosis I, homologous chromosomes pair; each homologue consists of _____ chromatids.
(116)   7. During meiosis II, the sister _____ of each _____ are separated from each other.
(112)   8. If the diploid chromosome number for a particular plant species is 18, the haploid number is
          _____.
(115)   9. Explain what is meant by the statement "each pair of homologues is assorted into gametes independently of the other pairs present in the cell."
(116)   10. Describe spermatogenesis in male animals.
(116)   11. Describe oogenesis in female animals.
(116)   12. Meiosis in the animal life cycle results in haploid _____; meiosis in the plant life cycle results in haploid _____.

Integrating and Applying Key Concepts

A few years ago, it was claimed that the actual cloning of a human being had been accomplished. Later, this claim was admitted to be fraudulent. If sometime in the future cloning of humans becomes possible, speculate about the effects of reproduction without sex on human populations.

Critical Thinking Exercise

1. The letters in the drawing below represent alleles on the chromosomes of a cell.

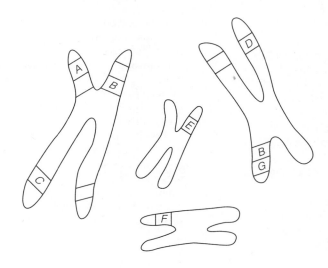

If the cell above underwent meiosis, which of the following combinations of alleles would most likely be present in the same gamete?

a. A, B, and C   b. A, B, and D   c. A and E   d. E and F   e. F and D

Answers

Answers to Interactive Exercises

ON SEXUAL AND ASEXUAL REPRODUCTION (8-I)
1. b; 2. a; 3. b; 4. b; 5. a; 6. b; 7. a; 8. b; 9. a; 10. b.

OVERVIEW OF MEIOSIS (8-II)
1. Meiosis; 2. T; 3. diploid; 4. T; 5. T.

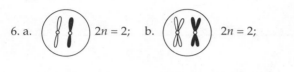

STAGES OF MEIOSIS (8-III)
1. anaphase II (H); 2. metaphase II (F); 3. metaphase I (A); 4. prophase II (B); 5. telophase II (C); 6. telophase I (G); 7. prophase I (E); 8. anaphase I (D).

GAMETE FORMATION (8-IV)
1. b; 2. c; 3. a; 4. c; 5. b; 6. b; 7. b; 8. b; 9. c; 10. b; 11. 2; 12. 5; 13. 4; 14. 1; 15. 3; 16. A; 17. B; 18. E; 19. D; 20. C.

MORE GENE SHUFFLINGS AT FERTILIZATION (8-V)
1. During prophase I of meiosis, crossing over and genetic recombination occur. During metaphase I of meiosis, the two members of each homologous chromosome assort independently of the other pairs.

Fertilization is a chance mix of different combinations of alleles from two different gametes.

MEIOSIS AND MITOSIS COMPARED (8-VI)
1. a. mitosis; b. mitosis; c. meiosis; d. meiosis; e. meiosis; f. meiosis; g. meiosis; h. mitosis; i. meiosis.
2. C; 3. F; 4. D; 5. A; 6. B; 7. E; 8. 4; 9. 8; 10. 4; 11. 8; 12. 2.

Answers to Self-Quiz
1. a; 2. a; 3. d; 4. b; 5. c; 6. b; 7. d; 8. a; 9. c; 10. b.

Critical Thinking Analysis

a. A, B, and C are present on the same chromatid. Remember that sister chromatids are identical unless crossing over occurs. Thus, these alleles will be in the same gamete, unless both chromatids undergo crossing over that separates the alleles. That event has a relatively low probability in a single cell.

b. A and G are on the same position on homologous chromosomes, and D occurs only on the chromosome that carries G. A will only occur on the same chromosome as D if a crossover occurs. Since B is present on both chromosomes, it will not be affected by any crossover.

c. A and E are on nonhomologous chromosomes. The probability that they will separate together in meiosis I is one in two.

d. E and F are at the same position on homologues. As long as meiosis I proceeds normally, they will not move into the same gamete.

e. F and D are on nonhomologous chromosomes. The probability that they will separate together in meiosis I is the same as for A and E, namely, one in two.

9

OBSERVABLE PATTERNS OF INHERITANCE

Interactive Exercises

MENDEL'S INSIGHTS INTO PATTERNS OF INHERITANCE (9-I, pp. 124–129)

Terms

The page-referenced terms are important; they were in boldface type in the chapter. Refer to the instructions given in Chapter 1, p. 1 of this workbook.

true-breeding (124): _____

genes (125):_____

alleles (125): _____

homozygous dominant (125):_____

homozygous recessive (125):_____

heterozygous (125):_____

genotype (125): _____

phenotype (125):_____

monohybrid crosses (126): _____

probability (126):_____

Punnett-square method (127):_____

theory of segregation (127): _____

testcross (127):_____

theory of independent assortment (129): _____

Short Answer

1. Describe how pea plants are fertilized in nature (p. 124):_____

2. Explain what is meant by a "true-breeding" pea plant; describe how Mendel planned to manipulate the reproduction of true-breeding garden pea plants to learn about heredity (p. 124):

_____

_____

_____

Matching

Choose the one most appropriate answer for each.

3. ___ genotype

4. ___ alleles

5. ___ heterozygous

6. ___ dominant allele

7. ___ phenotype

8. ___ genes

9. ___ recessive allele

10. ___ homozygous

11. ___ diploid cell

12. ___ locus

A. All the different molecular forms of a gene that exist
B. Particular location of a gene on a chromosome
C. Describes an individual having a pair of nonidentical alleles
D. Gene whose effect is masked by its partner
E. Refers to an individual's observable traits
F. Refers to the genes present in an individual organism
G. Gene whose effect "masks" the effect of its partner
H. Describes an individual for which two alleles of a pair are the same
I. Units of information about specific traits; passed from parents to offspring
J. Has a pair of genes for each trait, one on each of two homologous chromosomes

Fill-in-the-Blanks

A (13) _____ plant is one that is produced from two parents that have bred true for different forms of a single trait. In breeding experiments, (14) _____ is the symbol used for the parental generation; (15) _____ is the symbol used for first-generation offspring; (16) _____ is the symbol used for second-generation offspring. Because fertilization is a chance event, the rules of (17) _____ apply to genetics crosses. The separation of *A* and *a* as members of a pair of homologous chromosomes move to different gametes during meiosis is known as Mendel's theory of (18) _____. Crossing F_1 hybrids (possibly of unknown genotype) back to a plant known to be a true-breeding recessive plant is known as Mendel's

(19) _____. From this cross, a ratio of (20) _____ is expected. When F_1 offspring inherit two gene pairs, neither of which consists of identical alleles, the cross is known as a (21) _____ cross. "Gene pairs assorting into gametes independently of other gene pairs located on nonhomologous chromosomes" describes Mendel's theory of (22) _____ _____.

Problems

23. In garden pea plants, tall (T), is dominant over dwarf (t). In the cross $Tt \times tt$, the Tt parent would produce a gamete carrying T (tall) and a gamete carrying t (short) through segregation; the tt parent could only produce gametes carrying the t (short) gene. Using the Punnett-square method (refer to Figure 9.8 in the text), determine the genotype and phenotype probabilities of offspring from the above cross, $Tt \times tt$:

    Although the Punnett-square (checker board) method is a favored method for solving genetics problems, there is a quicker way. Six different outcomes are possible from monohybrid crosses. Studying the following *relationships* allows one to obtain the result of any monohybrid cross by *inspection*.

    (1) $AA \times AA$ = all AA (Each of the four blocks of the Punnett square would be AA.)
    (2) $aa \times aa$ = all aa
    (3) $AA \times aa$ = all Aa
    (4) $AA \times Aa = 1/2\ AA$; $1/2\ Aa$ (Two blocks of the Punnett square are AA, and two blocks are Aa.)
        or $Aa \times AA$
    (5) $aa \times Aa = 1/2\ aa$; $1/2\ Aa$
        or $Aa \times aa$
    (6) $Aa \times Aa = 1/4\ AA$; $1/2\ Aa$; $1/4\ aa$ (One block in the Punnett square is AA, two blocks are Aa, and one block is aa.)

    Using the gene symbols in exercise 23, apply the six Mendelian ratios listed above to solve the following monohybrid crosses by *inspection*. State results as genotype ratios.

24. $TT \times TT$ = _____

25. $Tt \times Tt$ = _____

26. $Tt \times tt$ = _____

27. $tt \times tt$ = _____

When working genetics problems dealing with two gene pairs, one can visualize the independent assortment of gene pairs located on nonhomologous chromosomes into gametes by use of a fork-line device. Assume that in man, pigmented eyes (B) are dominant (an eye color other than blue) over blue (b), and right-handedness (R) is dominant over left-handedness (r). To learn to solve a problem, cross the parents $BbRr \times BbRr$. A sixteen-block Punnett square is required with gametes from each parent arrayed on two sides of the Punnett square (refer to Figure 9.10 in the text). The gametes receive genes through independent assortment using a fork-line method:

$$B\ b\ R\ r \qquad \times \qquad B\ b\ R\ r$$
$$BR,\ Br,\ bR,\ br \qquad\qquad BR,\ Br,\ bR,\ br$$

28. Array the gametes above on two sides of the Punnett square; combine these haploid gametes to form diploid zygotes within the squares. In the blank spaces below, enter the probability ratios derived within the Punnett square for the phenotypes listed:

    a. ___ pigmented eyes, right-handed

    b. ___ pigmented eyes, left-handed

    c. ___ blue-eyed, right-handed

    d. ___ blue-eyed, left-handed

29. Albinos cannot form the pigments that normally produce skin, hair, and eye color, so albinos have white hair and pink eyes and skin (because the blood shows through). To be an albino, one must be homozygous recessive for the pair of genes that codes for the key enzyme in pigment production. Suppose a woman of normal pigmentation with an albino mother marries an albino man. State the possible kinds of pigmentation possible for this couple's children, and specify the ratio of each kind of child the couple is likely to have. Show the genotype(s) and state the phenotype(s).

    _____

    _____

30. In horses, black coat color is influenced by the dominant allele (B), and chestnut coat color is influenced by the recessive allele (b). Trotting gait is due to a dominant gene (T), pacing gait to the recessive allele (t). A homozygous black trotter is crossed to a chestnut pacer.

    a. What will be the appearance of the F_1 and F_2 generations? _____

    b. Which phenotype will be most common? _____

    c. Which genotype will be most common? _____

    d. Which of the potential offspring will be certain to breed true? _____

VARIATIONS ON MENDEL'S THEMES (9-II, pp. 130–132)

Terms

incomplete dominance (130): _____

codominance (130): _____

epistasis (132): _____

pleiotropy (130): _____

Complete the Table

1. Complete the following table by supplying the type of inheritance illustrated by each example. Choose from pleiotropy, multiple alleles, incomplete dominance, codominance, and gene interaction.

Type of Inheritance	Example
a.	pink-flowered snapdragons produced from red- and white-flowered parents
b.	AB type blood from a gene system of three alleles, *A*, *B*, and *O*
c.	a gene with three or more alleles such as the ABO blood typing alleles
d.	black, brown, or yellow fur of Labrador retrievers
e.	the multiple phenotypic effects of the gene causing human sickle cell anemia

Problems

Genes that are not always dominant or recessive may blend to produce a phenotype of a different appearance. This is termed *incomplete dominance*. In four o'clock plants, red flower color is determined by gene R and white flower color by R' while the heterozygous condition, RR', is pink. Determine the phenotypes and genotypes of the offspring from the following crosses:

2. $RR \times R'R' =$ _____

3. $R'R' \times R'R' =$ _____

4. $RR \times RR' =$ _____

5. $RR \times RR =$ _____

In other types of gene interactions, two alleles of a gene mask the expression of alleles of another gene, and some expected phenotypes never appear. *Epistasis* is the term given such interactions. Work the following problems on scratch paper to understand epistatic interactions.

In sweet peas, genes C and P are necessary for colored flowers. In the absence of either (_ _*pp* or *cc*_ _), or both (*ccpp*), the flowers are white. What will be the color of the offspring of the following crosses and in what proportions?

6. $CcPp \times ccpp =$ _____

7. $CcPP \times Ccpp =$ _____

8. $Ccpp \times ccPp =$ _____

The three genes I^A, I^B, and i produce proteins found on the surfaces of red blood cells that determine the four blood types in the ABO system, A, B, AB, and O. Genes I^A and I^B are both dominant over i but not over each other. They are codominant. Recognize that blood types A and B may be heterozygous or homozygous ($I^A I^A$, $I^A i$ or $I^B I^B$, $I^B i$) while blood type O is symbolized by ii. Indicate the genotypes and phenotypes of the offspring and their probabilities from the parental combinations in exercises 9–13.

9. $I^A i \times I^A I^B =$ _____

10. $I^B i \times I^A i =$ _____

11. $I^A I^A \times ii =$ _____

12. $ii \times ii =$ _____

13. $I^A I^B \times I^A I^B =$ _____

LESS PREDICTABLE VARIATIONS IN TRAITS (9-III, pp. 133–134)

Terms

continuous variation (134):_____

Choice

For questions 1–5, choose from the following primary contributing factors.

a. environment   b. gene interaction   c. a number of gene pairs affecting a trait

___ 1.   Height of human beings

___ 2.   Campodactyly, a human genetic disorder

___ 3.   Development of different water buttercup leaf shapes under and above water level

___ 4.   The range of eye colors in the human population

___ 5.   Heat-sensitive enzyme required for melanin production in Himalayan rabbits

Self-Quiz

___ 1. The best statement of Mendel's principle of independent assortment is that _____.
   a. one allele is always dominant to another
   b. hereditary units from the male and female parents are blended in the offspring
   c. the two hereditary units that influence a certain trait separate during gamete formation
   d. each hereditary unit is inherited separately from other hereditary units

___ 2. One of two or more alternative forms of a gene for a single trait is a(n) _____.
   a. chiasma
   b. allele
   c. autosome
   d. locus

___ 3. In the F_2 generation of a monohybrid cross involving complete dominance, the expected *phenotypic* ratio is _____.
   a. 3:1
   b. 1:1:1:1
   c. 1:2:1
   d. 1:1

___ 4. In the F_2 generation of a cross between a red-flowered four o'clock (homozygous) and a white-flowered four o'clock, the expected phenotypic ratio of the offspring is _____.
   a. 3/4 red, 1/4 white
   b. 100 percent red
   c. 1/4 red, 1/2 pink, 1/4 white
   d. 100 percent pink

___ 5. In a testcross, F_1 hybrids are crossed to an individual known to be _____ for the trait.
   a. heterozygous
   b. homozygous dominant
   c. homozygous
   d. homozygous recessive

___ 6. A man with type A blood could be the father of _____.
   a. a child with type A blood
   b. a child with type B blood
   c. a child with type O blood
   d. a child with type AB blood
   e. all of the above

___ 7. A single gene that affects several seemingly unrelated aspects of an individual's phenotype is said to be _____.
   a. pleiotropic
   b. epistatic
   c. mosaic
   d. continuous

___ 8. Suppose two individuals, each heterozygous for the same characteristic, are crossed. The characteristic involves complete dominance. The expected genotypic ratio of their progeny is _____.
   a. 1:2:1
   b. 1:1
   c. 100 percent of one genotype
   d. 3:1

___ 9. If the two homozygous classes in the F_1 generation of the cross in exercise 8 are allowed to mate, the observed genotypic ratio of the offspring will be _____.
   a. 1:1
   b. 1:2:1
   c. 100 percent of one genotype
   d. 3:1

___ 10. The skin color trait in humans exhibits _____.
   a. pleiotropy
   b. epistasis
   c. mosaic
   d. continuous variation

Chapter Objectives/Review Questions

Page	Objectives/Questions
(123)	1. What was the prevailing method of explaining the inheritance of traits before Mendel's work with pea plants?
(124)	2. Garden pea plants are naturally _____-fertilizing, but Mendel took steps to _____-fertilize them for his experiments.
(125)	3. Define allele; how many alleles are present in the genotypes *Tt*? *tt*? *TT*?
(127)	4. Be able to use the Punnett-square method of solving genetics problems.
(127)	5. Define the testcross and cite an example.
(127)	6. State the theory of segregation as formulated by Mendel.
(128)	7. Define dihybrid cross and distinguish it from a monohybrid cross.
(129)	8. State the theory of independent assortment as formulated by Mendel.
(130)	9. Distinguish between complete dominance, incomplete dominance, and codominance.
(131)	10. Define multiple allele system and cite an example.
(132)	11. Define epistasis and give an example.
(130)	12. Explain why sickle-cell anemia is a good example of pleiotropy.
(133)	13. Himalayan rabbits and water buttercups are good examples of environmental effects on _____.
(134)	14. List possible explanations for less predictable trait variations that are observed.

Integrating and Applying Key Concepts

Solve the following genetics problem:

In garden peas, one pair of alleles controls the height of the plant and a second pair of alleles controls flower color. The allele for tall (*D*) is dominant to the allele for dwarf (*d*), and the allele for purple (*P*) is dominant to the allele for white (*p*). A tall plant with purple flowers crossed with a tall plant with white flowers produces 3/8 tall purple, 3/8 tall white, 1/8 dwarf purple, and 1/8 dwarf white. What is the genotype of the parents?

Critical Thinking Exercise

1. Two black female mice were bred with the same brown male. In three litters, female A produced nine black offspring and eight brown offspring, while female B produced nineteen black offspring and no brown offspring. Assuming that coat color is determined by a single pair of genes and exhibits simple dominance in mice, which of the following is the most likely set of parental genotypes?
   a. The male is homozygous for brown, female A is heterozygous, and female B is homozygous for black.
   b. The male is heterozygous, female A is heterozygous, and female B is homozygous for black.
   c. The male is homozygous for brown, and female A and female B are both heterozygous for black.
   d. The male is heterozygous, and both females are homozygous for black.
   e. All three parents are heterozygous.

Answers

Answers to Interactive Exercises

MENDEL'S INSIGHTS INTO PATTERNS OF INHERITANCE (9-I)

1. Garden pea plants are self-fertilizing in nature.
2. True-breeding plants are those whose successive generations are exactly like the parents in one or more traits.

Mendel stopped pea plants from self-fertilizing by opening their flower buds and removing the stamens. He promoted cross-fertilization by brushing pollen from another plant onto the "castrated" flower bud.
3. F; 4. A; 5. C; 6. G; 7. E; 8. I; 9. D; 10. H; 11. J; 12. B; 13. monohybrid; 14. P; 15. F_1; 16. F_2; 17. probability; 18. segregation; 19. testcross; 20. 1:1; 21. dihybrid;

22. independent assortment; 23. genotype: 1/2 *Tt*; 1/2 *tt*; phenotype: 1/2 tall; 1/2 short; 24. All *TT*; 25. 1/4 *TT*; 1/2 *Tt*; 1/4 *tt* or 1 *TT*; 2 *Tt*; 1 *tt*; 26. 1/2 *Tt*; 1/2 *tt* or 1 *Tt*; 1 *tt*; 27. All *tt*.

28. Summary: a. 9/16 pigmented eyes, right-handed; b. 3/16 pigmented eyes, left-handed; c. 3/16 blue-eyed, right-handed; d. 1/16 blue-eyed, left-handed (note Punnett square below).

	BR	Br	bR	br
BR	BBRR	BBRr	BbRR	BbRr
Br	BBRr	BBrr	BbRr	Bbrr
bR	BbRR	BbRr	bbRR	bbRr
br	BbRr	Bbrr	bbRr	bbrr

29. Albino = *pp*, normal pigmentation = *PP* or *Pp*. The woman of normal pigmentation with an albino mother is genotype *Pp*; the woman received her recessive gene from her mother and her dominant gene (*P*) from her father. It is likely that half of the couple's children will be albinos (*pp*) and half will have normal pigmentation but be heterozygous (*Pp*).

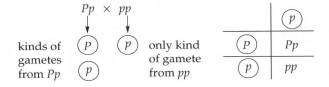

30. a. F$_1$: black trotter; F$_2$: nine black trotters, three black pacers, three chestnut trotters, one chestnut pacer; b. black trotter; c. *BbTt*; d. *bbtt*, chestnut pacers and BBTT, black trotters.

VARIATIONS ON MENDEL'S THEMES (9-II)

1. a. Incomplete dominance; b. Codominance; c. Multiple alleles; d. Epistasis; e. Pleiotropy. 2. phenotype: all pink, genotype: all *RR'*; 3. phenotype: all white, genotype: all *R'R'*; 4. phenotype: 1/2 red; 1/2 pink, genotype: 1/2 *RR*; 1/2 *RR'*; 5. phenotype: all red, genotype: all *RR*; 6. 1/4 color; 3/4 white; 7. 3/4 color; 1/4 white; 8. 1/4 color; 3/4 white; 9. genotypes: 1/4 $I^A I^A$; 1/4 $I^A I^B$; 1/4 $I^A i$; 1/4 $I^B i$, phenotypes: 1/2 A; 1/4 AB; 1/4 B; 10. genotypes: 1/4 $I^A I^B$; 1/4 $I^B i$; 1/4 $I^A i$; 1/4 *ii*, phenotypes: 1/4 AB; 1/4 B; 1/4 A; 1/4 O; 11. genotypes: all $I^A i$, phenotypes: all A; 12. genotypes: all *ii*, phenotypes: all O; 13. genotypes: 1/4 $I^A I^A$; 1/2 $I^A I^B$; 1/4 $I^B I^B$, phenotypes: 1/4 A; 1/2 AB; 1/4 B.

LESS PREDICTABLE VARIATIONS IN TRAITS (9-III)

1. c; 2. b; 3. a; 4. c; 5. a.

Answers to Self-Quiz

1. d; 2. b; 3. a; 4. c; 5. d; 6. e; 7. a; 8. a; 9. c; 10. d.

Answer to Integrating and Applying Key Concepts

DdPp × *Ddpp*

Critical Thinking Analysis

a. In this choice, female A is heterozygous and black; therefore, black is the dominant allele. About half the gametes of female A will carry the dominant allele and lead to black offspring, and about half will carry the recessive allele for brown fur and produce brown offspring in this mating with a brown homozygous male. Gametes of homozygous-black female B, on the other hand, will all carry the dominant black allele and produce black offspring in any mating. This set of parental genotypes predicts the observed distribution of offspring phenotypes.

b. This choice can be rejected, because the male and female A, brown and black, respectively, cannot both be heterozygous. With simple dominance, one genotype can't produce two different phenotypes.

c. In this choice, the black females are heterozygous; therefore, black is the dominant allele. About half of female B's gametes would be expected to carry the recessive allele for brown and produce brown offspring in a mating with a brown homozygous male. This choice is not impossible, but the observed outcome is very unlikely if female B is heterozygous.

d. In this choice, brown must be the dominant allele, because the male is heterozygous and brown. About half his gametes are predicted to carry the dominant brown allele and to produce brown offspring in a mating with homozygous black females. Female B produced no brown offspring. While not impossible, this deviation from the predicted outcome is very unlikely.

e. This choice, like choice (b), is impossible, because the parents have different phenotypes, and therefore can't have the same genotypes.

10

CHROMOSOMES AND HUMAN GENETICS

Interactive Exercises

EARLY STUDIES OF CHROMOSOMES (10-I, pp. 138–143)

Terms

The page-referenced terms are important; they were in boldface type in the chapter. Refer to the instructions given in Chapter 1, p. 1 of this workbook.

Y chromosome (138): _____

X chromosome (138): _____

sex chromosomes (138): _____

autosomes (138): _____

karyotype (138): _____

X-linked gene (141): _____

Y-linked genes (141): _____

linkage (143): _____

Matching

Choose the one most appropriate answer for each.

1. ___ SRY
2. ___ Autosomes
3. ___ Y-linked
4. ___ August Weismann
5. ___ linkage
6. ___ crossing over
7. ___ Walther Flemming
8. ___ karyotype
9. ___ sex chromosomes
10. ___ independent assortment

A. In humans, the X and Y chromosomes; concerned with gender determination
B. Situation in which the way one pair of chromosomes segregates into different gametes does not influence how a different pair of chromosomes segregates into the same gametes (compare with gene linkage)
C. Observed chromosomes in cells by 1882
D. The tendency of genes located on the same chromosome to travel together in inheritance
E. Genes carried on the smaller human sex chromosome
F. A cut-up, rearranged photograph of a cell's metaphase chromosomes
G. Any chromosome not concerned with gender determination
H. The master gene for sex determination; on the Y chromosome
I. Proposed that a special division process must reduce the chromosome number by half in gamete formation
J. Interrupts gene linkage

Completion

11. Complete the Punnett square below, which illustrates sex determination in humans; Y-bearing and X-bearing sperms combine randomly with eggs in fertilization.

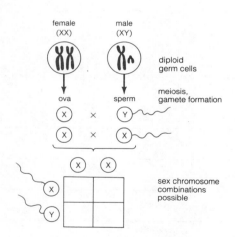

Dichotomous Choice

Circle one of two possible answers given between parentheses in each statement.

12. Male humans transmit their Y chromosome only to their (sons/daughters).
13. Male humans receive their X chromosome only from their (mothers/fathers).
14. Human mothers and fathers each provide an X chromosome for their (sons/daughters).
15. Crossing over occurs between (sister/nonsister) chromatids.
16. Morgan's experiments with fruit flies indicated a correlation between sex and eye color; eye color must be carried on the (X/Y) chromosome.

Problems

17. Which of the following represents a chromosome that has undergone crossover and recombination? It is assumed that the organism involved is heterozygous with the following genotype: $\left.\frac{A}{B}\right| \left|\frac{a}{b}\right.$ ————

a. $\left|\frac{A}{B}\right.$
b. $\frac{a}{b}\left|\right.$
c. $\frac{B}{A}\left|\right.$
d. $\left|\frac{a}{B}\right.$

18. Individuals with dominant gene *N* have the nail-patella syndrome and exhibit abnormally developed fingernails and absence of kneecaps; the recessive gene *n* is normal. Geneticists have learned that the genes determining the ABO blood groups and the nail-patella gene are linked, with both found on chromosome 9. A woman with blood type A and nail-patella syndrome marries a man with blood type O and normal fingernails and kneecaps. This couple has three children. One daughter has blood type A like her mother but has normal fingernails and kneecaps. A second daughter has blood type O with nail-patella syndrome. A son has blood type O and normal fingernails and kneecaps. The chromosomes of the parents are diagrammed at the right. Complete the exercise by writing the gene symbols of each child on his or her chromosomes (homologous chromosomes are represented by pairs of lines).

♀

| *A   n* |

| *O   N* |

×

♂

| *O   n* |

| *O   n* |

(Or *An/ON* × *On/On*)

a.   Daughter 1   Daughter 2   Son

_____ _____ _____

_____ _____ _____

b.   Explain how the son can have the genotype *On/On* when only one parent has a chromosome with genes *O* and *n* linked. _____

_____

HUMAN GENETICS (10-II, pp. 144–151)

Terms

pedigrees (144): _____

genetic disorder (145): _____

deletion (147): _____

duplication (148): _____

inversion (148): _____

translocation (148): _____

aneuploidy (148): _____

polyploidy (148): _____

nondisjunction (148): _____

Choice

For questions 1–10, choose from the following.

      a. autosomal recessive   b. X-linked   c. autosomal dominant

____ 1. Heterozygotes can remain undetected

____ 2. The trait appears in each generation

____ 3. The recessive phenotype shows up far more often in males than in females

____ 4. Both parents may be heterozygous normal

____ 5. A son cannot inherit the recessive allele from his father but his daughter can

____ 6. The trait is expressed in heterozygotes of either sex

____ 7. Heterozygous normal parents can expect that one fourth of their children will be affected by the disorder

____ 8. Individuals displaying this type of disorder will always be homozygous for the trait

____ 9. The trait is expressed in both the homozygote and the heterozygote

____ 10. Heterozygous females will transmit the recessive gene to half her sons and half her daughters

Complete the Table

11. Complete the table below by indicating whether the genetic disorder listed is due to inheritance that is autosomal recessive, autosomal dominant, or X-linked.

Inheritance	Genetic Disorder
a.	polydactyly
b.	Huntington's Disorder
c.	hemophilia A
d.	achondroplasia
e.	galactosemia
f.	progeria

Choice

For questions 12–21, choose from the following.

a. galactosemia   b. achondroplasia   c. Huntington's disorder   d. hemophilia A   e. progeria

___ 12. The aging disorder

___ 13. Homozygous dominant individuals are stillborn

___ 14. Affected individuals bleed for an abnormal time

___ 15. Progressive nervous system deterioration

___ 16. Heterozygotes cannot form cartilage during limb bone growth

___ 17. Newborns are homozygous recessive for a defective enzyme

___ 18. A mutated gene for "clotting factor VIII"

___ 19. Symptoms begin about age forty

___ 20. Detected by a high galactose level in urine; early symptoms are malnutrition, diarrhea, and vomiting

___ 21. Unusually high frequency among royal families of nineteenth-century Europe

Problems

22. The autosomal allele that causes albinism (*c*) is recessive to the allele for normal pigmentation (C). A normally pigmented woman whose father is an albino marries an albino man whose parents are normal. They have three children, two normal and one albino. Give the genotypes for each person listed.

23. Huntington's disorder is a rare form of autosomal dominant inheritance (*H*); the normal gene is (*h*). The disease causes progressive degeneration of the nervous system with onset exhibited near middle age. An apparently normal man in his early twenties learns that his father has recently been diagnosed as having Huntington's disorder. What are the chances that the son will develop Huntington's disorder?

24. A color-blind man and a woman with normal vision whose father was color-blind have a son. Color blindness, in this case, is caused by an X-linked recessive gene. If only male offspring are considered, what is the probability that their son is color-blind?

25. Hemophilia A is caused by an X-linked recessive gene. A woman who is seemingly normal but whose father was a hemophiliac marries a normal man.

a. What proportion of their sons will have hemophilia? _____

b. What proportion of their daughters will have hemophilia? _____

c. What proportion of their daughters will be carriers? _____

26. The pedigree below shows the pattern of inheritance of color blindness in a family (persons with the trait are indicated by black circles).

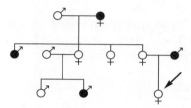

What is the chance that the third-generation female (indicated by the arrow) will have a color-blind son if she marries a normal male? A color-blind male? _____

Label-Match

27. On rare occasions, chromosomal structure becomes abnormally rearranged. Such changes may have profound effects on the phenotype of an organism. Identify the following diagrams of abnormal chromosome structure as a deletion, a duplication, an inversion, or a translocation. Complete the exercise by matching and entering the letter of the proper description in the parentheses following each label.

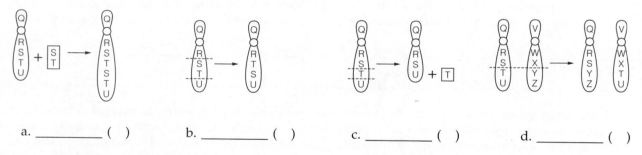

a. _____ ( )   b. _____ ( )   c. _____ ( )   d. _____ ( )

A. The loss of a chromosome segment
B. A gene sequence in excess of its normal amount in a chromosome
C. A chromosome segment that separated from the chromosome and then was inserted at the same place, but in reverse
D. The transfer of part of one chromosome to a nonhomologous chromosome

Complete the Table

28. Complete the table below to summarize the major mechanisms of chromosome number change in organisms.

Category of Change	Description
a. aneuploidy	
b. polyploidy	

Short Answer

29. If a nondisjunction occurs at anaphase I of the first meiotic division, what may the number of abnormal gametes (for the chromosome involved in the nondisjunction) be? (p. 148):_____

30. If a nondisjunction occurs at anaphase II of the second meiotic division, what may the number of abnormal gametes (for the chromosome involved in the nondisjunction) be? (p. 148):_____

31. Contrast the effects of polyploidy in plants and humans. (p. 148): _____

_____

Choice

For questions 32–41, choose from the following.

    a. Down syndrome   b. Turner syndrome   c. Klinefelter syndrome   d. XYY condition

____ 32. XXY male

____ 33. Ovaries nonfunctional and secondary sexual traits fail to develop at puberty

____ 34. Testes smaller than normal, sparse body hair, and some breast enlargement

____ 35. Could only be caused by a nondisjunction in males

____ 36. Older children smaller than normal with distinctive facial features; small skin fold over the inner corner of the eyelid

____ 37. XO female; often abort early, distorted female phenotype

____ 38. Males that tend to be taller than average; some mildly retarded but most are phenotypically normal

____ 39. Injections of testosterone reverse feminized traits but not the mental retardation

____ 40. Trisomy 21; skeleton develops more slowly than normal with slack muscles

____ 41. At one time these males were thought to be genetically predisposed to become criminals

Complete the Table

42. Complete the table below, which summarizes methods of dealing with the problems of human genetics. Choose from phenotypic treatments, genetic screening, genetic counseling, and prenatal diagnosis.

Method	Description
a.	detects genetic disorders before birth; may use karyotypes, biochemical tests, amniocentesis, and CVS
b.	parents at risk request emotional support and advice from clinical psychologists, geneticists, and social workers
c.	suppressing or minimizing symptoms of genetic disorders by surgical intervention, controlling diet or environment, or chemically modifying genes
d.	large-scale programs to detect affected persons or carriers in a population; early detection may allow introduction of preventive measures before symptoms develop

Self-Quiz

___ 1. All the genes located on a given chromosome compose a _____.
a. karyotype
b. genome
c. wild-type allele
d. linkage group

___ 2. Chromosomes other than those involved in sex determination are known as

_____.
a. nucleosomes
b. heterosomes
c. alleles
d. autosomes

___ 3. The farther apart two genes are on a chromosome, _____.
a. the less likely that crossing over and recombination will occur between them
b. the greater will be the frequency of crossing over and recombination between them
c. the more likely they are to be in two different linkage groups
d. the more likely they are to be segregated into different gametes when meiosis occurs

___ 4. Karyotype analysis is _____.
a. a means of detecting and reducing mutagenic agents

b. a surgical technique that separates chromosomes that have failed to segregate properly during meiosis II
c. a cut-up, rearranged photograph of a cell's metaphase chromosomes
d. a process that substitutes normal alleles for defective ones

___ 5. Which of the following did Morgan and his research group *not* do?
a. They isolated and kept under culture fruit flies with the sex-linked recessive white-eyed trait.
b. They developed the technique of amniocentesis.
c. They discovered X-linked genes.
d. Their work reinforced the concept that each gene is located on a specific chromosome.

___ 6. Red-green color blindness is a sex-linked recessive trait in humans. A color-blind woman and a man with normal vision have a son. What are the chances that the son is color-blind? If the parents ever have a daughter, what is the chance for each birth that the daughter will be color-blind? (Consider only the female offspring.)
a. 100 percent, 0 percent
b. 50 percent, 0 percent
c. 100 percent, 100 percent

d. 50 percent, 100 percent
e. none of the above

___ 7. Suppose a hemophilic male (sex-linked recessive allele) and a female carrier for the hemophilic trait have a nonhemophilic daughter with Turner syndrome. Nondisjunction could have occurred in _____.
   a. both parents
   b. neither parent
   c. the father only
   d. the mother only

___ 8. Nondisjunction involving the X chromosome occurs during oogenesis and produces two kinds of eggs, XX and O (no X chromosome). If normal Y sperm fertilize two types, which genotypes are possible?
   a. XX and XY
   b. XXY and YO
   c. XYY and XO
   d. XYY and YO

___ 9. A chromosome duplication is _____.
   a. a part of one chromosome inserted into a nonhomologous chromosome
   b. the loss of a chromosome region by irradiation, viral attack, or chemical action
   c. a segment that separated from a chromosome and then was inserted at the same place—but in reverse
   d. a repeated region of a chromosome

___ 10. Amniocentesis is _____.
   a. a surgical means of repairing deformities
   b. a form of chemotherapy that modifies or inhibits gene expression or the function of gene products
   c. used in prenatal diagnosis to detect chromosomal mutations and metabolic disorders in embryos
   d. a form of gene-replacement therapy

Chapter Objectives/Review Questions

This section lists general and detailed chapter objectives that can be used as review questions. You can make maximum use of these items by writing answers on a separate sheet of paper. Fill in answers where blanks are provided. To check for accuracy, compare your answers with information given in the chapter or glossary.

Page	Objectives/Questions
(138)	1. Define karyotype and state why it is useful; describe how karyotyping is done.
(141)	2. Distinguish between sex-linked genes and sex-determining genes.
(141)	3. Describe two significant contributions the Morgan research team made to our early understanding of genetics.
(138)	4. Explain how the meiotic segregation of sex chromosomes to gametes and subsequent random fertilization determines sex in many organisms.
(143)	5. The tendency of genes located on the same chromosome to end up together in the same gamete is called _____.
(143)	6. State the relationship between the probability of crossing over (and subsequent recombination) and the distance between two genes located on the same chromosome.
(144)	7. List reasons why it is difficult to study inheritance patterns in humans. Researchers can identify inheritance patterns and track genetic abnormalities through several generations by constructing _____ charts.
(145)	8. A genetic _____ is a rare or less common occurrence, whereas a _____ causes mild to severe medical problems.
(148)	9. When gametes or cells of an affected individual end up with one extra or one less than the parental number of chromosomes, it is known as _____; relate this to "trisomic" and "monosomic."

(149)   10. Practice sketching a diagram of nondisjunction similar to Figure 10.13 in the text.
(151)   11. Explain the procedures used in two types of prenatal diagnosis, amniocentesis and chorionic villi analysis; compare the risks.
(151)   12. List some benefits of genetic screening and genetic counseling to society.
(151)   13. Discuss some of the ethical considerations that might be associated with a decision of induced abortion.

Integrating and Applying Key Concepts

The parents of a young boy bring him to their doctor. They explain that the boy does not seem to be going through the same vocal developmental stages as his older brother. The doctor orders a common cytogenetics test to be done, and it reveals that the young boy's cells contain two X chromosomes and one Y chromosome. Describe the test that the doctor ordered and explain how and when such a genetic result—XXY—most logically occurred.

Critical Thinking Exercises

1. The offspring of one mated pair of mammals included three males, all of which showed an X-linked recessive trait. The fourth zygote was female. Which of the following statements could be made about her with the most confidence?
   a. Even if the father shows the trait, she will not.
   b. If the father does not show the trait, she will not.
   c. She will show the trait.
   d. If the mother shows the recessive trait, the daughter will also show the recessive trait.
   e. If the father shows the recessive trait, she will show the recessive trait.

Answers

Answers to Interactive Exercises

EARLY STUDIES OF CHROMOSOMES (10-I)
1. H; 2. G; 3. E; 4. I; 5. D; 6. J; 7. C; 8. F; 9. A; 10. B; 11. Two blocks of the Punnett square should be XX and two blocks should be XY; 12. sons; 13. mothers; 14. daughters; 15. nonsister; 16. X; 17. d; 18. a. Linked genes of the children: daughter, *On/An*; daughter, *ON/On*; son, *On/On*. b. The son has genotype *On/On* because a crossover must have occurred during meiosis in his mother to show the recombination *On*.

HUMAN GENETICS (10-II)
1. a; 2. c; 3. b; 4. a; 5. b; 6. c; 7. a; 8. a; 9. c; 10. b; 11. a. Autosomal dominant; b. Autosomal dominant; c. X-linked; d. Autosomal dominant; e. Autosomal recessive; f. Autosomal dominant; 12. e. 13. b; 14. d; 15. c; 16. b; 17. a; 18. d; 19. c; 20. a; 21. d. 22. The woman's mother is heterozygous normal, *Cc*; the woman is also heterozygous normal, *Cc*. The albino man, *cc*, has two heterozygous normal parents, *Cc*. The two normal children are heterozygous normal, *Cc*; the albino child is *cc*.
23. Assuming the father is heterozygous with Huntington's disorder and the mother is normal, the chances are 1/2 that the son will develop the disease.
24. If only male offspring are considered, the probability is 1/2 that the couple will have a color-blind son.
25. a. The probability is that half of the sons will have hemophilia; b. the probability is zero that a daughter will express hemophilia; c. the probability is that half of the daughters will be carriers. 26. If the woman marries a normal male, the chance that her son would be color-blind is 1/2. If she marries a color-blind male, the chance that her son would be color-blind is also 1/2.
27. a. duplication (B); b. inversion (C); c. deletion (A); d. translocation (D). 28. a. Aneuploidy is the abnormal condition in which gametes or cells of an affected individual end up with one extra or one less than the parental number of chromosomes; b. Polyploidy is the abnormal condition in which gametes or cells of an affected individual may end up with three or more of each type of chromosome characteristic of the parental stock. 29. a. All gametes will be abnormal; 30. Half the gametes will be abnormal; 31. About half of all flowering plant species are polyploids, but this condition is lethal for humans; 32. c; 33. b; 34. c; 35. d; 36. a; 37. b; 38. d; 39. c; 40. a; 41. d; 42. a. Prenatal diagnosis; b. Genetic counseling; c. Phenotypic treatments; d. Genetic screening.

Answers to Self-Quiz

1. d; 2. d; 3. b; 4. c; 5. b; 6. a; 7. d; 8. b; 9. d; 10. c.

Critical Thinking Analysis

A male mammal has one X chromosome from his mother and one Y chromosome from his father. These male offspring carry the recessive X-linked allele; thus, their mother is either heterozygous or homozygous for the allele, and no inference can be made about the father. The daughter will receive the father's only X chromosome and one of her mother's X chromosomes.

a. If the father carries the allele on his only X chromosome, the daughter will inherit it. She will be not show the recessive trait, however, if her mother is heterozygous and she happens to receive the mother's dominant X chromosome. If she receives another copy of the recessive allele from her mother, she will show the recessive trait.

b. If the father does not show the trait, he will transmit to the daughter his only X chromosome bearing the dominant allele. His daughter will not show the trait regardless of the mother's contribution to the daughter's genotype.

c. The daughter might or might not show the recessive trait, depending on the genotypes of the parents, which can be only partially inferred from the information about the sons.

d. If the mother shows the recessive trait, she must be homozygous and will contribute a recessive allele to the daughter's genotype. However, a father who does not show the trait will necessarily contribute the dominant allele and produce only daughters who do not show the trait.

e. A father who shows the trait will contribute the allele to all daughters, but the mother in this case could be heterozygous and contribute a dominant allele to her daughter. The result would be a dominant phenotype.

11

DNA STRUCTURE AND FUNCTION

DISCOVERY OF DNA FUNCTION
    Early Clues
    Confirmation of DNA Function

DNA STRUCTURE
    Components of DNA
    Patterns of Base Pairing

DNA REPLICATION

DNA REPAIR

ORGANIZATION OF DNA IN CHROMOSOMES

Interactive Exercises

DISCOVERY OF DNA FUNCTION (11-I, pp. 155–156)

Terms

The page-referenced terms are important; they were in boldface type in the chapter. Refer to the instructions given in Chapter 1, p. 1 in this workbook.

deoxyribonucleic acid, DNA (155) _deoxyribonucleic acid_____

bacteriophages (156) _a virus infect bacterial cell such as E coli._____

Complete the Table

1.   Complete the table below, which traces the discovery of DNA function.

Investigators	Year(s)	Contribution
a. Miescher	1868	
b.	1928	discovered the transforming principle in *Streptococcus pneumoniae*; live, harmless R cells were mixed with dead S cells, R cells became S cells
c. Avery (also MacLeod and McCarty)	1944	
d. Hershey and Chase	1952	

DNA STRUCTURE (11-II, pp. 157–158)

Terms

nucleotide (157) _the building blocks of nucleic acid_

adenine (A) (157)

guanine (G) (157)

thymine (T) (157)

cytosine (C) (157)

Short Answer

1. List the three parts of a nucleotide.

Labeling

Four nucleotides are illustrated below. In the blank, label each nitrogen-containing base correctly as guanine, thymine, cytosine, or adenine. In the parentheses following each blank, indicate whether that nucleotide base is a purine (pu) or a pyrimidine (py).

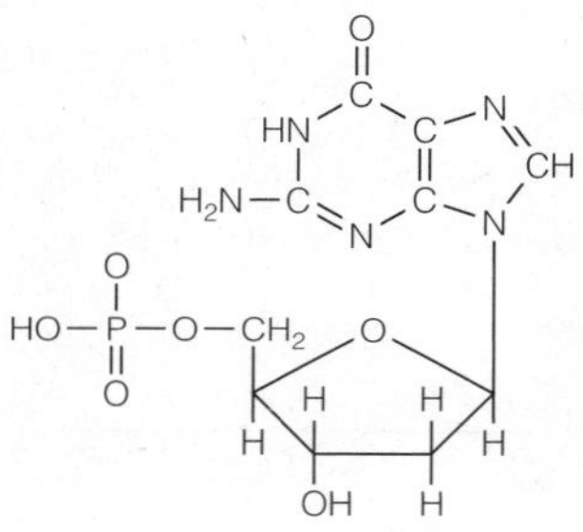

2._____ 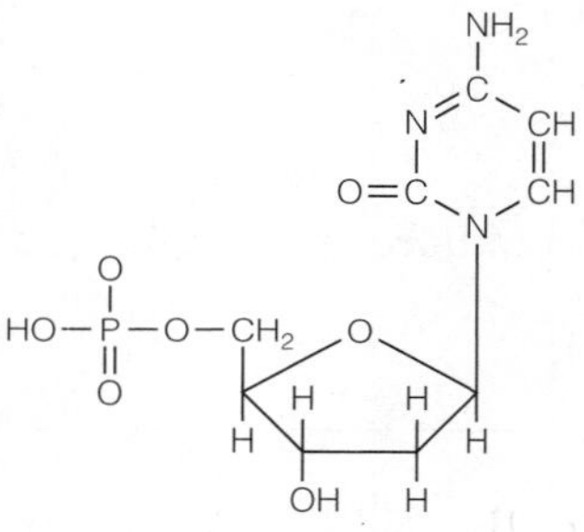 ( )   3._____ 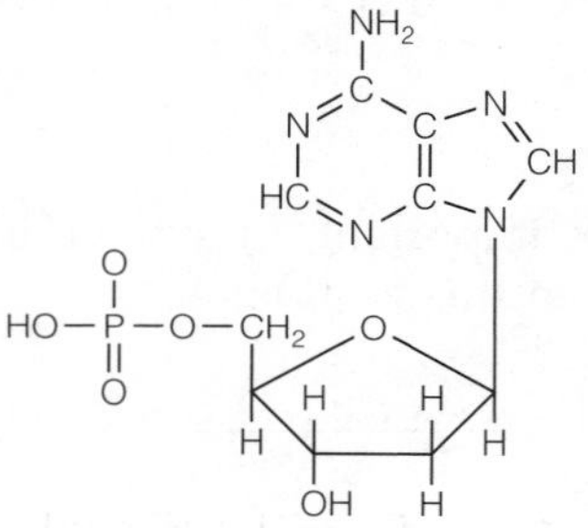 ( )   4._____ 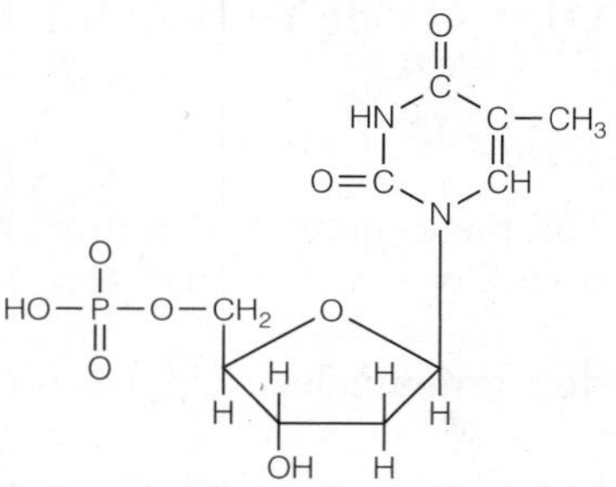 ( )   5._____ ( )

Label-Match

Identify each indicated part of the DNA illustration below. Choose from these answers: phosphate group, purine, pyrimidine, nucleotide, and deoxyribose. Complete the exercise by matching and entering the letter of the proper structure description in the parentheses following each label.

The following DNA memory devices may be helpful: Use pyrCUT to remember that the single-ring pyrimidines are cytosine, uracil, and thymine; use purAG to remember that the double-ring purines are adenine and guanine; pyrimidine is a long name for a narrow molecule; purine is a short name for a wide molecule; to recall the number of hydrogen bonds between DNA bases, remember that AT = 2 and CG = 3.

6. _____ ( )

7. _____ _____ ( )

8. _____ ( )

9. _____ ( )

10. _____ ( )

11. _____ ( )

12. _____ ( )

A. The pyrimidine is thymine because it has two hydrogen bonds.
B. A five-carbon sugar joined to two phosphate groups in the upright portion of the DNA ladder.
C. The purine is guanine because it has three hydrogen bonds.
D. The pyrimidine is cytosine because it has three hydrogen bonds.
E. The purine is adenine because it has two hydrogen bonds.
F. Composed of three smaller molecules: a phosphate group, five-carbon deoxyribose sugar, and a nitrogenous base (in this case, a pyrimidine).
G. A chemical group between two sugars in the upright portion of the DNA ladder.

True/False

If the statement is true, write a T in the blank. If the statement is false, make it correct by changing the underlined word(s) and writing the correct word(s) in the answer blank.

____T____ 13. DNA is composed of <u>four</u> different types of nucleotides.

____T____ 14. In the DNA of every species, the amount of adenine present always equals the amount of <u>thymine</u>, and the amount of cytosine always equals the amount of <u>guanine</u> (A = T and C = G).

_____ 15. In a nucleotide, the phosphate group is attached to the <u>nitrogen-containing base</u>, which is attached to the five-carbon sugar.

____F____ 16. Watson and Crick built their model of DNA in the early <u>1950s</u>.

____F____ 17. Guanine pairs with <u>cytosine</u> and adenine pairs with <u>thymine</u> by forming hydrogen bonds between them.

Fill-in-the-Blanks

Base (18)_____ between the two nucleotide strands in DNA is (19)_____ for all species (A-T; G-C). The base (20)_____ (determining which base follows the next in a nucleotide strand) is (21)_____ from species to species.

Short Answer

22. Explain why understanding the structure of DNA helps scientists understand how living organisms can have so much in common at the molecular level and yet be so diverse at the whole organism level (p. 158).

_____

_____

DNA REPLICATION (11-III, pp. 158–159)/DNA REPAIR (11-III, p. 159)

Terms

DNA replication (158)_____

DNA repair (159)_____

Labeling

1. The term *semiconservative replication* refers to the fact that each new DNA molecule resulting from the replication process is "half-old, half-new." In the illustration below, complete the replication required in the middle of the molecule by adding the required letters representing the missing nucleotide bases. Recall that ATP energy and the appropriate enzymes are actually required in order to complete this process.

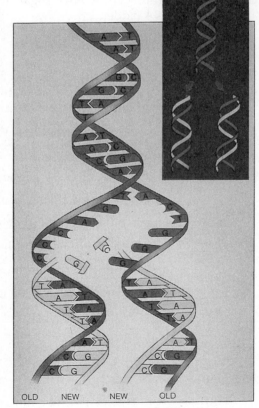

T - ___   ___ - A

G - ___   ___ - C

A - ___   ___ - T

C - ___   ___ - G

C - ___   ___ - G

C - ___   ___ - G

old new   new old

True/False

If the statement is true, write a T in the blank. If the statement is false, make it correct by changing the underlined word(s) and writing the correct word(s) in the answer blank.

_____ 2. The hydrogen bonding of adenine to <u>guanine</u> during replication is an example of complementary base pairing.

_____ 3. The replication of DNA is considered a <u>semiconservative</u> process because <u>the same four nucleotides are used again and again during replication</u>.

_____ 4. Each parent strand <u>remains intact</u> during replication, and a new companion strand is assembled on each of those parent strands.

_____ 5. Some of the enzymes associated with DNA assembly repair <u>errors</u> during the replication process.

ORGANIZATION OF DNA IN CHROMOSOMES (11-IV, p. 160)

Terms

histones (160)_____

nucleosome (160)_____

Fill-in-the-Blanks

Each chromosome has one (1)_____ molecule coursing through it. Eukaryotic DNA is complexed tightly with many (2)_____. Some (3)_____ proteins act as spools to wind up small pieces of (4)_____. A (5)_____ is a histone-DNA spool. The way the chromosome is packed is known to influence the activity of different (6)_____.

Labeling

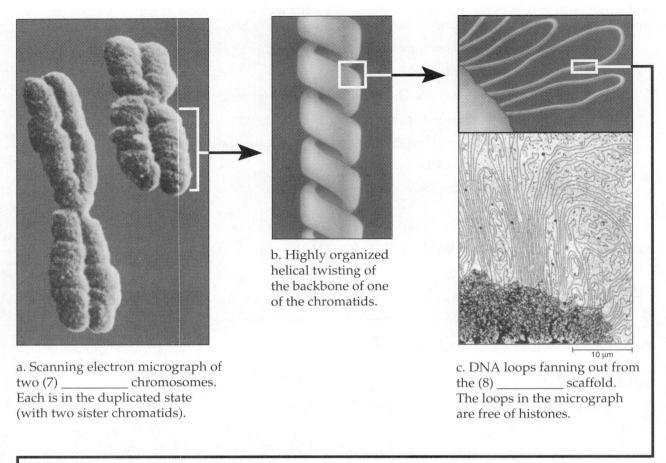

a. Scanning electron micrograph of two (7) _____ chromosomes. Each is in the duplicated state (with two sister chromatids).

b. Highly organized helical twisting of the backbone of one of the chromatids.

c. DNA loops fanning out from the (8) _____ scaffold. The loops in the micrograph are free of histones.

10 μm

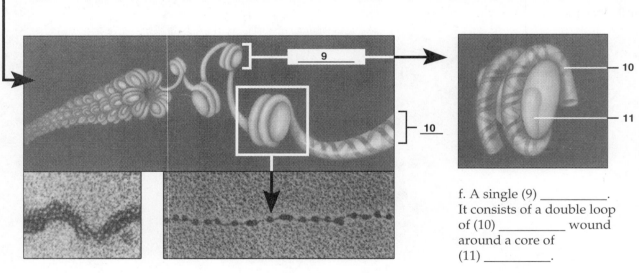

9

10

10

11

d. A cylindrical fiber (solenoid), 30-nm diameter. It results from DNA-protein interactions that lead torepeated coiling

e. In decondensed form, the proteins look like beads on a string (the "string" being DNA). Each "bead" is a (9) _____.

f. A single (9) _____. It consists of a double loop of (10) _____ wound around a core of (11) _____.

Crossword Puzzle: DNA Structure and Function

ACROSS

3. Developed the best x-ray diffraction images of DNA and deduced its basic dimensions.
5. DNA polymerases and DNA ligases _____ DNA.
8. Gaseous medium inhaled by Earth's organisms.
9. The molecule to which DNA passes its genetic code information (Chapter 12).
10. Built a model of DNA structure; won Nobel Prize.
11. Built a model of DNA structure; won Nobel Prize.
14. Not single, but _____.
15. Coiled structure, like a circular stairway.
18. Forms two hydrogen bonds with its mate.
19. Single-ringed nitrogenous base.
20. One of several kinds of proteins that bind tightly to DNA and organize it during its activities.
21. Duplication of hereditary material during the S phase of the cell cycle.
22. A new DNA strand is assembled on each of the two parent strands of DNA.
23. Forms three hydrogen bonds with its mate.
24. Five carbon sugar + phosphate group + nitrogenous base.

DOWN

1. In most organisms, the molecule that contains the genetic code.
2. _____ and Chase demonstrated that DNA contained the hereditary instructions for producing new bacteriophages.
4. A class of viruses that infect bacteria
6. A prefix that means "the same."
7. Showed that harmless cells could become pathogenic by acquiring a substance that was probably DNA (1944).
10. To move forward with one's belly close to the ground.
12. A double loop of DNA wound around a core of histone molecules.
13. Forms three hydrogen bonds with its mate.
16. Forms two hydrogen bonds with its mate.
17. Double-ringed nitrogenous base.

Self-Quiz

___1. Each DNA strand has a backbone that consists of alternating _____.
a. purines and pyrimidines
b. nitrogen-containing bases
c. hydrogen bonds
d. sugar and phosphate molecules

___2. In DNA, complementary base pairing occurs between _____.
a. cytosine and uracil
b. adenine and guanine
c. adenine and uracil
d. adenine and thymine

___3. Adenine and guanine are _____.
a. double-ringed purines
b. single-ringed purines
c. double-ringed pyrimidines
d. single-ringed pyrimidines

___4. Franklin used the technique known as _____ to determine many of the physical characteristics of DNA.
a. transformation
b. transmission electron microscopy
c. density-gradient centrifugation
d. x-ray diffraction

___5. The significance of Griffith's experiment that used two strains of pneumonia-causing bacteria is that _____.
a. the conserving nature of DNA replication was finally demonstrated
b. it demonstrated that harmless cells had become permanently transformed through a change in the bacterial hereditary system
c. it established that pure DNA extracted from disease-causing bacteria transformed harmless strains into "pathogenic strains"
d. it demonstrated that radioactively labeled bacteriophages transfer their DNA but not their protein coats to their host bacteria

___6. The significance of the experiments in which ^{32}P and ^{35}S were used is that _____.
a. the semiconservative nature of DNA replication was finally demonstrated
b. it demonstrated that harmless cells had become permanently transformed through a change in the bacterial hereditary system
c. it established that pure DNA extracted from disease-causing bacteria transformed harmless strains into "killer strains"
d. it demonstrated that radioactively labeled bacteriophages transfer their DNA but not their protein coats to their host bacteria

___7. Franklin's research contribution was essential in _____.
a. establishing the double-stranded nature of DNA
b. establishing the principle of base pairing
c. establishing most of the principal structural features of DNA
d. all of the above

___8. When Griffith injected mice with a mixture of dead pathogenic cells—encapsulated S cells and living, unencapsulated R cells of pneumonia bacteria—he discovered that _____.
a. the previously harmless strain had permanently inherited the capacity to build protective capsules
b. the dead mice teemed with living pathogenic (R) cells
c. the killer strain R was encased in a protective capsule
d. all of the above

___9. A single strand of DNA with the base-pairing sequence C-G-A-T-T-G is compatible only with the sequence _____.
a. C-G-A-T-T-G
b. G-C-T-A-A-G
c. T-A-G-C-C-T
d. G-C-T-A-A-C

___10. The nucleosome is a _____.
a. subunit of a nucleolus
b. coiled bead of histone-DNA
c. DNA packing arrangement within a chromosome
d. term synonymous with gene
e. both (b) and (c)

Chapter Objectives/Review Questions

This section lists general and detailed chapter objectives that can be used as review questions. You can make maximum use of these items by writing answers on a separate sheet of paper. Fill in answers where blanks are provided. To check for accuracy, compare your answers with information given in the chapter or glossary.

Page   *Objectives/Questions*

(155)   1. Before 1952, _____ molecules and _____ molecules were suspected of housing the genetic code.

(155–156) 2. Summarize the research carried out by Miescher, Griffith, Avery and colleagues, and Hershey and Chase; state the specific advances made by each in the understanding of genetics.

(156)   3. Viruses called _____ were used in early research efforts to discover the genetic material.

(156)   4. Summarize the specific research that demonstrated that DNA, not protein, governed inheritance.

(157)   5. DNA is composed of double-ring molecules known as _____ and single-ring molecules known as _____; the two purines are _____ and _____, while the two pyrimidines are _____ and _____.

(158)   6. Draw the basic shape of a deoxyribose molecule and show how a phosphate group is joined to it when forming a nucleotide.

(158)   7. Show how each nucleotide base would be joined to the sugar-phosphate combination drawn in objective 6.

(157)   8. List the pieces of information about DNA structure that Rosalind Franklin discovered through her x-ray diffraction research.

(158)   9. The two scientists who assembled the clues to DNA structure and produced the first model were _____ and _____.

(158–159) 10. Explain what is meant by the pairing of nitrogen-containing bases (base pairing), and explain the mechanism that causes bases of one DNA strand to join with bases of the other strand.

(158)   11. Assume that the two parent strands of DNA have been separated and that the base sequence on one parent strand is A-T-T-C-G-C; the base sequence that will complement that parent strand is _____.

(158)   12. Describe how double-stranded DNA replicates from stockpiles of nucleotides.

(159)   13. Explain what is meant by "each parent strand is conserved in each new DNA molecule."

(159)   14. During DNA replication, enzymes called DNA _____ assemble new DNA strands.

(160)   15. The basic histone-DNA packing unit of the chromosome is the _____.

(160)   16. List possible reasons for the highly organized packing of nucleoprotein into chromosomes.

Integrating and Applying Key Concepts

Review the stages of mitosis and meiosis, as well as the process of fertilization. Include what has now been learned about DNA replication and the relationship of DNA to a chromosome. As you cover the stages, be sure each cell receives the proper number of DNA threads.

Critical Thinking Exercises

1. If each nucleotide in DNA paired only with itself (A with A, C with C, and so on), which of the following statements would have the greatest validity?
   a. A pairs with T and C pairs with G.
   b. DNA could not replicate.
   c. DNA could not carry genetic information.
   d. The two strands of each DNA molecule would be identical.
   e. The diameter of the DNA molecule would be variable.

Answers

Answers to Interactive Exercises

DISCOVERY OF DNA FUNCTION (11-I)
1. a. Miescher: identified "nuclein" from nuclei of pus cells and fish sperm; discovered DNA; b. Griffith: discovered the transforming principle in *Streptococcus pneumoniae*; live, harmless R cells were mixed with dead S cells; R cells became S cells; c. Avery: reported that the transforming substance in Griffith's bacteria experiments was probably DNA, the substance of heredity; d. Hershey and Chase: worked with radioactive sulfur (protein) and phosphorus (DNA) labels; T4 bacteriophage and *E. coli* demonstrated that labeled phosphorus was in bacteriophage DNA and contained hereditary instructions for new bacteriophages.

DNA STRUCTURE (11-II)
1. A five-carbon sugar called deoxyribose, a phosphate group, and one of the four nitrogen-containing bases; 2. guanine (purine); 3. cytosine (pyrimidine); 4. adenine (purine); 5. thymine (pyrimidine); 6. deoxyribose (B); 7. phosphate group (G); 8. purine (C); 9. pyrimidine (A); 10. purine (E); 11. pyrimidine (D); 12. nucleotide (F); 13. T; 14. T; 15. F, sugar; 16. T; 17. T; 18. pairing; 19. constant; 20. sequence; 21. different; 22. Living organisms have so many diverse body structures and behave in different ways because the many different habitats of Earth have selected those genotypes most able to survive in those habitats. The remaining genotypes have perished. The directions that code for the building of those body structures and that enable the specific successful behaviors reside in DNA or in a few cases, RNA. All living organisms follow the same rules for base pairing between the two nucleotide strands in DNA; adenine always pairs with thymine in undamaged DNA and cytosine always pairs with guanine. All living organisms must extract energy from food molecules and the reactions of glycolysis occur in virtually all of Earth's species. That means that similar enzyme sequences enable similar metabolic pathways to occur. While virtually all living organisms on Earth use the same code and the same enzymes during replication, transcription, and translation, the particular array of proteins being formed differs from individual to individual, even in the same species, according to the sequences of nitrogenous bases that make up an individual's chromosome(s), and therein lies the key to the enormous diversity of life on Earth: No two individuals have the exact same array of proteins in their phenotypes.

DNA REPLICATION/DNA REPAIR (11-III)
1.
T - A	T - A
G - C	G - C
A - T	A - T
C - G	C - G
C - G	C - G
C - G	C - G
old new	new old

2. Thymine (during replication) or uracil (during transcription); 3. it is a conserving process because each "new" DNA molecule contains one "old" strand from the parent cell attached to a strand of "new" complementary nucleotides that were assembled from stockpiles in the cell; 4. T; 5. T.

ORGANIZATION OF DNA IN CHROMOSOMES (11-IV)
1. DNA; 2. histones; 3. histone; 4. DNA; 5. nucleosome; 6. genes; 7. metaphase; 8. protein; 9. nucleosome; 10. DNA; 11. histones.

```
D   H       F R A N K L I N   B     R E P A I R
N   E   G                     A           S
A I R   R N A           C R I C K   W A T S O N
    S   I       N       R     T
C   H   F   D O U B L E   H E L I X       A   P
Y   E   F       C       E     R           D   U
T H Y M I N E   L       P Y R I M I D I N E   R
O       T       E       O             N       I
S       H I S T O N E   R E P L I C A T I O N
I           S           H             N       E
N       S E M I C O N S E R V A T I V E   E
E       M               G
    G U A N I N E   N U C L E O T I D E
```

Answers to Self-Quiz

1. d; 2. d; 3. a; 4. d; 5. b; 6. d; 7. c; 8. d; 9. d; 10. e.

Critical Thinking Analysis

a. This statement simply denies the conditions of the question. While this is the actual observed pairing pattern, the question concerns the consequences of a different pattern.

b. The essence of the replication process is that base pairing is specific. It does not matter which base pairs with which. As long as each kind of base pairs with only one other kind, each strand will direct the synthesis of a complementary strand, and the double-stranded molecule will be replicated.

c. The genetic information is encoded in the sequence of nucleotides along the strand. This is not affected by the pattern of pairing between the two strands.

d. This statement is true to an extent. If you started at one end of the double-stranded molecule and followed both strands, they would have identical sequences of nucleotides. However, remember that the two ends of a single strand of DNA are different and that the two strands in a double helix run in opposite directions. If you separated the two strands, placed them with their similar ends together, and read their sequences, you would find that the sequences were opposite each other.

e. A and G are larger than C and T. Thus, A-A and G-G pairs would create wide spots in the double-stranded molecule, and C-C and T-T pairs would be narrow.

12

FROM DNA TO PROTEINS

Interactive Exercises

TRANSCRIPTION AND TRANSLATION: AN OVERVIEW (12-I, p. 163)

Terms

The page-referenced terms are important; they were in boldface type in the chapter. Refer to the instructions given in Chapter 1, p. 1 in this workbook.

base sequence (163)_____

transcription (163) _____

_____

translation (163)_____

_____

RNA, ribonucleic acid (163)_____

_____

mRNA (163)_____

rRNA (163)_____

tRNA (163)_____

Fill-in-the-Blanks

Which base follows the next in a strand of DNA is referred to as the base (1)_____. A region of DNA that calls for the assembly of specific amino acids into a polypeptide chain is a (2)_____. The two steps from genes to proteins are called (3)_____ and (4)_____. In (5)_____, single-stranded molecules of RNA are assembled on DNA templates in the nucleus. In (6)_____, the RNA molecules are shipped from the nucleus into the cytoplasm, where they are used as templates for assembling (7)_____ chains. Following translation, one or more chains become (8)_____ into the three-dimensional shape of protein molecules. Proteins have (9)_____ and (10)_____ roles in cells, including control of DNA.

Complete the Table

11. Three types of RNA are transcribed from DNA in the nucleus (from genes that code only for RNA). Complete the following table, which summarizes information about these molecules.

RNA Molecule	Abbreviation	Description/Function
a. Ribosomal RNA		
b. Messenger RNA		
c. Transfer RNA		

TRANSCRIPTION OF DNA INTO RNA (12-II, pp. 164–166)

Terms

uracil (164)_____

promoter (164)_____

exons (165)_____

introns (165)_____

Short Answer

1. List three ways in which a molecule of RNA is structurally different from a molecule of DNA.

_____

_____

2. Cite two similarities in DNA replication and transcription._____

_____

3. What are the three key ways in which transcription differs from DNA replication?

_____

Sequence

Arrange the steps of transcription in correct chronological sequence. Write the letter of the first step next to 4, the letter of the second step next to 5, and so on.

4.___   A. The RNA strand grows along exposed bases until RNA polymerase meets a DNA base sequence that signals "stop."
5.___   B. RNA polymerase binds with the DNA promoter region to open up a local region of the DNA double helix.
6.___   C. An RNA polymerase enzyme locates the DNA bases of the promoter region of one DNA strand by recognizing DNA-associated proteins near a promoter.
7.___   D. RNA is released from the DNA template as a free, single-stranded transcript.
8.___   E. RNA polymerase moves stepwise along exposed nucleotides of one DNA strand; as it moves, the DNA double helix keeps unwinding.

Completion

9. Suppose the line below represents the DNA strand that will act as a template for the production of mRNA through the process of transcription. Fill in the blanks below the DNA strand with the sequence of complementary bases that will represent the message carried from DNA to the ribosome in the cytoplasm.

A   T   C   G   C   A   A   T   C   G   G   A   T   A   G   G   C   A

—   —   —   —   —   —   —   —   —   —   —   —   —   —   —   —   —   —

(transcribed single strand of mRNA)

Label-Match

Newly transcribed mRNA contains more genetic information than is necessary to code for a chain of amino acids. Before the mRNA leaves the nucleus for its ribosome destination, an editing process occurs as certain portions of nonessential information are snipped out. Identify each indicated part of the illustration below; use abbreviations for the nucleic acids. Complete the exercise by matching and entering the letter of the description in the parentheses following each label.

10._____ ( )

11._____ ( )

12._____ ( )

13._____ ( )

14._____ ( )

15._____ _____ _____ ( )

A. The actual coding portions of mRNA
B. Noncoding internal portions of the newly transcribed mRNA
C. Presence of cap and tail, introns snipped out and exons spliced together
D. Acquiring of a poly-A tail by the maturing mRNA transcript
E. The region of the DNA template strand to be copied
F. Reception of a nucleotide cap by the 5′ end of mRNA (the first synthesized)

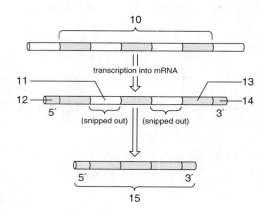

TRANSLATION OF mRNA (12-III, pp. 166–169)

Terms

genetic code (166)_____

_____

codon (166) _____

anticodon (167)_____

Matching

1.___ codon
2.___ three at a time
3.___ sixty-one
4.___ the genetic code
5.___ release factors
6.___ ribosome
7.___ anticodon
8.___ the "stop" codons

A. Composed of two subunits, the small subunit with P and A amino acid binding sites as well as a binding site for mRNA
B. Reading frame of the nucleotide bases in mRNA
C. Detach(es) protein and mRNA from the ribosome
D. UAA, UAG, UGA
E. A sequence of three nucleotide bases that can pair with a specific mRNA codon
F. Name for each base triplet in mRNA
G. The number of codons that actually specify amino acids
H. Term for how the nucleotide sequences of DNA and then mRNA correspond to the amino acid sequence of a polypeptide chain

Complete the Table

9. Complete the following table, which distinguishes the stages of translation.

Translation Stage	Description
a.	Special initiator tRNA loads onto small ribosomal subunit and recognizes AUG; small subunit binds with mRNA, and large ribosomal subunit joins small one.
b.	Amino acids are strung together in sequence dictated by mRNA codons as the mRNA strand passes through the two ribosomal subunits; two tRNAs interact at P and A sites.
c.	mRNA "stop" codon signals the end of the polypeptide chain; release factors detach the ribosome and polypeptide chain from the mRNA.

Completion

10. Given the following DNA sequence, deduce the composition of the mRNA transcript:

TAC   AAG   ATA   ACA   TTA   TTT   CCT   ACC   GTC   ATC

___   ___   ___   ___   ___   ___   ___   ___   ___   ___

(mRNA transcript)

11. Deduce the composition of the tRNA anticodons that would pair with the above specific mRNA codons as these tRNAs deliver the amino acids (identified below) to the P and A binding sites of the small ribosomal subunit.

___   ___   ___   ___   ___   ___   ___   ___   ___   ___

(tRNA anticodons)

12. From the mRNA transcript in exercise 10, use Figure 12.5 of the text to deduce the amino acid composition of the polypeptide sequence.

___   ___   ___   ___   ___   ___   ___   ___   ___   ___

(amino acids)

Label-Match

A summary of the flow of genetic information in protein synthesis is useful as an overview. Identify the indicated parts of the illustration on the next page by filling in the blanks with the names of the appropriate structures or functions. Choose from the following: DNA, mRNA, tRNA, polypeptide, rRNA subunits, intron, exon, mature mRNA transcript, new mRNA transcript, anticodon, amino acids, ribosome-mRNA complex. Complete the exercise by matching and entering the letter of the description in the parentheses following each label.

13._____ ( )

14._____ _____ _____ ( )

15._____ ( )

16._____ ( )

17._____ _____ _____ ( )

18._____ ( )

19._____ _____ ( )

20._____ ( )

21._____ ( )

22._____ _____ ( )

23._____ ( )

24._____-_____ _____ ( )

25._____ ( )

A. Coding portion of mRNA that will translate into proteins
B. Carries a modified form of the genetic code from DNA in the nucleus to the cytoplasm
C. Transports amino acids to the ribosome and mRNA
D. The building blocks of polypeptides
E. Noncoding portions of newly transcribed mRNA
F. tRNA after delivering its amino acid to the ribosome-mRNA complex
G. Join when translation is initiated
H. Holds the genetic code for protein production
I. Place where translation occurs
J. Includes introns and exons
K. A sequence of three bases that can pair with a specific mRNA codon
L. Snipping out of introns, only exons remaining
M. May serve as a functional protein (enzyme) or a structural protein

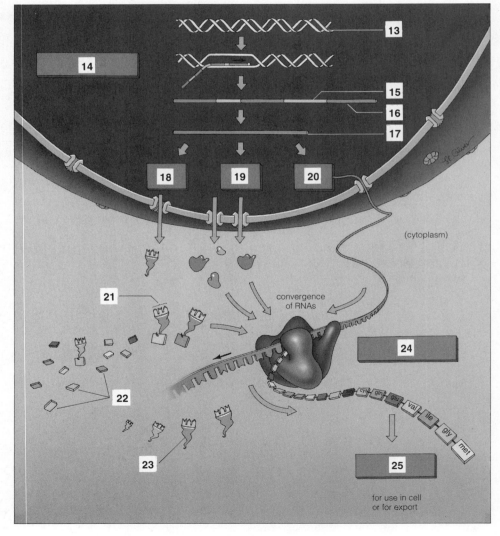

Fill-in-the-Blanks

The order of (26)_____ _____ in a protein is specified by a sequence of nucleotide bases. The genetic code is read in units of (27)_____ nucleotides; each unit of three codes for (28)_____ amino acid(s). In the table that showed which triplet specified a particular amino acid, the triplet code was incorporated in (29)_____ molecules. Each of these triplets is referred to as a(n) (30)_____ . (31)_____ alone carries the instructions for assembling a particular sequence of amino acids from the DNA to the ribosomes in the cytoplasm, where (32)_____ of the polypeptide occurs. (33)_____ RNA acts as a shuttle molecule as each type brings its particular (34)_____ _____ to the ribosome where it is to be incorporated into the growing (35)_____. A(n) (36)_____ is a triplet on mRNA that forms hydrogen bonds with a(n) (37)_____, which is a triplet on one end of tRNA.

MUTATION AND PROTEIN SYNTHESIS (12-IV, pp. 170–171)
THE NATURE OF GENE CONTROL (pp. 172–176)

Terms

gene mutations (170)_____

regulatory proteins (172)_____

operator (172) _____

operon (172)_____

repressors (172)_____

cell differentiation (174)_____

hormones (174)_____

cancer (175) _____

Short Answer

1. Define the term *mutagen* and cite several examples. _____

_____

Fill-in-the-Blanks

In addition to changes in chromosomes (crossing over, recombination, deletion, addition, translocation, and inversion), changes can also occur in the structure of DNA; these modifications are referred to as gene mutations. Complete the following exercise on types of spontaneous gene mutations.

Viruses, ultraviolet radiation, and certain chemicals are examples of environmental agents called (2)_____ that may enter cells and damage strands of DNA. If A becomes paired with C instead of T during DNA replication, this spontaneous mutation is a base-pair (3)_____. Sickle-cell anemia is a genetic disease whose cause has been traced to a single DNA base pair; the result is that one (4)_____ _____ is substituted for another in the beta chain of (5)_____. Some DNA regions "jump" to new DNA locations and often inactivate the genes in their new environment; such (6)_____ elements may give rise to observable changes in the phenotype of an organism.

Complete the Table

7. All of the diploid cells in an organism possess the same genes, and every cell utilizes most of the same genes; yet specialized cells must activate only certain genes. Some agents of gene control have been discovered. Transcriptional controls are the most common. Complete the following table to summarize the agents of gene control.

Agents of Gene Control	Method of Gene Control
a. Repressor protein	
b. Hormones	major agents of vertebrate gene control; signaling molecules that move through the bloodstream to affect gene expression in target cells
c. Promoter	
d.	short DNA base sequences between promoter and the start of a gene; a binding site for control agents

Label-Match

Escherichia coli, a bacterial cell living in mammalian digestive tracts, is able to exert a negative type of gene control over lactose metabolism. Use the numbered blanks to identify each part of the illustration on the next page. Use abbreviations for nucleic acids. Choose from the following:

     lactose   regulator gene   promoter   mRNA transcript   enzyme genes   lactose operon

     lactose enzymes   repressor-lactose complex   repressor protein   RNA polymerase   operator

Complete the exercise by matching and entering the letter of the proper function description in the parentheses following each label.

8._____ _____ ( )

9._____ _____ ( )

10._____ _____ ( )

11._____ ( )

12._____ ( )

13._____ _____ ( )

14._____ _____ ( )

15._____-_____ _____ ( )

16._____ ( )

17._____ _____ ( )

18._____ _____ ( )

A. Includes promoter, operator, and the lactose-metabolizing enzymes
B. Short DNA base sequence between promoter and the beginning of a gene
C. The nutrient molecule in the lactose operon
D. Major enzyme that catalyzes transcription
E. Binds to operator and overlaps promoter; this prevents RNA polymerase from binding to DNA and initiating transcription
F. Prevents repressor from binding to the operator
G. Genes that produce lactose-metabolizing enzymes
H. Catalyze the digestion of lactose
I. Carries genetic instructions to ribosomes for production of lactose enzymes
J. Specific base sequence that signals the beginning of a gene
K. Gene that contains coding for production of repressor protein

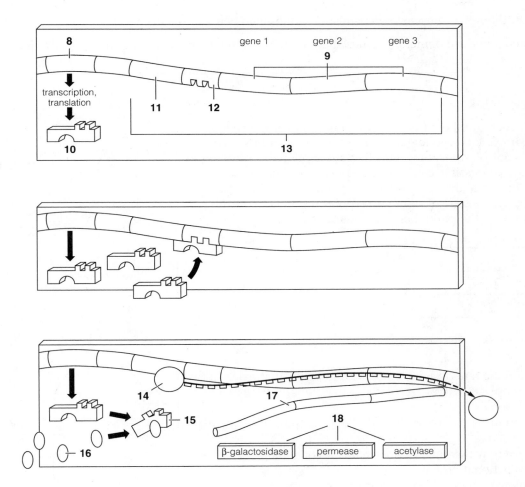

Fill-in-the-Blanks

(19)_____ _____ is a species of bacterium that lives in mammalian digestive tracts and provided some of the first clues about gene control. A(n) (20)_____ is any group of genes together with its promoter and operator sequence. Promoter and operator provide (21)_____ _____. The (22)_____ codes for the formation of mRNA, which assembles a repressor protein. The affinity of the (23)_____ for RNA polymerase dictates the rate at which a particular operon will be transcribed. Repressor protein allows (24)_____ _____ over the lactose operon. Repressor binds with operator and overlaps promoter when lactose concentrations are (25)_____. This blocks (26)_____ _____ from the genes that will process lactose. This (27) [choose one] ( ) blocks, ( ) promotes production of lactose-processing enzymes. When lactose is present, lactose molecules bind with the (28)_____ _____. Thus, repressor cannot bind to (29)_____, and RNA polymerase has access to the lactose-processing genes. This gene control works well because lactose-degrading enzymes are not produced unless they are (30)_____.

31. Although a complex organism such as a human being arises from a single cell, the zygote, differentiation occurs in development. Define differentiation and relate it to a definition of selective gene expression._____

Label-Match

Identify each numbered part of the illustration at the right, which shows eukaryotic gene control. Choose from translational control, transport controls, transcriptional controls, transcript processing controls, and post-translational controls. Complete the exercise by correctly matching and entering the letter of the corresponding gene control description in the parentheses following each label.

32._____ _____ ( )

33._____ _____ _____ ( )

34._____ _____ ( )

35._____ _____ ( )

36._____ - _____ _____ ( )

A. Govern the rates at which mRNA transcripts that reach the cytoplasm will be translated into polypeptide chains at ribosomes
B. Govern modification of the initial mRNA transcripts in the nucleus
C. Govern how the polypeptide chains become modified into functional proteins
D. Dictate which mature mRNA transcripts will be shipped out of the nucleus and into the cytoplasm for translation
E. Influence when and to what degree a particular gene will be transcribed (if at all)

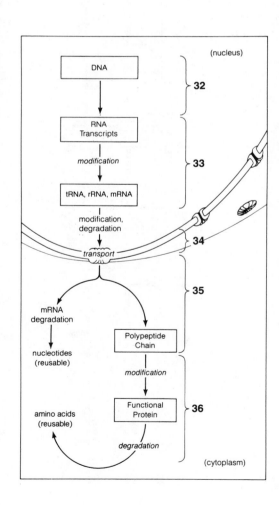

Fill-in-the-Blanks

All diploid cells in our bodies contain copies of the same (37)_____. These genetically identical cells become structurally and functionally distinct from one another through a process called (38)_____, which arises through (39)_____ gene expression in different cells. Cells depend on (40)_____, which govern transcription, translation, and enzyme activity. Controls that operate during transcription and transcript processing utilize (41)_____ proteins, especially (42)_____ that are turned on and off by the addition and removal of phosphate. A (43)_____ body is a condensed X chromosome. X chromosome inactivation produces adult human females who are (44)_____ for X-linked traits. This effect is shown in human females as patches of skin that lack normal sweat glands, a disorder known as (45)_____ _____ _____ and provides evidence for (46)_____ gene expression.

True/False

If the statement is true, write a T in the blank. If the statement is false, make it correct by changing the underlined word(s) and writing the correct word(s) in the answer blank.

_____ 47. When cells become cancerous, cell populations <u>decrease</u> to very <u>low</u> densities and <u>stop</u> dividing.

_____ 48. <u>All</u> abnormal growths and massings of new tissue in any region of the body are called tumors.

_____ 49. Malignant tumors have cells that <u>migrate and divide</u> in other organs.

_____ 50. Oncogenes are genes that <u>combat</u> cancerous transformations.

_____ 51. Proto-oncogenes <u>rarely</u> trigger cancer.

_____ 52. The normal expression of proto-oncogenes is vital, even though their <u>normal</u> expression may be lethal.

Self-Quiz

___1. Transcription _____.
  a. occurs on the surface of the ribosome
  b. is the final process in the assembly of a protein
  c. occurs during the synthesis of any type of RNA by use of a DNA template
  d. is catalyzed by DNA polymerase

___2. _____ carry(ies) amino acids to ribosomes, where amino acids are linked into the primary structure of a polypeptide.
  a. mRNA
  b. tRNA
  c. Introns
  d. rRNA

___3. Transfer RNA differs from other types of RNA because it _____.
  a. transfers genetic instructions from cell nucleus to cytoplasm
  b. specifies the amino acid sequence of a particular protein
  c. carries an amino acid at one end
  d. contains codons

___4. _____ dominates the process of transcription.
  a. RNA polymerase
  b. DNA polymerase
  c. Phenylketonuria
  d. Transfer RNA

___5. _____ and _____ are found in
RNA but not in DNA.
a. Deoxyribose; thymine
b. Deoxyribose; uracil
c. Uracil; ribose
d. Thymine; ribose

___6. Each "word" in the mRNA language con-
sists of _____ letters.
a. three
b. four
c. five
d. more than five

___7. If each nucleotide codes for only one
amino acid, how many different types of
amino acids could be selected?
a. four
b. sixteen
c. twenty
d. sixty-four

___8. The genetic code is composed of
_____ codons.
a. three
b. twenty
c. sixteen
d. sixty-four

___9. _____ binds to operator whenever
lactose concentrations are low.
a. Operon
b. Repressor

c. Promoter
d. Operator

___10. Any gene or group of genes together with
its promoter and operator sequence is a(n)
_____.
a. repressor
b. operator
c. promoter
d. operon

___11. The operon model explains the regulation
of _____ in prokaryotes.
a. replication
b. transcription
c. induction
d. selective gene expression

___12. In multicelled eukaryotes, cell differentia-
tion occurs as a result of _____.
a. growth
b. selective gene expression
c. repressor molecules
d. the death of certain cells

___13. _____ controls govern the rates at
which mRNA transcripts that reach the
cytoplasm will be translated into polypeptide chains at the ribosomes.
a. Transport
b. Transcript processing
c. Translational
d. Transcriptional

Chapter Objectives/Review Questions

This section lists general and detailed chapter objectives that can be used as review questions. You can make maximum use of these items by writing answers on a separate sheet of paper. Fill in answers where blanks are provided. To check for accuracy, compare your answers with information given in the chapter or glossary.

Page   *Objectives/Questions*

(163–164) 1. State how RNA differs from DNA in structure and function, and indicate what features RNA has in common with DNA.

(163)   2. _____ RNA combines with certain proteins to form the ribosome; _____ RNA carries genetic information for protein construction from the nucleus to the cytoplasm; _____ RNA picks up specific amino acids and moves them to the area of mRNA and the ribosome.

(164)   3. Describe the process of transcription and indicate three ways in which it differs from replication.

(166)   4. What RNA code would be formed from the following DNA code: TAC-CTC-GTT-CCC-GAA?

(163–166) 5. State the relationship between the DNA genetic code and the order of amino acids in a protein chain.

(166)   6. Scrutinize Figure 12.5 in the text and decide whether the genetic code in this instance applies to DNA, mRNA, or tRNA.

(166–169)   7. Describe how the three types of RNA participate in the process of translation.

(171)   8. Explain how the DNA message TAC-CTC-GTT-CCC-GAA would be used to code for a segment of protein, and state what its amino acid sequence would be.

(172)   9. The negative control of _____ protein prevents the enzymes of transcription from binding to DNA; the positive control of _____ protein enhances the binding of RNA polymerases to DNA.

(172)   10. Gene expression is controlled through regulatory _____, hormones, and other molecules that interact with DNA, _____, and the protein products of _____.

(172)   11. The cells of *E. coli* manage to produce enzymes to degrade lactose when those molecules are _____ and to stop production of lactose-degrading enzymes when lactose is _____.

(174)   12. Explain how selective gene expression relates to cell differentiation in multicelled eukaryotes.

(174)   13. Explain how X-chromosome inactivation provides evidence for selective gene expression; use the example of anhidrotic ectodermal dysplasia.

(176)   14. Describe the relationship of proto-oncogenes, environmental irritants, and oncogenes.

Integrating and Applying Key Concepts

Genes code for specific polypeptide sequences. Not every substance in living cells is a polypeptide. Explain how genes might be involved in the production of a storage starch (such as glycogen) that is constructed from simple sugars.

Critical Thinking Exercises

1. The existence of introns was first inferred from a technique called hybridization. DNA and mRNA were isolated from cells, mixed and heated, then cooled to allow base pairing between the two kinds of molecules. When the resulting DNA-RNA "hybrid" double-stranded molecules were examined with electron-microscopy, the DNA strands were looped away from the mRNA strands.

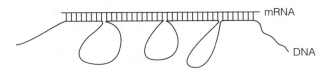

The DNA loops were thought to be nucleotide sequences that were not complementary to any part of the mRNA. Thus, the mRNA was made of sequences that occurred discontinuously along the DNA. Which of the following assumptions were these investigators most likely making?

   a. Radioactivity causes mutations.
   b. Heating DNA changes the sequence of its monomers.
   c. The DNA loops were nucleotide sequences that were not complementary to any part of the RNA.
   d. Only nucleic acids with complementary sequences of monomers can stick together by base pairing.
   e. Only nucleic acids with the same sequence of monomers can stick together by base pairing.

Answers

Answers to Interactive Exercises

TRANSCRIPTION AND TRANSLATION: AN OVERVIEW (12-I)

1. sequence; 2. gene; 3. transcription (translation); 4. translation (transcription); 5. transcription; 6. translation; 7. protein; 8. folded; 9. structural (functional); 10. functional (structural); 11. a. Ribosomal RNA; rRNA; RNA molecule that associates with certain proteins to form the ribosome, the "workbench" on which polypeptide chains are assembled; b. Messenger RNA; mRNA; RNA molecule that moves to the cytoplasm, complexes with the ribosome where translation will result in polypeptide chains; c. Transfer RNA; tRNA; RNA molecule that moves into the cytoplasm, picks up a specific amino acid, and moves it to the ribosome where tRNA pairs with a specific mRNA code word for that amino acid.

TRANSCRIPTION OF DNA INTO RNA (12-II)

1. RNA molecules are single-stranded, while DNA has two strands; uracil substitutes in RNA molecules for thymine in DNA molecules; ribose sugar is found in RNA, while DNA has deoxyribose sugar; 2. Both DNA replication and transcription follow base-pairing rules; nucleotides are added to a growing RNA strand one at a time as in DNA replication; 3. Only one region of one DNA strand serves as a template for transcription; transcription requires different enzymes (three types of RNA polymerase); the results of transcription are single-stranded RNA molecules, but replication results in DNA, a double-stranded molecule; 4. C; 5. B; 6. E; 7. A; 8. D; 9. U-A-G-C-G-U-U-A-G-C-C-U-A-U-C-C-G-U; 10. DNA (E); 11. introns (B); 12. cap (F); 13. exons (A); 14. tail (D); 15. mature mRNA transcript (C).

TRANSLATION OF mRNA (12-III)

1. F; 2. B; 3. G; 4. H; 5. C; 6. A; 7. E; 8. D; 9. a. Initiation; b. Chain elongation; c. Chain termination. 10. mRNA transcript: AUG UUC UAU UGU AAU AAA GGA UGG CAG UAG; 11. tRNA anticodons: UAC AAG AUA ACA UUA UUU CCU ACC GUC AUC; 12. amino acids: met phe tyr cys asn lys gly try gln stop (start); 13. DNA (H); 14. new mRNA transcript (J); 15. intron (E); 16. exon (A); 17. mature mRNA transcript (L); 18. tRNAs (C); 19. rRNA subunits (G); 20. mRNA (B); 21. anticodon (K); 22. amino acids (D); 23. tRNA (F); 24. ribosome-mRNA complex (I); 25. polypeptide (M); 26. amino acids; 27. three; 28. one; 29. mRNA; 30. codon; 31. mRNA; 32. translation (synthesis); 33. Transfer; 34. amino acid; 35. protein (polypeptide); 36. codon; 37. anticodon.

MUTATION AND PROTEIN SYNTHESIS (12-IV)/ THE NATURE OF GENE CONTROL

1. Mutagens are environmental agents that attack a DNA molecule and modify its structure. Viruses, ultraviolet radiation, and certain chemicals are examples; 2. mutagens; 3. substitution; 4. amino acid; 5. hemoglobin; 6. transposable; 7. a. Repressor protein; prevent transcription enzymes (RNA polymerases) from binding to DNA; this is negative transcription control; b. Hormones; major agents of vertebrate gene control; signaling molecules that move through the bloodstream to affect gene expression in target cells; c. Promoter; specific base sequences on DNA that serve as binding sites for control agents; before RNA assembly can occur on DNA, the enzymes must bind with the promoter site; d. Operators; short DNA base sequences between promoter and the start of a gene; a binding site for control agents; 8. regulator gene (K); 9. enzyme genes (G); 10. repressor protein (E); 11. promoter (J); 12. operator (B); 13. lactose operon (A); 14. RNA polymerase (D); 15. repressor-lactose complex (F); 16. lactose (C); 17. mRNA transcript (I); 18. lactose enzymes (H); 19. *Escherichia coli*; 20. operon; 21. transcription controls; 22. regulator; 23. promoter; 24. negative control; 25. low; 26. RNA polymerase (mRNA transcription); 27. blocks; 28. repressor protein; 29. operator; 30. needed (required); 31. All cells in the body descend from the same zygote; as cells divide to form the body, they become specialized in composition, structure, and function—they differentiate through selective gene expression. 32. transcriptional control (E); 33. transcript processing control (B); 34. transport control (D); 35. translational control (A); 36. post-translational control (C); 37. DNA (genes); 38. differentiation; 39. selective; 40. controls; 41. regulatory; 42. activators; 43. Barr; 44. mosaic; 45. anhidrotic ectodermal dysplasia; 46. selective; 47. F, increase, high, start; 48. T; 49. T; 50. F, bring about; 51. T; 52. F, abnormal.

Answers to Self-Quiz

1. c; 2. b; 3. c; 4. a; 5. c; 6. a; 7. a; 8. d; 9. b; 10. d; 11. b; 12. b; 13. c.

Critical Thinking Analysis

a. Radioactivity does cause mutations, but in order to make inferences about the sequences of nucleotides, it must be assumed that any incident radioactivity does not change the sequences during the course of the experiment.

b. As in (a), it is necessary to assume the opposite. If heating changed the sequences of nucleotides, it would be impossible to infer the sequence of native DNA from the behavior of heated DNA.

c. This is the interpretation, not an assumption that was necessary in order to make the interpretation.

d. The conclusion is that the single-stranded loops are noncomplementary regions and the double-stranded regions are complementary. If noncomplementary regions could stick together, by base pairing between occasional complementary single bases, for example, the stated conclusion could not be reached.

e. If this assumption were made, the conclusion would have to be that the hybridized segments were identical sequences. This would be contradictory to the conclusion that the hybridization occurred between the mRNA and the portions of the DNA on which the RNA was transcribed.

13

RECOMBINANT DNA AND GENETIC ENGINEERING

RECOMBINANT DNA TECHNOLOGY
    Producing Restriction Fragments
    Amplifying Specific Fragments
    Focus on Science: Applications of RFLPs
    Locating Genetically Altered Host Cells
    Expressing the Gene of Interest

APPLICATIONS OF THE NEW TECHNOLOGY
    Uses in Basic Research
    Genetic Modification of Organisms: Some
      Examples
    Focus on Bioethics: Some Implications of
      Human Gene Therapy

Interactive Exercises

RECOMBINANT DNA TECHNOLOGY (13-I, pp. 179–183)

Terms

The page-referenced terms are important; they were in boldface type in the chapter. Refer to the instructions given in Chapter 1, p. 1 of this workbook.

recombinant DNA technology (179) _____

plasmid (179)_____

restriction enzymes (180)_____

DNA ligase (180)_____

DNA library (180)_____

cloned DNA (180)_____

PCR, "polymerase chain reaction" (181)_____

DNA polymerase (181) _____

DNA fingerprint (182)_____

DNA probes (182)_____

nucleic acid hybridization (182)_____

cDNA (183)_____

reverse transcription (183)_____

Short Answer

1. Define recombinant DNA technology._____

_____

2. Describe and distinguish between the bacterial chromosome and plasmids present in a bacterial cell.

_____

_____

True/False

If the statement is true, write a T in the blank. If the statement is false, make it correct by changing the underlined word(s) and writing the correct word(s) in the answer blank.

_____ 3. Plasmids are <u>organelles on the surfaces of which amino acids are assembled into</u>

 <u>polypeptides</u>.

_____ 4. <u>Gene transfer</u> and <u>recombination</u> are common in nature.

Complete the Table

5. Complete the table below, which summarizes some of the basic tools and procedures used in recombinant DNA technology.

Tool/Procedure	Definition and Role in Recombinant DNA Technology
a. Restriction enzymes	
b. DNA ligase	
c. DNA library	
d. Cloned DNA	
e. Reverse transcriptase	
f. cDNA	
g. PCR	
h. cDNA probe	

Matching

Match the steps in the formation of a DNA library with the parts of the illustration below.

6.___
7.___
8.___
9.___
10.___
11.___

A. Joining of chromosomal and plasmid DNA using DNA ligase
B. Restriction enzyme cuts chromosomal DNA at specific recognition sites
C. Cut plasmid DNA
D. Recombinant plasmids containing cloned library
E. Fragments of chromosomal DNA
F. Same restriction enzyme is used to cut plasmids

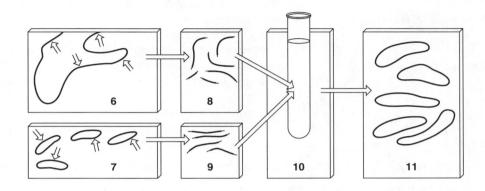

True/False

A genetic engineer used restriction enzymes to prepare fragments of DNA from two different species that were then mixed. Four of these fragments are illustrated below. Fragments (a) and (c) are from one species, (b) and (d) from the other species. Answer exercises 12–16. If the statement is true, write a T in the blank. If the statement is false, explain why.

TACA	TTCA	CGTA	ATGT
a	b	c	d

_____ 12. Some of the fragments represent sticky ends.

_____ 13. The same restriction enzyme was used to cut fragments (b), (c), and (d).

_____ 14. Different restriction enzymes were used to cut fragments (a) and (d).

_____ 15. Fragment (a) will base-pair with fragment (d) but not with fragment (c).

_____ 16. The same restriction enzyme was used to cut the different locations in the DNA of the two species shown.

Matching

Select the most appropriate answer.

17.___ polymerase chain reaction (PCR)

18.___ DNA ligase

19.___ cDNA

20.___ DNA library

21.___ cloned DNA

22.___ DNA fingerprint

23.___ recombinant DNA technology

24.___ plasmids

25.___ restriction enzymes

26.___ genome

A. All the DNA in a haploid set of chromosomes
B. Process by which a gene is split into two strands and then copied over and over (most common type of gene amplification) by enzymes
C. Caused by a unique array of RFLPs inherited from each parent
D. Connects DNA fragments
E. Small circular DNA molecules that carry only a few genes
F. DNA assembled through use of reverse transcriptase and coding on mRNA
G. Cuts DNA molecules
H. A collection of DNA fragments produced by restriction enzymes and incorporated into plasmids
I. Multiple, identical copies of DNA fragments from an original chromosome
J. Method of genetic engineering

Fill-in-the-Blanks

Each person has a genetic (27)_____: a unique pattern of RFLPs that can be used to map the human genome, apprehend criminals, and resolve cases of disputed paternity and maternity. Bacterial host cells lack the proper (28)_____ _____ to translate cloned genes unless the (29)_____ have been cut out.

APPLICATIONS OF THE NEW TECHNOLOGY (13-II, pp. 184–186)

Terms

human genome project (184) _____

genetic engineering (184)_____

gene therapy (186)_____

Short Answer

1. Explain the goal of the human genome project. (p. 184) _____

_____

In Exercises 2–8, summarize the results of the given experimentation dealing with genetic modifications of plants and animals.

2. Plants such as carrots that are regenerated from cultured cells (p. 184): _____

_____

3. The bacterium, *Agrobacterium tumefaciens* (p. 185):_____

_____

4. Cotton plants (p. 185):_____

_____

5. Introduction of the rat and human somatotropin gene into fertilized mouse eggs (p. 186):

_____

_____

6. "Ice-minus" bacteria and strawberry plants (p. 184):_____

_____

7. Genetically modified lymphoblasts and Severe Combined Immune Deficiency (p. 178):

_____

_____

Fill-in-the-Blanks

Determination of the order and identity of nucleotides in DNA is called (8)_____. Bacterial strains
used in (9)_____ _____ may be initially harmless, but there is concern about danger to humans
and the environment. A strain of bacteria, *Pseudomonas syringae*, lives on stems and leaves of crop plants and
makes them susceptible to (10)_____. The harmful gene is called the "ice-forming" gene, and the bacteria are known as (11)_____ - _____ bacteria; genetic engineers were able to remove the harmful
gene and test the modified bacterium on strawberry plants with no adverse effects. Inserting one or more
genes into the body cells of an organism for the purpose of correcting genetic defects is known as
(12)_____ _____.

Crossword Puzzle: Terms from Chapters 12 and 13

ACROSS

1. _____ DNA (multiple, identical copies of DNA fragments) is produced by repeated replications and cell divisions.
4. The process of transferring DNA code to any of three forms of RNA.
9. A collection of DNA fragments produced by (see # 11) _____ enzymes that have been incorporated into plasmids.
10. This is a portion of an operon that produces a repressor protein.
11. _____ enzymes cut apart foreign DNA that has been injected into the bacterial cell; the short, single-stranded ends of a DNA fragment are "sticky."
12. The place where a protein is assembled.
14. A small circlet of DNA or RNA that contains only a few genes; found in many bacteria.
15. DNA _____ uses short DNA sequences assembled from radioactively labeled nucleotides to express a gene of interest.
17. DNA _____ uses short nucleotide sequences as "primers" to begin replication.
18. _____ transcription assembles a DNA strand on mRNA and produces a "hybrid" DNA/RNA molecule.
20. These are spliced together to form the *mature* mRNA transcript.
21. _____ DNA technology splices together DNA cut from different species and inserts the product into bacteria or other rapidly dividing cells.

22. The part of an operon to which RNA polymerase attaches.
23. That which attaches by hydrogen bonds to a triplet of mRNA.
24. Different kinds of RNA, enzymes, cofactors, and raw materials collaborate to construct a specific protein.

DOWN

2. DNA _____ seals base-pairings during replication.
3. Base-pairing between nucleotide sequences from different sources is called nucleic acid _____.
5. An intervening sequence of RNA.
6. A DNA _____ is a unique array of RFLPs inherited from each parent in a Mendelian pattern.
7. Gene _____ transfers one or more normal genes into the body cells of an organism to correct a genetic defect.
8. The common method of amplifying DNA.
13. The "on-off" switch of an operon; the repressor protein may bind to this.
16. A group of genes that govern transcription of a series of genes that code for enzymes in a single metabolic pathway.
19. A modified form of the genetic code that directs the translation of a protein.

Self-Quiz

___1. Small circular molecules of DNA in bacteria are called _____.
   a. plasmids
   b. desmids
   c. pili
   d. F particles
   e. transferins

___2. Base-pairing between nucleotide sequences from different sources is called _____.
   a. nucleic acid hybridization
   b. reverse replication
   c. heterocloning
   d. plasmid formation

___3. Enzymes used to cut genes in recombinant DNA research are _____.
   a. ligases
   b. restriction enzymes
   c. transcriptases
   d. DNA polymerases
   e. replicases

___4. The total DNA in a haploid set of chromosomes of a species is its _____.
   a. plasmid
   b. enzyme potential
   c. genome
   d. DNA library
   e. none of the above

___5. An enzyme that heals random base-pairing of chromosomal fragments and plasmids is _____.
   a. reverse transcriptase
   b. DNA polymerase
   c. cDNA
   d. DNA ligase

___6. A DNA library is _____.
   a. a collection of DNA fragments produced by restriction enzymes and incorporated into plasmids

   b. cDNA plus the required restriction enzymes
   c. mRNA-cDNA
   d. composed of mature mRNA transcripts

___7. Amplification results in _____.
   a. plasmid integration
   b. bacterial conjugation
   c. cloned DNA
   d. production of DNA ligase

___8. Any DNA molecule that is copied from mRNA is known as _____.
   a. cloned DNA
   b. cDNA
   c. DNA ligase
   d. hybrid DNA

___9. The most commonly used method of DNA amplification is _____.
   a. polymerase chain reaction
   b. gene expression
   c. genome mapping
   d. RFLPs

___10. Restriction fragment length polymorphisms are valuable because _____.

   a. they reduce the risks of genetic engineering
   b. they provide an easy way to sequence the human genome
   c. they allow fragmenting DNA without enzymes
   d. they provide DNA fragment sizes unique to each person

Chapter Objectives/Review Questions

This section lists general and detailed chapter objectives that can be used as review questions. You can make maximum use of these items by writing answers on a separate sheet of paper. Fill in answers where blanks are provided. To check for accuracy compare your answers with information given in the chapter or glossary.

Page   *Objectives/Questions*

(179)   1. List the means by which natural genetic experiments occur.
(179)   2. Define recombinant DNA technology.
(179)   3. _____ are small, circular, self-replicating molecules of DNA or RNA within a bacterial cell.
(180)   4. Some bacteria produce _____ enzymes that cut apart DNA molecules injected into the cell by viruses; such DNA fragments or "_____ ends" often have staggered cuts capable of base-pairing with other DNA molecules cut by the same _____ enzymes.
(180)   5. Base-pairing between chromosomal fragments and cut plasmids is made permanent by DNA _____.
(180)   6. Be able to explain what a DNA library is; review the steps used in creating such a library.
(181)   7. Describe the major method of DNA amplification.
(180)   8. Multiple, identical copies of DNA fragments produced by restriction enzymes are known as _____ DNA.
(183)   9. A special viral enzyme, _____ _____, presides over the process by which mRNA is transcribed into DNA.
(183)   10. Define cDNA.
(181)   11. Polymerase chain reaction is the most commonly used method of DNA _____.
(182)   12. Explain how gel electrophoresis is used to sequence DNA.
(183)   13. How is a cDNA probe used to identify a desired gene carried by a modified host cell?
(183)   14. Why do researchers prefer to work with cDNA when working with human genes?
(184)   15. Tell about the human genome project and its implications.
(182)   16. List some practical genetic uses of RFLPs.
(186)   17. Define gene therapy and eugenic engineering.

Integrating and Applying Key Concepts

How could scientists guarantee that *Escherichia coli*, the human intestinal bacterium, will not be transformed into a severely pathogenic form and released into the environment if researchers use the bacterium in recombinant DNA experiments?

Critical Thinking Exercises

1. You isolate an mRNA of interest from a eukaryotic cell and use it to prepare a cDNA. You incorporate the cDNA into the genome of a prokaryote. The cell grows well in culture but fails to synthesize the protein coded by the original mRNA. You have independent evidence that the cDNA was successfully inserted into the prokaryotic genome and that it replicates properly. Which of the following hypotheses best explains the failure?

    a. The prokaryotic cell had no enzymes to edit the introns out of its RNA transcripts.
    b. The cDNA was complementary, not identical, to the mRNA.
    c. There were no introns in the mRNA, and thus the transcripts from the cDNA were incomplete.
    d. The promoter was not transcribed into the original mRNA.
    e. The mRNA had a poly-A "tail" that produced a confused message.

Answers

Answers to Interactive Exercises

RECOMBINANT DNA TECHNOLOGY (13-I)

1. DNA from different species can be cut, spliced together, and then inserted into bacteria or other types of rapidly dividing cells, which multiply the recombinant DNA molecules in quantity. Genes can be isolated, modified, and reinserted into the organism (or transplanted to a different one). 2. The bacterial chromosome, a circular DNA molecule, contains all the genes necessary for normal growth and development. Plasmids, small, circular molecules of "extra" DNA, carry only a few genes and are self-replicating. 3. F, small circles of DNA in bacteria; 4. T; 5. a. bacterial enzymes that cut apart DNA molecules injected into the cell by viruses; several hundred have been identified; b. a replication enzyme that joins the short fragments of DNA; c. a collection of DNA fragments produced by restriction enzymes and incorporated into plasmids; d. after a DNA library is inserted into a host cell's cloning vector (often a plasmid), repeated replications and divisions of the host cells produce multiple, identical copies of DNA fragments, or cloned DNA; e. a viral enzyme that allows mRNA to be transcribed into DNA; f. any DNA molecule "copied" from mRNA; g. most common method of DNA amplification; DNA containing a gene of interest is split into two single strands that enzymes convert back to double-stranded forms; h. a nucleic acid hybridization technique used to identify bacterial colonies harboring the DNA (gene) of interest; a short nucleotide sequence is assembled from radioactively labeled subunits (part of the sequence must be complementary to that of the desired gene). 6. B; 7. F; 8. E; 9. C; 10. A; 11. D; 12. T; 13. The same restriction enzyme would have cut the DNA wherever the same specific very short nucleotide sequence appears and would have produced fragments with identical end sequences—which these do *not* have. 14. (a) and (d) have complementary nucleotide sequences and thus were probably cut by the same restriction enzyme. 15. T;

16. Neither (a) and (c) nor (b) and (d) is totally complementary to the other. 17. B; 18. D; 19. F; 20. H; 21. I; 22. C; 23. J; 24. E; 25. G; 26. A; 27. fingerprint; 28. splicing enzymes; 29. introns.

APPLICATIONS OF THE NEW TECHNOLOGY (13-II)

1. Researchers are working to sequence the estimated 3 billion nucleotides present in human chromosomes. 2. Such culturing of cells apparently increases mutation rates and provides genetic novelties; researchers have successfully inserted genes into cultured cells. 3. The plasmid of *Agrobacterium* can be used as a vector to introduce desired genes into cultured plant cells; *Agrobacterium* was used to deliver a firefly gene into cultured tobacco plant cells. 4. Certain cotton plants have been genetically engineered for resistance to worm attacks. 5. In separate experiments the rat and human somatotropin genes became integrated into the mouse DNA. The mice grew much larger than their normal littermates. 6. Bacteria that have specific proteins on their cell surfaces facilitate ice crystal formation on whatever substrate the bacteria are located; bacteria without the ability to synthesize those proteins ("ice-minus") have been genetically engineered and were sprayed upon strawberry plants. Nothing bad happened. 7. Researchers have been able to introduce the ADA gene into some of the SCID sufferers' lymphoblasts and stimulate them to divide. About a billion copies of the genetically modified cells have been injected into the bloodstream of a girl with SCID and she is doing well. 8. sequencing; 9. genetic engineering; 10. frost; 11. "ice-minus"; 12. gene therapy

Answers to Self-Quiz

1. a; 2. a; 3. b; 4. c; 5. d; 6. a; 7. c; 8. b; 9. a; 10. d.

Answer to Crossword Puzzle

Critical Thinking Analysis

a. This is a true statement; prokaryotes have neither introns nor the enzymes to remove them. However, this is irrelevant. The mRNA had already been edited when it was used to produce the cDNA. No introns are present to be edited.

b. The original cDNA strand was complementary—it must be in order to be transcribed as a new copy of the mRNA.

c. It is true that the original introns had been edited before the mRNA was used. Thus, the new transcripts would be incomplete relative to the original eukaryotic DNA, but they are complete in terms of the information for the sequence of amino acids in the protein.

d. This is probably true, and it would mean that the cDNA contained no promoter. Unless the cDNA happened to be inserted into the prokaryotic genome adjacent to a promoter, it could not be transcribed, and the protein would not be synthesized in the cell.

e. The mRNA probably did have a poly-A tail, and it would copy as a poly-T sequence in the cDNA, leading eventually to poly-A tails on the new transcripts. This would not be a confused message, however, since the original message also had the poly-A tail and was translated readily. Furthermore, the stop translation signal must occur before the tail is reached, making the tail irrelevant to the information content of the message.

14

MICROEVOLUTION

Interactive Exercises

EMERGENCE OF EVOLUTIONARY THOUGHT (14-I, pp. 191–195)

The page-referenced terms are important; they were in boldface type in the chapter. Refer to the instructions given in Chapter 1, p. 1 in this workbook.

Terms

fossils (192)_____

evolution (192)_____

uniformitarianism (194)_____

Short Answer

1. State what a fourteenth-century scholar would have understood by the word "species."

   _____

   _____

2. Briefly state the prevailing view that most fourteenth- and fifteenth-century scholars held regarding the origins of the different kinds of living things. _____

   _____

3. Global explorations of the sixteenth century "expanded" the natural world of European scholars. It soon became apparent that the diversity of organisms in the world was overwhelming. What was the crucial question that scholars then began to ask regarding the origin of species?

   _____

   _____

4. During the eighteenth century, anatomists began to ask questions about organisms they were studying. What questions did they ask when they carefully compared human arms, whale flippers, and bat wings?_____

   _____

5. Why would it be disturbing to people of the eighteenth century to learn that humans had body parts (the coccyx) that looked exactly like the bones of a tail?_____

   _____

6. Define the word fossil._____

   _____

7. Summarize the work and ideas of the mid-eighteenth-century geologists; how did these ideas challenge the prevailing beliefs about the origin of species?_____

   _____

8. If one can accept evidence that there have been changes in the lines of descent of species that have happened during long periods of time, then that person also is likely to believe that species _____.

9. Several key players and events in the life of Charles Darwin led him to his conclusions about natural selection and evolution. Summarize these influences by completing the table that follows.

Event/Person	Importance to Synthesis of Evolutionary Theory
a. John Henslow	
b.	British ship that carried Darwin to South America as naturalist for that voyage
c.	English geologist who wrote *Principles of Geology*, a book read by Darwin that argued that very slow and gradual processes now molding the earth's surface had also been at work in the past
d. Thomas Malthus	
e.	English naturalist who arrived at the same conclusion as Darwin regarding evolution; papers released jointly by him and Darwin first announced their thinking to the world

True/False

If the statement is true, write a T in the blank. If the statement is false, make it correct by changing the underlined word(s) and writing the correct word(s) in the answer blank.

_____ 10. There is <u>considerable</u> evidence indicating that species have remained unchanged since the time of their <u>creation</u>.

_____ 11. One expects to find the most recent fossil organisms in the <u>upper</u> layers of Earth's <u>sedimentary</u> rocks.

_____ 12. Malthus believed that the <u>food supply of a population</u> increases faster than the <u>population</u> increases.

_____ 13. <u>Lyell's</u> ideas on overpopulation and <u>Malthus's</u> ideas on slow geologic change provoked Darwin's thinking on how organisms change with time.

_____ 14. When resources become scarce, competition for similar resources promotes greater <u>specialization</u> among the competitors.

_____ 15. <u>Wallace's letter to Charles Darwin</u> was the principal early stimulus that caused Darwin to recognize natural selection as an important process that brings about evolutionary change in populations.

_____ 16. *Archaeopteryx* was a unique fossil that supplied a "missing link" for the idea that a group of reptiles evolved into birds.

_____ 17. The members of a natural population show <u>little</u> variation in their traits, and much of this variation is passed on through generations.

_____ 18. A population in nature might change because nature would select individuals with <u>advantageous</u> traits and eliminate others.

MICROEVOLUTIONARY PROCESSES (14-II, pp. 196–200)

Terms

population (196)_____

allele (197)_____

allele frequencies (197) _____

genetic equilibrium (197)_____

microevolution (198)_____

mutation (198) _____

lethal mutation (198) _____

neutral mutation (198) _____

gene flow (198) _____

genetic drift (198) _____

founder effect (198) _____

bottleneck (199) _____

endangered species (199)_____

natural selection (200) _____

Short Answer

1. A central idea forming the basis of evolutionary processes is the statement that "individual organisms do not evolve, populations do." Why cannot an individual organism evolve?

_____

_____

2. What is meant by the genetic variation that exists in populations? _____

_____

3. In populations, genetic variation arises in individuals through five events. List these five sources of variation._____

_____

4. List the conditions that must be met before genetic equilibrium (or nonevolution) will occur (any order). _____

_____

5. For the following situation, assume that the conditions listed in exercise 4 do exist; therefore, there should be no change in gene frequency, generation after generation. Consider a population of hamsters in which dominant gene B produces black coat color and recessive gene b produces gray coat color (two alleles are responsible for color). The dominant gene has a frequency of 80 percent (or 0.80). It would follow that the frequency of the recessive gene is 20 percent (or 0.20). From this, the assumption is made that 80 percent of all sperm have gene B and 80 percent of all eggs have gene B. Also, 20 percent of all sperm carry gene b and 20 percent of all eggs carry gene b. (See pages 196–197 in the text.)

a. Calculate the probabilities of all possible matings in the Punnett square.
b. Summarize the genotype frequencies of the F_1 generation:

Genotypes	Phenotypes
___BB	
___Bb	___% black
___bb	___% gray

Sperm

	0.80 B	0.20 b
Eggs 0.80 B	BB	Bb
0.20 b	Bb	bb

c. Further assume that the individuals of the F_1 generation produce another generation and the assumptions of the Hardy-Weinberg principle still hold. What are the frequencies of the sperm produced?

Parents (F_1)	B sperm	b sperm
___BB	___	___
___Bb	___	___
___bb	___	___
Totals =	___	___

The egg frequencies can be similarly calculated. Note that the gamete frequencies of the F_2 generation are the same as the gamete frequencies of the last generation. Phenotype percentage also remains the same. Thus, the gene frequencies did not change between the F_1 and the F_2 generations. Again, given the assumptions of the Hardy-Weinberg equilibrium, gene frequencies do not change generation after generation.

6. In a population, 81 percent of the organisms are homozygous dominant, and 1 percent are homozygous recessive. Find the following.

a. the percentage of heterozygotes_____

b. the frequency of the dominant allele_____

c. the frequency of the recessive allele_____

7. In a population of 200 individuals, determine the following for a particular locus if p = 0.80.

   a. the number of homozygous dominant individuals_____

   b. the number of homozygous recessive individuals _____

   c. the number of heterozygous individuals if p = 0.80_____

8. If the percentage of gene *D* is 70 percent in a gene pool, find the percentage of gene *d*.

   _____

   _____

9. If the frequency of gene *R* in a population is 0.60, what percentage of the individuals are heterozygous *Rr*?

   _____

   _____

10. Distinguish between the two extreme cases of genetic drift: the founder effect and bottlenecks.

   _____

   _____

Labeling

The traits of individuals in a population are often classified as morphological, physiological, or behavioral. In the list of traits below, enter M in the blank if the trait is morphological, P if the trait is physiological, and B if the trait is behavioral.

11.___ Frogs have a three-chambered heart.

12.___ The active transport of Na^+ and K^+ ions is unequal.

13.___ Humans and orangutans possess an opposable thumb.

14.___ An organism periodically seeks food.

15.___ Some animals have a body temperature that fluctuates with the environmental temperature.

16.___ The platypus is a strange mammal; it has a bill like a duck and lays eggs.

17.___ Some vertebrates exhibit greater parental protection of offspring than others.

18.___ During short photoperiods, the pituitary gland releases small quantities of gonadotropins.

19.___ Red grouse defend large, multipurpose territories where they forage, mate, nest, and rear young.

20.___ Lampreys have an elongated cylindrical body without scales.

Matching

Select the single best answer.

21.___ gene flow

22.___ allele frequencies

23.___ neutral mutations

24.___ microevolution

25.___ lethal mutation

26.___ alleles

27.___ mutation

28.___ genetic drift

29.___ genetic equilibrium

30.___ natural selection

A. A heritable change in DNA
B. Zero evolution
C. Different molecular forms of a gene
D. Change in allele frequencies as individuals leave or enter a population
E. Change or stabilization of allele frequencies due to differences in survival and reproduction among variant members of a population
F. Random fluctuation in allele frequencies over time due to chance
G. Change in which expression of mutated gene always leads to the death of the individual
H. The abundance of each kind of allele in the entire population
I. Neither harmful nor helpful to the individual
J. Changes in allele frequencies brought about by mutation, genetic drift, gene flow, and natural selection

Fill-in-the-Blanks

A(n) (31)_____ is a group of individuals of the same species that occupy a given area at a specific time. The (32)_____-_____ principle allows researchers to establish a theoretical reference point (baseline) against which changes in allele frequency can be measured. Variation can be expressed in terms of (33)_____ _____, which means the relative abundance of different alleles carried by the individuals in that population. The stability of allele ratios that would occur if all individuals had equal probability of surviving and reproducing is called (34)_____ _____. Over time, allele frequencies tend to change through infrequent but inevitable (35)_____, which are the original source of genetic variation. Random fluctuation in allele frequencies over time due to chance occurrence alone is called (36)_____ _____; it is more pronounced in small populations than in large ones. (37)_____ flow associated with immigration and/or emigration also changes allele frequencies. (38)_____ _____ is the differential survival and reproduction of individuals of a population that differ in one or more traits. (39)_____ _____ is the most important microevolutionary process.

EVIDENCE OF NATURAL SELECTION (14-III, pp. 201–203)

Terms

stabilizing selection (201)_____

directional selection (202)_____

disruptive selection (202)_____

balanced polymorphism (202)_____

sexual dimorphism (203) _____

sexual selection (203)_____

Complete the Table

1. Complete the following table, which defines three types of natural selection effects.

Effect	Characteristics	Example
a. Stabilizing selection		
b. Directional selection		
c. Disruptive selection		

Labeling

2. Identify the three curves below as stabilizing selection, directional selection, or disruptive selection.

  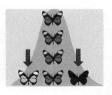

a._____   b._____   c._____

Fill-in-the-Blanks

When individuals of different phenotypes in a population differ in their ability to survive and reproduce, their alleles are subject to (3)_____ selection. (4)_____ selection favors the most common phenotypes in the population. (5)_____ selection occurs when a specific change in the environment causes a heritable trait to occur with increasing frequency and the whole population tends to shift in a parallel direction. (6)_____ selection favors the development of two or more distinct polymorphic varieties such that they become increasingly represented in a population and the population splits into different phenotypic variations. (7)_____ provide an excellent example of stabilizing selection, because they have existed essentially unchanged for hundreds of millions of years. (8) _____ - _____ _____, a genetic disorder that produces an abnormal form of hemoglobin, is an example of (9) _____ selection that maintains a high frequency of the harmful (10) _____ along with the normal gene. Such a genetic juggling act is called a (11) _____ _____. Differences in appearance between males and females of a species are known as (12)_____ _____. Among birds and mammals, females act as agents of (13)_____ when they choose their mates. (14)_____ selection is based on any trait that gives the individual a competitive edge in mating and producing offspring. The more colorful, showier, and larger appearance of male pheasants when compared with females is the result of (15)_____ selection.

SPECIATION (14-IV, pp. 204–205)

species (204)_____

reproductive isolation (204) _____

divergence (204)_____

speciation (204)_____

polyploidy (204)_____

Short Answer

1. Describe the process of divergence. _____

_____

_____

_____

_____

2. The illustration below depicts the divergence of one species into two as time passes. Answer the questions below the illustration.

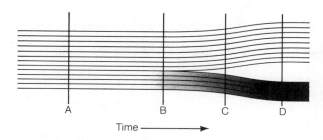

Time ————————▶

a. In what areas is there distinctly one species? _____

b. Between what letters does divergence begin?_____

c. Does this divergence appear to be slow or rapid? _____

d. What letter represents the time when divergence is probably complete?_____

3. Study the illustration below, which shows the possible course of wheat evolution; similar genomes are designated by the same letter. Answer the questions below the diagram.

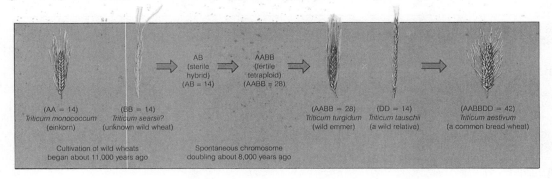

a. How many pairs of chromosomes are found in *Triticum monococcum*? _____

b. How are the *T. monococcum* chromosomes designated?_____

c. How many pairs of chromosomes are found in the unknown wild wheat?_____

d. How are the unknown wild wheat's chromosomes designated?_____

e. Why is the hybrid designated AB sterile? _____

f. What cellular event must have occurred to create the *T. turgidum* genome?_____

g. Describe the genome of the plant arising from the cross of *T. turgidum* and *T. tauschii* (not shown on the diagram)._____

h. What cellular event must have occurred to create the *T. aestivum* genome? _____

i. What is the source of the A genome in *T. aestivum*? The source of the B genome in *T. aestivum*? The source of the D genome in *T. aestivum*?_____

Complete the Table

4. Complete the table below to summarize reproductive isolating mechanisms.

Reproductive Isolating Mechanism	Description/Example
a. Structural isolation	
b.	Incompatibilities between the sperm of one species and the egg (or female reproductive system) of another may prevent fertilization; example: two species of sea urchin.
c. Hybrid inviability and infertility	
d.	Behavior such as complex courtship rituals can be a strong isolating mechanism; example: courtship rituals of many bird species.
e.	Differences in reproductive timing may serve as an isolating mechanism; example: two closely related species of cicadas.

Matching

Select the single best answer.

5. ___ geographic isolation

6. ___ isolation within a population

7. ___ polyploidy

8. ___ lineage

9. ___ gradualism

10. ___ punctuation

A. A single line of descent (a twig or branch on the "family tree")

B. Speciation resulting from barriers forming between two populations that normally interbreed

C. Slow divergence over long time spans

D. Rapid changes during speciation; quick spurt of changes when first diverging from parental lineage

E. Offspring receiving three or more of each parental chromosome due to meiotic nondisjunction

F. Formation of genetic barriers in a single population, followed by speciation

Fill-in-the-Blanks

(11)_____ is the process whereby species are formed. A (12)_____ is composed of one or more populations of individuals who can interbreed under natural conditions and produce fertile, reproductively isolated offspring. (13)_____ can be described as a process in which differences in alleles accumulate between populations. Any aspect of structure, function, or behavior that prevents interbreeding is a (14)_____ _____ mechanism. Physical barriers that prevent gene flow as in the case of large rivers changing course or forests giving way to grasslands are (15)_____ barriers. Wheat is an example of a species formed by (16)_____ and (17)_____.

Self-Quiz

___1. Changes in allele frequencies brought about by mutation, genetic drift, gene flow, and natural selection are called _____ processes.
a. genetic equilibrium
b. microevolution
c. founder effect
d. independent assortment

For questions 2–4, choose from the following answers:
a. stabilizing selection
b. genetic drift
c. directional selection
d. balanced polymorphism

___2. Increasing resistance of cockroaches to pesticides is an example of _____.

___3. The founder effect is a special case of _____.

___4. Different alleles that code for different forms of hemoglobin have led to _____ in the midst of the malarial belt that extends through West and Central Africa.

___5. The idea that any population tends to outgrow its resources and that its members must compete for what is available belonged to _____.
a. Malthus
b. Darwin
c. Lyell
d. Henslow

___6. In Lyell's book *Principles of Geology*, he suggested that _____.
a. tail bones in a human have no place in a perfectly designed body
b. perhaps species originated in more than one place and have been modified over time
c. the force for change in organisms was a built-in drive for perfection, up the Chain of Being
d. gradual processes now molding the earth's surface had also been at work in the past

For questions 7–8, choose from the following answers:

    a. balanced polymorphism
    b. polyploidy
    c. genetic equilibrium
    d. sexual selection

___7. Hardy and Weinberg invented an ideal population that was in _____ and used it as a baseline against which to measure evolution in real populations.

___8. Many domestic varieties of fruits, vegetables, and grains have evolved larger vegetative and reproductive structures through _____.

For questions 9–10, choose from the following answers:

    a. stabilizing selection
    b. divergence
    c. a reproductive isolating mechanism
    d. polyploidy

___9. Two species of sage plants with differently shaped floral parts that prevent pollination by the same pollinator serve as an example of _____.

___10. _____ occurs when isolated populations accumulate allele frequency differences between them over time.

Chapter Objectives/Review Questions

This section lists general and detailed chapter objectives that can be used as review questions. You can make maximum use of these items by writing answers on a separate sheet of paper. Fill in answers where blanks are provided. To check for accuracy, compare your answers with information given in the chapter or glossary.

Page   Objectives/Questions

(192) 1. It was the study of _____ that raised the following question: If all species were created at the same time in the same place, why were certain species found in only some parts of the world and not others?

(192) 2. Give examples of the type of evidence found by comparative anatomists that suggested living things may have changed with time.

(192–195) 3. Be able to state the significance of the following: Henslow, H.M.S. *Beagle*, Lyell, Darwin, Malthus, Wallace, and *Archaeopteryx*.

(194–195) 4. What conclusion was Darwin led to when he considered the various species of finches living on the separate islands of the Galapagos?

(200) 5. Outline the key statements of the theory of natural selection.

(197) 6. Review the five categories through which genetic variation occurs among individuals.

(197) 7. The abundance of each kind of allele in the whole population is referred to as allele _____.

(197) 8. A point at which allele frequencies for a trait remain stable through the generations is called genetic _____.

(197) 9. Be able to list the five conditions that must exist before conditions for the Hardy-Weinberg principle are met.

(198) 10. Changes in allele frequencies brought about by mutation, genetic drift, gene flow, and natural selection are called _____.

(198) 11. Random fluctuations in allele frequencies over time due to chance are called _____.

(201) 12. _____ _____ is defined as the change or stabilization of allele frequencies due to differences in survival and reproduction among variant members of a population.

(198–199) 13. Distinguish the founder effect from a bottleneck.

(201) 14. When the most common phenotypes in a population are favored, _____ selection is operating.

(202) 15. When both ends of the phenotypic range are favored and intermediate forms are selected against, _____ selection is occurring.

(203)   16. _____ selection is based on any trait that gives an individual a competitive edge in mating and producing offspring.

(204)   17. What part does divergence play in speciation?

(204)   18. Discuss the role that polyploidy plays in speciation.

Integrating and Applying Key Concepts

Can you imagine any way in which directional selection may have occurred or may be occurring in humans? Which factors do you suppose are the driving forces that sustain the trend? Do you think the trend could be reversed? If so, by what factor(s)?

Critical Thinking Exercises

1. It is often stated that most new mutations are recessive. Which of the following assumptions would be most important in arguing for this statement?
   a. Most individuals are heterozygous for most of their genes.
   b. Most individuals are homozygous for most of their genes.
   c. Most existing alleles are recessive.
   d. Most existing alleles are dominant.
   e. Mutations are rare events.

Answers

Answers to Interactive Exercises

EMERGENCE OF EVOLUTIONARY THOUGHT (14-I)

1. A species was a kind of being created by a creator. Naturally, species would not change over time because species existed according to the Creator's design.
2. A Great Chain of Being was visualized that extended from the lowest form to humans and on to spiritual beings. Each kind of being had a separate place in the divine order of things and had not changed since creation; each was a link in the chain. Once all the links were discovered, named, and described, the meaning of life would be revealed. 3. Where did the great diversity of organisms fit into the Great Chain? 4. Why are these different structures made of the same materials? Why do these organs have similar locations in the body? Why do these structures form and develop in similar ways in the animal embryo? 5. This was disturbing because the human body was thought to be perfectly designed and should not have parts of a tail. 6. Fossils are the recognizable remains or body impressions of organisms that lived in the past. 7. Geologists began mapping the horizontal layers of sedimentary rocks they felt had been slowly deposited. Different layers held different kinds of fossils that lived in the past. These findings challenged the traditional view of a recent creation. 8. evolve
9. a. John Henslow: Cambridge botanist who perceived and respected Darwin's real interests in nature; arranged that Darwin be offered the position of ship's naturalist aboard H.M.S. *Beagle*. b. H.M.S. *Beagle*: British ship that carried Darwin to South America as naturalist for that voyage. c. Charles Lyell: English geologist who wrote *Principles of Geology*, a book read by Darwin that argued that very slow and gradual processes now molding the earth's surface had also been at work in the past. d. Thomas Malthus: English clergyman and economist whose essay, read by Darwin, suggested that any population tends to outgrow its resources and that its members must compete for what is available. e. Alfred Wallace: English naturalist who arrived at the same conclusion as Darwin regarding evolution; papers released jointly by him and Darwin first announced their thinking to the world. 10. F, no, appearance; 11. T ; 12. F, population, food supply of a population; 13. F, Malthus's ideas on overpopulation, Lyell's ideas on slow geologic change; 14. T; 15. F, at least 20 years before, the ideas of Lyell and Malthus and Darwin's five-year voyage around the world provided earlier stimuli; 16. T; 17. F, great; 18. T.

MICROEVOLUTIONARY PROCESSES (14-II)

1. Evolution is a change in *allele frequency*, which is a characteristic of a population, not of an individual.
2. Some number of the genes in one individual have a slightly different molecular structure than their counterpart in another individual. 3. Gene mutation, abnormal changes in chromosome structure or number, crossing over and genetic recombination at meiosis, independent assortment of chromosomes at meiosis, and fertilization between genetically different gametes 4. No mutation; very large population; isolation from other populations of the same species; all members survive, mate, and reproduce (no selection); and random mating 5. a. 0.64 *BB*, 0.16 *Bb*, 0.16 *Bb*, and 0.04 *bb*. b. genotypes: 0.64 *BB*, 0.32 *Bb*, and 0.04 *bb*; phenotypes: 96% black, 4% gray.

c. Parents	B sperm	b sperm
0.64 BB	0.64	0
0.32 Bb	0.16	0.16
0.04 bb	0	0.04
Totals =	0.80	0.20

6. Find (b) first, then (c), and finally (a). a. $2pq = 2 \times (0.9) \times (0.1) = 2 \times (0.09) = 0.18 = 18\%$, which is the percentage of heterozygotes; b. $p^2 = 0.81$, $p = \sqrt{0.81} = 0.9$ = the frequency of the dominant allele; c. $p + q = 1$, $q = 1 - 0.9 = 0.1$ = the frequency of the recessive allele. 7. a. homozygous dominant = $p^2 \times 200 = (0.8)2 \times 200 = 0.64 \times 200 = 128$ individuals; b. homozygous recessive = $q^2 \times 200 = (0.2)2 \times 200 = (0.04) \times (200) = 8$ individuals; c. heterozygotes = $2pq \times 200 = 2 \times 0.8 \times 0.2 \times 200 = 0.32 \times 200 = 64$ individuals. Check: $128 + 8 + 64 = 200$. 8. If $p = 0.70$, since $p + q = 1$, $0.70 + q = 1$, then $q = 0.30$, or 30 percent. 9. If $p = 0.60$, since $p + q = 1$, $0.60 + q = 1$, then $q = 0.40$; thus, $2pq = 0.48$, or 48 percent. 10. In the founder effect, a few individuals leave a population and establish a new one; by chance, allele frequencies will differ from the original population. In bottlenecks, disease, starvation, or some other stressful situation nearly eliminates a population; relative allele frequencies are randomly changed. 11. M; 12. P; 13. M; 14. B; 15. P; 16. M; 17. B; 18. P; 19. B; 20. M; 21. D; 22. H; 23. I; 24. J; 25. G; 26. C; 27. A; 28. F; 29. B; 30. E; 31. population; 32. Hardy-Weinberg; 33. gene frequency; 34. genetic equilibrium; 35. mutations; 36. genetic drift; 37. Gene; 38. Natural selection; 39. Natural selection.

EVIDENCE OF NATURAL SELECTION (14-III)
1. a. the most common phenotypes are favored; average human birth weight of 7 pounds; b. allele frequencies shift in a steady, consistent direction in response to a new environment or a directional change in an old one; light to dark forms of peppered moths; c. forms at both ends of the phenotypic range are favored, and intermediate forms are selected against; finches on the Galapagos Islands. 2. a. directional; b. disruptive; c. stabilizing; 3. natural; 4. Stabilizing; 5. Directional; 6. Disruptive; 7. Horsetails; 8. Sickle-cell anemia; 9. stabilizing; 10. allele; 11. balanced polymorphism; 12. sexual dimorphism; 13. selection; 14. Sexual; 15. sexual.

SPECIATION (14-IV)
1. Divergence is a process that blocks gene flow in some degree between local population units. When divergence

becomes great enough, members of two populations will not be able to breed successfully; they become two separate species. 2. a. A-B, some of B-C; b. B and C; c. slow; d. D. 3. a. 7 pairs; b. AA; c. 7 pairs; d. BB; e. The chromosomes fail to pair in meiosis; f. Fertilization of nonreduced gametes produced a fertile tetraploid; g. a sterile hybrid, ABD; h. Fertilization of nonreduced gametes resulted in a fertile hexaploid; i. *Triticum monococcum*, the unknown wild wheat, *T. tauschii*. 4. a. Differences in reproductive structures or other body parts may prevent members of two populations from interbreeding; example: two species of sage plants and their pollinators; b. Isolation of gametes; c. Even when fertilization occurs between the gametes of different species, the resulting embryo usually dies due to physical or chemical incompatibilities; example: sometimes hybrids such as the mule and zebroid are vigorous but sterile; d. Behavioral isolation; e. Isolation in time. 5. B; 6. F; 7. E; 8. A; 9. C; 10. D; 11. Speciation; 12. species; 13. Divergence; 14. reproductive isolating; 15. geographic; 16. polyploidy (hybridization); 17. hybridization (polyploidy).

Answers to Self-Quiz

1. b; 2. c; 3. b; 4. d; 5. a; 6. d; 7. c; 8. b; 9. c; 10. b.

Critical Thinking Analysis

a. Because the statement concerns the pattern of expression of new mutations, the frequencies and pattern of occurrence of existing alleles are largely irrelevant.

b. This assumption is subject to the same argument as (a).

c. If most existing alleles are recessive, they already code for a protein that is inactive in some way. Therefore, any changes in the amino acid sequence of the protein are likely to simply continue their recessive status and not change the phenotype.

d. If most existing alleles are dominant, they code for active proteins or enzymes. Changes in such proteins would be most likely to inactivate them. These would represent new phenotypes and would be recessive to the active alleles.

e. The statement concerns the qualitative effects of mutations, not their frequency.

15

LIFE'S ORIGINS AND MACROEVOLUTION

EVIDENCE OF MACROEVOLUTION
    Fossils—Evidence of Ancient Life
    The Changing Geologic Stage
    Comparative Morphology
    Comparative Biochemistry
    Branchings, Extinctions, and Adaptive
      Radiations
MACROEVOLUTION AND EARTH HISTORY
    Origin of Life
    The Archean and Proterozoic Eras

First Half of the Paleozoic Era
Second Half of the Paleozoic Era
The Mesozoic Era
The Cenozoic Era
Focus on the Environment—Plumes, Global
    Impacts, and the Dinosaurs
ORGANIZING THE EVIDENCE—CLASSIFICATION
SCHEMES

Interactive Exercises

EVIDENCE OF MACROEVOLUTION (15-I, pp. 210–219)

Terms

The page-referenced terms are important; they were in boldface type in the chapter. Refer to the instructions given in Chapter 1, p. 1 in this workbook.

macroevolution (210)_____

lineage (210)_____

fossil (211)_____

geologic time scale (211) _____

comparative morphology (213) _____

homologous structures (214)_____

morphological divergence (214)_____

morphological convergence (214) _____

analogous structures (214) _____

molecular clock (215)_____

DNA-DNA hybridization (215)_____

Short Answer

1. How do microevolution and macroevolution deal with the process and the patterns of evolution?

_____

_____

2. State the simple fact that we can use to interpret the evidence of large-scale patterns and trends.

_____

_____

3. Cite some examples of fossils, the most direct evidence available that organisms lived long ago.

_____

_____

4. What conditions favor fossil formation?_____

_____

Sequence

Earth history has been divided into five great eras, which are based on four abrupt transitions in the fossil record. Arrange the eras in correct chronological sequence from the oldest to the youngest.

5. ___   A. Mesozoic
                B. Cenozoic
6. ___   C. Proterozoic
7. ___   D. Archean
                E. Paleozoic
8. ___

9. ___

Labeling

Evidence for macroevolution comes from comparative morphology and comparative biochemistry. For each of the following items, place an M in the blank if the evidence comes from comparative morphology and a B if the evidence comes from comparative biochemistry.

10. ___ Using neutral mutations to date the divergence of two species from a common ancestor

11. ___ Similar embryonic stages, such as aortic arches, that persist in different vertebrates

12. ___ Similarities in vertebrate forelimbs when comparing the wings of pterosaurs, birds, and bats and porpoise flippers

13. ___ Comparison of the proportional development changes in a chimpanzee skull and a human skull

14. ___ The amino acid sequence in the cytochrome *c* of humans precisely matching the sequence in chimpanzees

15. ___ Distantly related vertebrates, such as sharks, penguins, and porpoises, showing a similarity to one another in their proportion, position, and function of body parts

16. ___ Establishing a rough measure of evolutionary distance between two organisms by use of DNA hybridization studies that demonstrate the extent to which the DNA from one species base-pairs with that of another

17. ___ The striking resemblances between certain cacti of North America and euphorbs of Africa, even though these two plant families are rather distant

18. ___ Storks and new-world vultures being more closely related to each other than flamingos and ibises are to each other, according to DNA hybridization studies

19. ___ Regulatory genes controlling the rate of growth of different body parts and being able to produce large differences in the development of two very similar embryos

Fill-in-the-Blanks

The establishment of large-scale patterns, trends, and rates of change among groups of species is known as (20)_____. Most of the information about the history of life on Earth comes from (21)_____. Earth history can be divided into five great eras; beginning with the oldest, they are the (22)_____, the (23) _____, the (24)_____, the (25)_____, and the (26) _____. These eras are based on abrupt (27)_____ in the (28)_____ record. Evidence for macroevolution comes from detailed comparisons of body form and structural patterns of major taxa; this is known as (29)_____ _____. It is believed that variations among genetically similar adult vertebrates come about by mutations of (30)_____ _____. The wings of birds and bats show genetic relationships and are termed (31)_____ structures. Such departure from a common ancestral form is known as morphological (32)_____. Penguin wings and shark fins perform similar functions in a similar environment; these structures are said to be (33)_____ and demonstrate morphological (34)_____. Divergence between two species can be dated using accumulation rates of neutral (35)_____. DNA (36)_____ is a method of converting DNA of different species to single-stranded forms and allowing them to recombine as a measure of similarity.

EVIDENCE OF MACROEVOLUTION (cont.) (15-II, pp. 216–219)

Branchings, Extinctions, and Adaptive Radiations

Terms

evolutionary tree (216) _____

gradual model (216) _____

punctuation model (216) _____

mass extinction (216) _____

adaptive radiation (217) _____

adaptive zone (217) _____

Sequence

Refer to Figure 15.12 in the text. Study of the geologic record reveals that, as the major events in the evolution of the earth and its organisms occurred, there were periodic major extinctions of organisms followed by major radiations of organisms. Arrange the letters of the extinctions and radiations listed below in the approximate order in which they occurred, from youngest to oldest.

1. ___
2. ___
3. ___
4. ___
5. ___
6. ___
7. ___
8. ___
9. ___
10. ___
11. ___
12. ___
13. ___
14. ___

A. Pangea, worldwide ocean forms; shallow seas squeezed out. Major radiations of reptiles, gymnosperms.

B. Glaciations as Gondwana crosses South Pole. Mass extinction of many marine organisms.

C. Mass extinction of many marine invertebrates, most fishes.

D. Pangea breakup begins. Rich marine communities. Major radiations of dinosaurs.

E. Asteroid impact? Mass extinction of all dinosaurs and many marine organisms.

F. Recovery, radiations of marine invertebrates, fishes, dinosaurs. Gymnosperms the dominant land plants. Origin of mammals.

G. Major glaciations. Modern humans emerge and begin what may be the greatest mass extinction of all time on land, starting with Ice Age hunters.

H. Unprecedented mountain building as continents rupture, drift, collide. Major climatic shifts; vast grasslands emerge. Major radiations of flowering plants, insects, birds, mammals. Origin of earliest human forms.

I. Gondwana moves south. Major radiations of marine invertebrates, early fishes.

J. Laurasia forms. Gondwana moves north. Vast swamplands, early vascular plants. Radiation of fishes continues. Origin of amphibians.

K. Pangea breakup continues; broad inland seas form. Major radiations of marine invertebrates, fishes, insects, dinosaurs; origin of angiosperms (flowering plants).

L. Asteroid impact? Mass extinction of many organisms in seas, some on land; dinosaurs, mammals survive.

M. Mass extinction. Nearly all species in seas and on land perish.

N. Tethys sea forms. Recurring glaciations. Major radiations of insects, amphibians. Spore-bearing plants dominant; gymnosperms present; origin of reptiles.

Chronology of Events

Refer to Figure 15.12 in the text. From the list of evolutionary events on the previous page (A-N), select the events that occurred in a particular era by circling the appropriate letters.

15. Cenozoic: A - B - C - D - E - F - G - H - I - J - K - L - M - N

16. Cenozoic–Mesozoic border: A - B - C - D - E - F - G - H - I - J - K - L - M - N

17. Mesozoic: A - B - C - D - E - F - G - H - I - J - K - L - M - N

18. Mesozoic–Paleozoic border: A - B - C - D - E - F - G - H - I - J - K - L - M - N

19. Paleozoic: A - B - C - D - E - F - G - H - I - J - K - L - M - N

MACROEVOLUTION AND EARTH HISTORY (15-III, pp. 220–230)

Terms

stromatolites (224)_____

Archean (224)_____

Proterozoic (224) _____

Paleozoic (224)_____

Mesozoic (226)_____

Cenozoic (230)_____

Complete the Table

1. Refer to Figure 15.12 in the text. To review some of the important events that occurred in the geologic past, complete the table below by entering the geologic era (Archean, Proterozoic, Paleozoic, Mesozoic, and Cenozoic) and the *approximate* time in millions of years since the time of the events.

Era	Time	Events
a.		A few reptile lineages give rise to mammals and the dinosaurs.
b.		Formation of Earth's crust, early atmosphere, oceans; chemical evolution leading to the origin of life.
c.		Origin of amphibians.
d.		Origin of animals with hard parts.
e.		Rocks 3.5 billion years old contain fossils of well-developed prokaryotic cells that probably lived in tidal mud flats.
f.		Flowering plants emerge, gymnosperms begin their decline.
g.		In the Carboniferous, there were major radiations of insects and amphibians; gymnosperms present, origin of reptiles.
h.		Oxygen accumulated in the atmosphere.
i.		Before the close of this era, the first photosynthetic bacteria had evolved.
j.		Insects, amphibians, and early reptiles flourished in the swamp forests of the Permian.
k.		Most of the major animal phyla evolved in rather short order.
l.		Dinosaurs ruled.
m.		Origin of aerobic metabolism; origin of protistans, algae, fungi, animals.
n.		Grasslands emerge and serve as new adaptive zones for plant-eating mammals and their predators.
o.		Humans destroy habitats and many species.
p.		The first ice age initiated the first global mass extinction; reef life everywhere collapsed.
q.		The invasion of land begins; small stalked plants establish themselves along muddy margins, and the lobe-finned fishes ancestral to amphibians move onto land.

Matching

Several probable stages in the physical and chemical evolution of life have been identified. Match the numbered statements and conditions below to the four listed categories. Letters may be used more than once.

A. Early Earth and its atmosphere   C. Self-replicating systems

B. Synthesis of biological molecules   D. The first plasma membrane

2. ___ Suppose a clay template attracted nucleotides.

3. ___ Farther out from the center, Earth was forming along with other planets.

4. ___ We also know that lightning, hot volcanic ash, and even shock waves have enough energy to drive the construction of biological molecules under abiotic conditions.

5. ___ During the 300 million years after the first rains began, organic compounds accumulated in the shallow waters of Earth. The interacting molecules DNA, RNA, and proteins were formed.

6. ___ This first atmosphere had very little free oxygen—a condition that favored the origin of life.

7. ___ Sidney Fox heated amino acids under dry conditions to form protein chains, which he placed in hot water. The cooled chains self-assembled into small, stable spheres.

8. ___ All the components found in biological molecules were present on the early earth.

9. ___ Early on, water vapor must have been released from the breakdown of rocks during volcanic eruptions, but it would have evaporated in the intense heat blanketing the crust.

10. ___ Clay crystals at the bottom of tidal flats and estuaries may have been the first template for protein synthesis.

11. ___ Stanley Miller mixed hydrogen, methane, ammonia, and water in a reaction mixture. He recirculated the mixture and bombarded it with a spark discharge to simulate lightning.

Matching

Choose the single best answer.

12. ___ Fox's experiments that subjected amino acids to dry heat

13. ___ instructions coded in DNA (or RNA)

14. ___ lightning or intense heat associated with volcanoes

15. ___ liquid water

16. ___ Miller's sparking chamber experiments

17. ___ molecular oxygen, O_2

A. Demonstrated that amino acids and other small organic compounds formed within a week
B. Without this, life as we know it could not have originated on Earth
C. Earth's original atmosphere lacked this
D. Cellular reproduction requires this
E. Demonstrated that selectively-permeable spheres resembling cell membranes formed
F. Most likely provided the energy needed for the synthesis of organic compounds

Sequence

Put into proper order the events that led to the first living cells. 1 = earliest event, 7 = most recent event.

18. ___ DNA formation

19. ___ Enzymes and other proteins were formed

20. ___ Formation of biological molecules under anaerobic conditions

21. ___ Formation of lipid spheres

22. ___ Membrane-bound proto-cells appeared

23. ___ Protein-RNA systems were formed

24. ___ Self-replicating systems evolved

Fill-in-the-Blanks

About (25) [choose one] ( ) 10, ( ) 4.6, ( ) 3.6, ( ) 3.2 billion years ago, the cloud that formed our solar system had flattened out into a slowly rotating disk. By (26) [choose one] ( ) 4, ( ) 3.8, ( ) 3.2, ( ) 2 billion years ago, the earth was hurtling through space as a thin-crusted inferno. In time, the crust cooled and rain fell, stripping mineral salts from parched rocks, and the oceans were formed. Rocks 3.5 billion years old contain fossils of (27)_____ _____ that probably existed in tidal mud flats. Little or no free (28)_____ was in the atmosphere, so (29)_____, an anaerobic pathway, must have been the first metabolic pathway to develop. Between (30) [choose one] ( ) 3.2, ( ) 2.5, ( ) 1 billion and 700 million years ago, (31)_____ _____ dominated the shallow seas. Mound-shaped bacterial cell mats found in rock formations more than 2 billion years old are called (32)_____. Oxygen accumulated in the (33)_____ atmosphere. This prevented further synthesis of (34)_____ _____ and opened up new worldwide (35)_____ _____ in which organisms such as multicelled plants, fungi, and animals evolved that carried on (36)_____ respiration. By 1.2 billion years ago, the green (37)_____ had evolved. The Paleozoic era is subdivided into the Cambrian, Ordovician, Silurian, Devonian, Carboniferous, and Permian periods. Nearly all animal phyla had evolved by the end of the (38)_____ period; (39)_____ were the dominant animals. Gondwana drifted southward and seas flooded land to open new, shallow marine environments for flourishing reef organisms during the (40)_____ period. In the area of the South Pole, glaciers formed on Gondwana, and many forms of reef life became extinct. Gondwana drifted northward during the Silurian and the Devonian. Reef organisms recovered as a major radiation of (41)_____ fishes (with armor plates and massive jaws) occurred. (42)_____ - _____ fishes that were ancestral to amphibians began to invade the land as small stalked plants began to evolve in muddy land–water margins. Another global (43)_____ _____ occurred at the Devonian–Carboniferous boundary. During the (44)_____ period, major adaptive radiations of insects and treelike coal-forming plants occurred in swamp forests. When Pangea was formed by collisions of all land masses, nearly all land and sea species perished in the greatest known mass extinction as the (45)_____ period drew to a close. Great changes in geology and climate occurred, but the small reptilian ancestors of dinosaurs, birds, snakes, and lizards survived.

The Mesozoic era is subdivided into Triassic, Jurassic, and Cretaceous periods and lasted about 175 million years. The major radiation of reptiles had ended abruptly before the Mesozoic, but divergences in a few lineages now gave rise to (46)_____ and reptiles, including dinosaurs. During a (47)_____ _____ at the Triassic–Jurassic boundary, many marine organisms were lost. (48)_____ ruled the earth during the Jurassic and early Cretaceous periods. There is physical evidence that Earth's collision with a(n) (49)_____ may have brought about dinosaur extinction. In the early Cretaceous period, (50)_____ _____ emerged and began a major radiation amid the decline of gymnosperms that continues today. When the (51)_____ era began, major mountain ranges were formed, the great land masses underwent major renovations, and climate changes took place that greatly altered the course of evolution. New environments promoted a great diversification of (52) _____.

ORGANIZING THE EVIDENCE—CLASSIFICATION SCHEMES (15-IV, pp. 230–231)

Terms

phylogeny (230)_____

classification system (230)_____

Short Answer

1. Give some reasons why the knowledge of phylogeny is very valuable to researchers._____

_____

2. Arrange the following jumbled taxonomic categories in proper order, with the most inclusive listed first: family, class, species, kingdom, genus, phylum (or division), and order._____

_____

3. Arrange the following jumbled Latin and/or Greek names (with their accompanying categories) in proper order, with the most inclusive listed first: genus: *Archibaccharis*; kingdom: Plantae; order: Asterales; species: *lineariloba*; class: Dicotyledonae; division: Anthophyta; family: Asteraceae.

_____

_____

Matching

Choose the one best answer for each.

4. ___ Monera
5. ___ Protista
6. ___ Fungi
7. ___ Plantae
8. ___ Animalia

A. Multicelled heterotrophs that feed by extracellular digestion and absorption
B. Single-celled prokaryotes, some autotrophs, others heterotrophs
C. Diverse multicelled heterotrophs, including predators and parasites
D. Multicelled photosynthetic autotrophs
E. Diverse single-celled eukaryotes, some photosynthetic autotrophs, many heterotrophs

Fill-in-the-Blanks

(9)_____ is the process of categorizing phylogenetic information into a retrieval system consisting of many hierarchical levels, or ranks. Identifying organisms and assigning names to them comprise the activity called (10)_____.

Matching

Choose the single most appropriate answer

11. ___ black bear
12. ___ Whittaker
13. ___ binomial system
14. ___ *Pinus strobus* and *Pinus banksiana*
15. ___ species
16. ___ genus
17. ___ family
18. ___ taxa
19. ___ species of the same lineage
20. ___ Linnaeus

A. A group of similar species
B. The taxon most often studied
C. Groups of organisms evolving independently of other such organisms
D. Developed the 5-kingdom system of phylogenetic classification
E. Classification level that includes similar genera
F. Originated the modern practice of providing two scientific names for organisms
G. An example of a common name
H. Species belonging to one genus
I. Species connected by genetic contact in the space of the environment and through time
J. System of assigning a two-part Latin name to species

Self-Quiz

For questions 1–8, choose from the following answers:
   a. Archean
   b. Cenozoic
   c. Mesozoic
   d. Paleozoic
   e. Proterozoic

___1. Dinosaurs and gymnosperms were the dominant forms of life during the _____ era.

___2. The Alps, Andes, Himalayas, and Cascade Range were born during major reorganization of land masses early in the _____ era.

___3. The composition of Earth's atmosphere changed during the _____ era from anaerobic to aerobic.

___4. Invertebrates, primitive plants, and primitive vertebrates were the principal groups of organisms on Earth during the _____ era.

___5. Before the close of the _____ era, the first photosynthetic bacteria had evolved.

___6. The _____ era ended with the greatest of all extinctions, the Permian extinction.

___7. Late in the _____ era, flowering plants arose and underwent a major radiation.

___8. The _____ era included adaptive zones into which plant-eating mammals and their predators radiated.

For questions 9–11, choose from the following answers.
   a. convergence
   b. divergence
   c. adaptive radiation

___9. Penguins and porpoises serve as examples of _____.

___10. The wings of pterosaurs, birds, and bats serve as examples of _____.

___11. Evolution of the major groups of mammals following dinosaur extinction occurred by _____.

For questions 12–15, choose from the following answers:
    a. DNA–DNA hybridization
    b. punctuation
    c. gradualism
    d. mass extinction

___12. A rough measure of the evolutionary distance between two species is provided by _____.

___13. The vertebrate invasion of the land at the end of the Devonian was followed by rapid evolution; this is an example of _____.

___14. The disappearance of dinosaurs from the earth is an example of _____.

___15. Fossil foraminiferans (tiny shelled protists) show a constant change of form (over lengthy time periods) within one lineage; this is an example of _____.

For questions 16–20, choose from the following answers:
    a. homologous
    b. analogous
    c. macroevolution
    d. plate tectonics
    e. regulatory genes

___16. _____ structures resemble one another due to common descent.

___17. _____ is the study of the slabs floating on the earth's underlying mantle.

___18. Examples of _____ structures are the shark fin and the penguin wing.

___19. _____ refers to changes in groups of species.

___20. _____ control the rate of growth in different body parts.

___21. Phylogeny is _____.
    a. identifying organisms and assigning names to them
    b. producing a "retrieval system" consisting of several levels
    c. an evolutionary history of organisms, both living and extinct
    d. a study of the adaptive responses of various organisms

___22. All species that are connected by genetic contact in the space of the environment and through time represent a _____.
    a. population
    b. kingdom
    c. phylum
    d. lineage

Chapter Objectives/Review Questions

This section lists general and detailed chapter objectives that can be used as review questions. You can make maximum use of these items by writing answers on a separate sheet of paper. Fill in answers where blanks are provided. To check for accuracy, compare your answers with information given in the chapter or glossary.

Page	Objectives/Questions
(210)	1. _____ refers to the large-scale patterns, trends, and rates of change among groups of species.
(210)	2. Evolution proceeds by _____ of organisms that already exist.
(210)	3. Another term for a line of descent is a _____.
(211)	4. Arrange the great eras of Earth's history in the correct order from oldest to youngest.
(213)	5. What type of macroevolutionary evidence is collected by workers in comparative morphology?
(213)	6. Make a general statement about embryos of different organisms within the same major group.

(214)   7. Structures that are similar due to descent from a common ancestor are _____ structures.
(214)   8. Relate the term *morphological divergence* to homologous structures.
(214)   9. Similar body parts used for similar function in evolutionarily remote lineages are said to be _____ to one another; cite an example of morphological convergence.
(215)   10. Comparative _____ is a field that examines genes and gene products to determine evolutionary relationships.
(220–223) 11. Be able to generally "tell the story" of the origin of life; include statements relating to the following: primitive Earth and its atmosphere, synthesis of biological molecules, self-replicating systems, the first plasma membranes.
(220–223) 12. Describe the contributions of Stanley Miller and Sidney Fox to studies of life's origin.
(212)   13. Describe the essentials of plate tectonic theory.
(216)   14. Discuss the gradualistic model and the punctuational model in relation to speciation and rates of change.
(216–217) 15. Two trends have dominated life's history: mass _____ and adaptive _____.
(217)   16. What is an adaptive zone?
(224–230) 17. Be able to generally discuss the important geological and biological events occurring throughout the Archean, Proterozoic, Paleozoic, Mesozoic, and Cenozoic eras.
(230)   18. Identifying organisms and assigning names to them is a field of systematic inquiry known as _____.
(210, 216) 19. What is meant by "all species of the same lineage"?
(230)   20. Explain the purpose of applying a genus name and a species name to an organism.
(230–231) 21. Be able to arrange the series of ever more inclusive taxa of a classification system in proper order: kingdom, phylum (or division), class, order, family, genus, species.
(231)   22. Single-celled prokaryotes belong to kingdom _____; diverse, single-celled eukaryotes are placed in kingdom _____; kingdom _____ contains multicelled heterotrophs that feed by extracellular digestion; multicelled photosynthetic autotrophs are in kingdom _____; diverse multicelled heterotrophs that include parasites are classified in kingdom _____.

Integrating and Applying Key Concepts

Imagine that in the next decade three more Chernobyl-type disasters happen, the oceans acquire critical levels of carcinogenic pesticides that work their way up the food chains, and the ozone layer shrinks dramatically in the upper atmosphere. Describe the macroevolutionary events that you believe might happen.

As Earth becomes increasingly loaded with carbon dioxide and various industrial waste products, how do you think living forms on Earth will evolve to cope with these changes?

Critical Thinking Exercises

1. Which of the following would best test the hypothesis that two populations of very similar squirrels, isolated from contact with each other, are different species?
   a. Examine many individuals to determine whether there are any morphological differences between the two populations.
   b. Examine DNA from many individuals to assess genetic differences between the two populations.
   c. Examine the environments in the two locations to determine whether the same selection pressures would be operating.
   d. Capture live individuals from both populations and determine whether they can crossbreed and produce viable, fertile offspring.
   e. Look for hybrids between the two populations in the wild.

2. The table below presents some of the characteristics of twenty-six species of American coniferous trees. Classify the species into genera and families, using these observations. Do not be concerned about names; simply gather like species into groups representing genera, and like genera into groups representing families. Make as many groups as you think the observations justify.

No.	Cones				Leaves				Remarks
No.	Texture	Shape	Grouping	Position	Type	Position	Number	Cross-section	Remarks
1	wo	cy	so	pe	ne	sp	5	rd	Needles ridged and blunt
2	le	ov	cl	tr	sc	op			
3	le	ob	cl	pe	sc	op			Leaf tips curved inward, cones toothed
4	le	ov	cl	tr	sc	op			
5	pa	cy	so	er	ne	sp	1	fl	Blunt needles with two ribs
6	wo	ov	so	pe	ne	sp	2	rd	Needles ridged, sharp-tipped
7	wo	ov	so	tr	sc	sp			
8	pa	cy	so	pe	ne	sp	1	sq	Needles sharp-tipped, ridged
9	wo	cy	so	pe	ne	sp	5	rd	Needles stiff, sharp-tipped, twisted
10	wo	ob	so	pe	ne	sp	3	rd	Needles stiff
11	wo	cy	so	pe	ne	sp	2	ob	Needles stout and twisted
12	le	ov	so	tr	sc	op			Leaves with toothed edges
13	pa	cy	so	pe	ne	sp	1	fl	Needles flexible with rounded tip
14	wo	ob	so	pe	ne	sp	5	rd	Needles stiff
15	le	ov	cl	tr	sc	op			
16	wo	ob	so	tr	sc	sp			Scales sharp-pointed
17	wo	ob	cl	pe	sc	op			Scales pointed
18	pa	cy	so	pe	ne	sp	1	sq	Needles sharp-tipped and ridged
19	pa	cy	so	er	ne	sp	1	fl	Needles blunt and flexible
20	pa	cy	so	er	ne	sp	1	fl	Needles blunt and flexible
21	wo	ob	cl	pe	sc	op			Scales triangular and sharp
22	wo	ob	so	pe	ne	sp	3	rd	Needles stiff
23	wo	ob	so	pe	ne	sp	1	rd	Needles stiff, sharp-tipped, curved
24	le	ov	cl	tr	ne	wh			Needles awl-shaped
25	wo	ob	so	pe	ne	op	2	fl	Needles flexible, round tips, grooved
26	wo	ov	so	pe	ne	sp	5	rd	Needles ridged, twigs flexible

Key to abbreviations is at the top of p. 176

Abbreviations:
Cone texture: le = leathery, pa = papery, wo = woody
Cone shape: ob = oblong, ov = ovoid (almost spherical), cy = cylindrical
Cone grouping: cl = clustered, so = solitary
Cone position: er = erect (pointing upward from twig), pe = pendulous (pointing downward from twig),
    tr = transverse (pointing horizontally from twig)
Leaf type: ne = needlelike, sc = scalelike
Leaf position: (refers to arrangement of leaves along the twig) op = opposite, sp = spiral, wh = whorls
Leaf number: refers to the number of needles in a bundle sharing one attachment to the twig; not applicable
    to scalelike or awl-shaped leaves
Leaf cross-section: fl = flat, ob = oblong, rd = round, sq = square

Answers

Answers to Interactive Exercises

EVIDENCE OF MACROEVOLUTION (15-I)
1. Microevolution deals primarily with the process of evolution (the origin and success of given genotypes and their associated phenotypes); macroevolution deals with large-scale patterns of diversity through time.
2. Evolution proceeds by modifications of organisms that already exist.
3. Fossilized skeletons, shells, leaves, seeds, and tracks.
4. To be preserved as a fossil, body parts or impressions must be buried (burial medium varies) before they decompose; there must be no disturbance of this material.
5. D; 6. C; 7. E; 8. A; 9. B; 10. B; 11. M; 12. M; 13. M; 14. B; 15. M; 16. B; 17. M; 18. B; 19. M; 20. macroevolution; 21. fossils; 22. Archean; 23. Proterozoic; 24. Paleozoic; 25. Mesozoic; 26. Cenozoic; 27. changes; 28. fossil; 29. comparative morphology; 30. regulatory genes; 31. homologous; 32. divergence; 33. analogous; 34. convergence; 35. mutations; 36. hybridization.

EVIDENCE OF MACROEVOLUTION (CONT.) (15-II)
1. G; 2. H; 3. E; 4. K; 5. D; 6. L; 7. F; 8. M; 9. A; 10. N; 11. C; 12. J; 13. B; 14. I; 15. G, H; 16. E; 17. D, F, K, L; 18. M; 19. A, B, C, I, J, N

MACROEVOLUTION AND EARTH HISTORY (15-III)
1. a. Mesozoic, 240–205; b. Archean, 4,600–3,800; c. Paleozoic, 435–360; d. Paleozoic, 550–500; e. Archean, 4,600–3,800; f. Mesozoic, 135–65; g. Paleozoic, 360–280; h. Proterozoic, 2,500–570; i. Archean, 3,800–2,500; j. Paleozoic, 290–240; k. Paleozoic, 550–500; l. Mesozoic, 181–65; m. Proterozoic, 2,500–570; n. Cenozoic, 65–1.65; o. Cenozoic, 1.65–present; p. Paleozoic, 440–435; q. Paleozoic, 435–360;
2. C; 3. A; 4. B; 5. C (B); 6. A; 7. D (B); 8. B; 9. A; 10. B; 11. B; 12. E; 13. D; 14. F; 15. B; 16. A; 17. C; 18. 4; 19. 5; 20. 1; 21. 3(2); 22. 7; 23. 2(3); 24. 6; 25. 4.6; 26. 3.8; 27. prokaryotic cells (anaerobic bacteria); 28. oxygen (O_2); 29. fermentation; 30. 2.5; 31. photosynthetic bacteria; 32. stromatolites; 33. Proterozoic; 34. organic compounds (chemicals); 35. adaptive zones; 36. aerobic; 37. algae; 38. Cambrian; 39. trilobites; 40. Ordovician; 41. predatory; 42. Lobe-finned; 43. mass extinction; 44. Carboniferous; 45. Permian; 46. mammals; 47. mass extinction; 48. Dinosaurs (Reptiles); 49. asteroid; 50. flowering plants; 51. Cenozoic; 52. species.

ORGANIZING THE EVIDENCE—CLASSIFICATION SCHEMES (15-IV)
1. Researchers arrive at generalizations after comparing similar attributes among a variety of organisms. Inferences about one organism can be based on observations or experiments with another organism that is closely related and is apt to have similar responses.
2. kingdom, phylum (or division), class, order, family, genus, species.
3. kingdom: Plantae; division: Anthophyta; class: Dicotyledonae; order: Asterales; family: Asteraceae; genus: *Archibaccharis*; species: *lineariloba*.
4. B; 5. E; 6. A; 7. D; 8. C.
9. Classification; 10. taxonomy; 11. G; 12. D; 13. J; 14. H; 15. B; 16. A; 17. E; 18. C; 19. I; 20. F.

Answers to Self-Quiz

1. c; 2. b; 3. e; 4. d; 5. a; 6. d; 7. c; 8. b; 9. a; 10. b; 11. c, b; 12. a; 13. b; 14. d; 15. c; 16. a; 17. d; 18. b; 19. c; 20. e; 21. c; 22. d.

Critical Thinking Analysis

1. a. Morphological differences are not a reliable guide to separating species. Most species have a variety of characteristics within the species, and many species are very similar.
    b. Even DNA sequences are not necessarily species-specific. Individual humans can be identified on the basis of their DNA sequences. There is a great deal of genetic diversity within species.
    c. This might lead to predictions that certain characteristics will be similar in the two populations, but it does not define species.
    d. Reproductive isolation is the factor that separates two species. If the two populations interbreed, they

are classifiable as a single species. If they do not, it may be that they are two separate species, or it may be that they will not reproduce in captivity.

e. Because the two populations are spatially isolated, there can be no hybrids.

2. The most efficient way to approach this problem is to photocopy the table and cut the copy into strips with one species (one row) on each slip. Then choose a characteristic that you think is the most important and arrange the strips in groups according to that characteristic. Then rearrange the strips within each group according to another characteristic.

Eventually, you will perceive certain patterns. The final scheme you devise depends on which characteristic you choose as being more important than others. With the limited information given in the table, it might be very difficult to duplicate the commonly accepted version below. The importance of this exercise lies in working with a variety of observations to make a rational classification, not in "learning" the current version of conifer classification. The following table is not the right classification or a fact itself; it is better than yours only in that it is based on more facts.

Family	Species	Common Name	Number
Cupressaceae	Calocedrus decurrens	Incense cedar	3
	Cupressus arizonica	Arizona cypress	21
	C. macrocarpa	Monterey cypress	17
	Juniperus communis	Common juniper	24
	J. deppeana	Alligator juniper	4
	J. osteosperma	Utah juniper	12
	J. scopulorum	Rocky Mountain juniper	15
	Thuja plicata	Giant cedar	2
Taxodiaceae	Sequoia sempervirens	Coastal redwood	16
	Sequoiadendron gigantium	Giant sequoia	7
Pinaceae	Abies concolor	White fir	19
	A. lasiocarpa	Subalpine fir	20
	A. magnifica	Red fir	5
	Picea engelmannii	Engelmann spruce	18
	P. pungens	Blue spruce	8
	Pinus contorta	Lodgepole pine	11
	P. edulis	Two-needle pinyon	6
	P. flexilis	Limber pine	26
	P. lambertiana	Sugar pine	9
	P. longaeva	Bristlecone pine	1
	P. monophylla	One-needle pinyon	23
	P. ponderosa	Ponderosa pine	22
	P. sabiniana	Digger pine	10
	P. torreyana	Torrey pine	14
	Pseudotsuga menziesii	Douglas fir	13
	Tsuga heterophylla	Western hemlock	25

16

HUMAN EVOLUTION

THE MAMMALIAN HERITAGE
    Primate Classification
    From Primate to Human: Key Evolutionary
    Trends
PRIMATE ORIGINS

THE HOMINIDS
    Australopiths
    Stone Tools and Early *Homo*
    Homo erectus
    Homo sapiens

Interactive Exercises

THE MAMMALIAN HERITAGE/PRIMATE ORIGINS (16-I, pp. 235–239)

Terms

The page-referenced terms are important; they were in boldface type in the chapter. Refer to the instructions given in Chapter 1, p. 1 of this workbook.

mammal (235) _____

incisor (235) _____

canine (235) _____

premolar (235)_____

molar (235)_____

primate (236) _____

hominoid (236)_____

hominid (236) _____

bipedalism (236) _____

culture (238) _____

dryopiths (239)_____

Fill-in-the-Blanks

(1)_____ are warm-blooded animals with hair that began their evolution more than 200 million years ago. Humans are (2)_____; members of this order that originated about 65 million years ago rely less on their sense of (3)_____ and more on daytime vision. Their (4)_____ became larger and more complex; changes in the skull led to the development of (5)_____: the collection of behavior patterns of a social group, passed from generation to generation by learning and by symbolic behavior, some of which is language. Modification in the (6)_____ led to increased dexterity and manipulative skills that set the stage for the development of tools and technology.

Animals leaping from branch to branch benefit greatly from (a) excellent depth perception and (b) a (7) _____ that can guess accurately the distances required to reach the far limb and issue quick commands to the right muscles to achieve their goal; those who guess inaccurately may not live long enough to pass their genes on to the next generation.

Changes in primate (8)_____ indicate that there was a shift from eating insects to fruit and leaves and on to a mixed diet. Primates evolved from ancestral mammals more than (9)_____ million years ago, during the Paleocene. The first primates resembled small (10)_____ or tree shrews; they foraged at night for (11)_____, seeds, buds, and eggs on the forest floor, and they could clumsily climb trees searching for safety and sleep.

During the (12)_____ era, birds, mammals, and flowering plants rose to dominate Earth's assemblage of organisms. Humans, apes, monkeys, and prosimians are all (13)_____; they have excellent (14)_____ perception as a result of their forward-directed eyes, and their fingers and toes are adapted for (15)_____ instead of running. About (16)_____ million years ago, the continents began to assume their current positions, and climates became cooler and (17)_____. Under these conditions, forests began to give way to (18)_____; perhaps speciation was favored as subpopulations of apes became reproductively isolated within the shrinking stands of trees. The "tree apes," or (19)_____, originated during this time; eventually, they ranged throughout Africa, Europe, and (20)_____. Between (21)_____ million and five million years ago, there were several notable divergences; one gave rise to the gorilla and a second one resulted in the separation of the (22) _____ lineage from the early hominids, including the ancestors of humans.

Matching

Fill in the blanks of the evolutionary diagram below with the appropriate letter from the choice list.

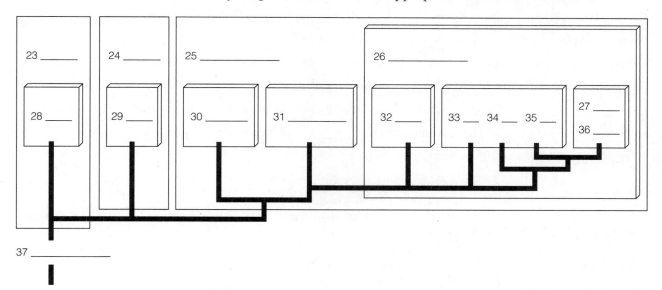

A. Anthropoids
B. Chimpanzee
C. Gibbon, siamang
D. Gorilla
E. Hominids

F. Hominoids
G. Humans and
   their most
   recent ancestors
H. Lemurs, lorises

I. New World monkeys
   (spider monkeys, etc.)
J. Old World monkeys
   (baboons, etc.)
K. Orangutan

L. Prosimians
M. Tarsiers
N. Tarsioids
O. Rodentlike primate
   of the Paleocene

THE HOMINIDS (16-II, pp. 240–243)

Terms

plasticity (240) _____

australopiths (241) _____

early *Homo* (242) _____

Homo erectus (242) _____

Homo sapiens (242) _____

Fill-in-the-Blanks

The hominid group (which includes all species on the genetic path leading to humans since the time when

that path diverged from the path leading to the (1)_____) emerged somewhere around the

Miocene–Pliocene boundary approximately (2)_____ million years ago. (3)_____-million-year-old

fossils of humanlike forms have been discovered in Africa, and they all were bipedal and omnivorous and

had an expanded brain of approximately (4) _____ ccs. Lucy was one of the earliest (5)_____, a

collection of transitional forms that combined ape and human features. They were fully bipedal, with essen-

tially human bodies and ape-shaped heads. The oldest fossils of the genus *Homo*, makers of stone tools, date from approximately (6)_____ million years ago. Between 1.5 million and 300,000 years ago, during the Pleistocene periods of glaciation, there were also intermittent periods of warming; during these interglacial times, a larger-brained human species, (7)_____ _____, migrated out of Africa and into China, Southeast Asia, and Europe. The (8)_____ were a distinct hominid population that appeared about 130,000 years ago in Europe and Asia; their cranial capacity was indistinguishable from our own, and they had a complex culture. By 30,000 years ago, there was only one remaining hominid species: (9)_____ _____ _____. These hominids learned to store and share food among themselves, to anticipate seasonal migrations, and to plan community (10) _____. They created exquisite tools and (11) _____ treasures at Lascaux and elsewhere.

Choice

For questions 12–14, enter the letter(s) of the feature(s) that first developed in or characterized the creatures listed. Choose from these possibilities:

a. bipedalism

b. brain volume of ≈ 1,000 ccs

c. thumb and forefingers could touch each other

d. speech and language

e. shaping and use of stone tools

___ 12. tree-dwelling primates

___ 13. *Australopithecus afarensis*, "Lucy"

___ 14. *Homo sapiens*

Identification/Matching

Suppose you are a student who wants to be chosen to accompany a paleontologist who has spent forty years teaching and roaming the world in search of human ancestors. Above the desk in her office, she keeps reconstructions of the seven skulls shown on the next page, and you have heard that she chooses the graduate students who accompany her on her summer safaris on the basis of their ability on a short matching quiz. The quiz is presented on the next page, but *before* you begin, perhaps you would like to review what is known about hominid fossils.

Without consulting your text (or any other) while you are taking the quiz, match up all seven skulls with *all* applicable letters. Sixteen correctly placed letters win you a place on the expedition. Fourteen correct answers put you on a waiting list. Thirteen or fewer and she suggests that you may wish to investigate dinosaur fossils instead of human ancestry. Which will it be for you?

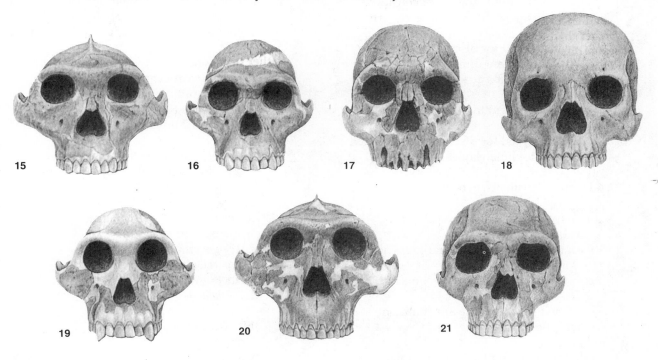

A. gracile forms of australopiths
B. robust forms of australopiths
C. fashioned stone tools and used them first
D. has a chin
E. first controlled use of fire
F. lived 4 million years ago
G. from a lineage that lived from approximately 3 million years ago until about 1.25 million years ago
H. lived 2 million years ago
I. lived 1.5 million years ago until at least 100,000 years ago
J. most recent hominid to appear

15. _____

16. _____

17. _____

18. _____

19. _____

20. _____

21. _____

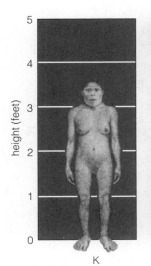
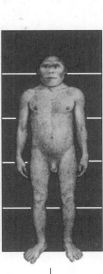
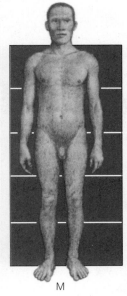

Self-Quiz

___1. Which of the following is not considered to have been a key character in early primate evolution?
a. eyes adapted for discerning color and shape in a three-dimensional field
b. body and limbs adapted for tree climbing
c. bipedalism and increased cranial capacity
d. eyes adapted for discerning movement in a three-dimensional field

___2. Primitive primates generally live _____.
a. in tropical and subtropical forest canopies
b. in temperate savanna and grassland habitats
c. near rivers, lakes, and streams in the East African Rift Valley
d. in caves where there are abundant supplies of insects

___3. All the ancestral placental mammals apparently arose from ancestral forms of a group _____.
a. that includes omnivorous shrews and moles
b. that includes dogs, cats, and seals
c. that includes mice and beavers
d. that includes the koala ("teddy bear") and flying phalanger (resembles the flying squirrel)

___4. The hominid evolutionary line stems from a divergence (fork in a phylogenetic tree) from the ape line that apparently occurred _____.
a. somewhere between 8 million and 5 million years ago
b. about 3 million years ago
c. during the Pliocene epoch
d. less than 2 million years ago

___5. _____ was an Oligocene anthropoid that probably predated the divergence leading to Old World monkeys and the apes, with dentition more like that of dryopiths and less like that of the Paleocene primates with rodentlike teeth.
a. *Aegyptopithecus*
b. *Australopithecus*
c. *Homo erectus*
d. *Plesiadapis*

___6. Johanson's Lucy was a(n) _____.
a. dryopith
b. australopith
c. Miocene
d. prosimian

___7. A hominid of Europe and Asia that became extinct nearly 30,000 years ago was _____.
a. a dryopith
b. *Australopithecus*
c. *Homo erectus*
d. Neandertal

Matching

8. ___ anthropoids
9. ___ australopiths
10. ___ Cenozoic
11. ___ dentition
12. ___ dryopiths
13. ___ hominids
14. ___ hominoids
15. ___ Miocene
16. ___ Neandertal
17. ___ plasticity
18. ___ Pliocene
19. ___ primates
20. ___ prosimians

A. A group that includes apes and humans
B. A population of *Homo sapiens* that lived from at least 100,000 to as recently as 30,000 years ago; tool users and artisans
C. The ability to adapt to a wide range of demands
D. Organisms in a suborder that includes New World and Old World monkeys, apes, and humans
E. An era that began 65 to 63 million years ago; characterized by the evolution of birds, mammals, and flowering plants
F. The type, number, and size of teeth
G. A group that includes humans and their most recent ancestors
H. An epoch of the Cenozoic era lasting from 25 million to 5 million years ago; characterized by the appearance of primitive apes, whales, and grazing animals of the grasslands
I. Organisms in a suborder that includes tree shrews, lemurs, and others
J. An epoch of the Cenozoic era lasting from 5 million to 2 million years ago; characterized by the appearance of distinctly modern plants and animals
K. A group that includes prosimians, tarsioids, and anthropoids
L. Bipedal organisms living from about 4 million to 1 million years ago, with essentially human bodies and ape-shaped heads; brains no larger than those of chimpanzees
M. Transitional apelike forms that could climb about in trees and walk on the ground

Chapter Objectives/Review Questions

This section lists general and detailed chapter objectives that can be used as review questions. You can make maximum use of these items by writing answers on a separate sheet of paper. Fill in answers where blanks are provided. To check for accuracy, compare your answers with information given in the chapter or glossary.

Page	Objectives/Questions
(236–238)	1. Describe the general physical features and behavioral patterns attributed to early primates.
(239)	2. Trace primate evolutionary development through the Cenozoic era. Describe how Earth's climates were changing as primates changed and adapted. Be specific about times of major divergence.
(239–240)	3. State which anatomical features underwent the greatest changes along the evolutionary line from early anthropoids to humans.
(236, 239)	4. Where do prosimian survivors dwell today on Earth?
(236–238)	5. Five key characters of primate evolution are _____ _____ _____ _____ _____.
(236)	6. Beginning with the primates most closely related to humans, list the main groups of primates in order by decreasing closeness of relationship to humans.
(242–243)	7. Explain how you think *Homo sapiens sapiens* arose. Make sure your theory incorporates existing paleontological (fossil), biochemical, and morphological data.

Integrating and Applying Key Concepts

Suppose someone told you that some time between 12 million and 6 million years ago dryopiths were forced by larger predatory members of the cat family to flee the forests and take up residence in estuarine, riverine, and sea coastal habitats where they could take refuge in the nearby water to evade the tigers. Those that, through mutations, became naked, developed an upright stance, developed subcutaneous fat deposits as insulation, and developed a bridged nose that had advantages in watery habitats (features that other dryopiths that remained inland never developed) survived and expanded their populations. As time went on, predation by the big cats and competition with other animals for available food caused most of the terrestrial dryopiths to become extinct, but the water-habitat varieties survived as scattered remnant populations, adapting to easily available shellfish and fish, wild rice and oats, and various tubers, nuts, and fruits. It was in these aquatic habitats that the first food-getting tools (baskets, nets, and pebble tools) were developed, as well as the first words that signified different kinds of food. How does such a story fit with current speculations about and evidence of human origins? How could such a story be shown to be true or false?

Crossword Puzzle: Primate Evolution

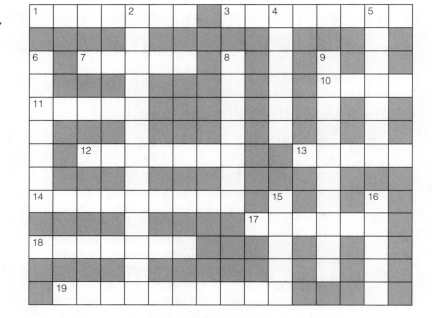

ACROSS
1. A group that includes prosimians, tarsioids, and anthropoids
3. A group that includes apes and humans
7. _____ evolution is a time of rapid branchings and adaptive radiation
10. A kind of tall, showy flower; usually lavender or purple; blue flag
11. A cheek tooth that crushes and grinds
12. All the behavior patterns of a social group passed by learning and language from generation to generation
13. Pay great homage to; worship
14. Forest apes that lived 13 million years ago in Africa, Europe, and southern Asia
17. Pointed teeth that enable the tearing of flesh
18. *Homo* _____ was the first hominid to control and use fire for heating and cooking
19. The ability to adapt to a variety of agents of change in an animal's environment

DOWN
2. Southern ape-humans, the fossils of which have dates from 3.7 to 1.25 million years ago
4. A vertebrate with hair
5. A flat chisel or conelike tooth that nips or cuts food
6. A group that includes modern humans and their direct-line ancestors since divergence from the ape line
8. The specific (species) name of modern humans
9. Habitual two-legged method of locomotion
15. _____ *Homo* used simple stone tools
16. Hard structures that can provide clues about what an animal typically eats

Critical Thinking Exercise

The brain of the members of the human lineage enlarged dramatically as the genus *Homo* evolved. Which of the following is the strongest evolutionary explanation for this phenomenon?

a. The lineage developed large brains in order to produce humans.

b. The prehuman ancestors developed large brains because they needed them.

c. The early humans developed large brains in order to use their new bipedal locomotion, stereoscopic vision, and opposable thumb.

d. Early humans developed large brains because they had begun to develop culture.

e. Because early humans practiced cooperative food gathering, fishing and hunting, individuals with larger brains were more successful and left more offspring.

Answers

Answers to Interactive Exercises

THE MAMMALIAN HERITAGE/PRIMATE ORIGINS (16-I)

1. Mammals; 2. primates; 3. smell; 4. brains; 5. culture; 6. hands; 7. brain; 8. teeth; 9. 60; 10. rodents; 11. insects; 12. Cenozoic; 13. primates; 14. depth; 15. grasping; 16. 23 to 20; 17. drier; 18. grasslands (savannas); 19. dryopiths; 20. Asia; 21. 10; 22. chimpanzee; 23. L; 24. N; 25. A; 26. F; 27. E; 28. H; 29. M; 30. I; 31. J; 32. C; 33. K; 34. D; 35. B; 36. G; 37. O.

THE HOMINIDS (16-II)

1. apes; 2. five; 3. Four; 4. 400; 5. australopiths; 6. 2.5; 7. *Homo erectus*; 8. Neandertals; 9. *Homo sapiens sapiens*; 10. hunts; 11. artistic; 12. c; 13. a; 14. d; 15. B, G, H; 16. A, G, L; 17. C, H; 18. D, J, M; 19. A, F, K; 20. B, G, H; 21. E, I.

Answers to Self-Quiz

1. c; 2. a; 3. a; 4. a; 5. a; 6. b; 7. d; 8. D; 9. L; 10. E; 11. F; 12. M; 13. G; 14. A; 15. H; 16. B; 17. C; 18. J; 19. K; 20. I.

Answers to Crossword Puzzle

Answers appear on next page.

Critical Thinking Analysis

a. Evolution has no predetermined outcome or "goal." It proceeds according to the dictates of selective pressures applied by the environment and to the historical accidents of mutation, genetic drift, and extinction.

b. Evolution does not proceed according to the "needs" of the population. Selection is a pragmatic force that favors whatever accidental developments reproduce more rapidly.

c. Again, this statement implies a predetermined direction or purpose for evolution. Evolution has none.

d. This is a sort of evolutionary explanation. It identifies a force that favors mutations of a certain type. However, it is a very general and vague statement, and it fails to link large brains directly with reproductive success.

e. This is a clear, specific evolutionary hypothesis. Cooperative food gathering, fishing, and hunting would be more successful for individuals who could communicate and do abstract planning of future movements instead of simply reacting as individuals to the immediate situation. Assuming that these abilities are improved by increasing brain size, they would constitute a strong selective force for larger brains.

Crossword puzzle (answers shown):

Across:
1. PRIMATE
3. HOMINOID
7. BUSHY
10. IRIS
11. MOLAR
12. CULTURE
13. ADORE
14. DRYOPITHS
17. CANINE
18. ERECTUS
19. PLASTICITY

Down:
2. MUTATION
4. MAMMAL
5. INCISOR
6. HOMININI
8. SAPIENS
9. BIPEDALISM
13. ARBOREAL
15. ELSMERE (ELSMT...)
16. TEETH

17

VIRUSES, BACTERIA, AND PROTISTANS

Interactive Exercises

VIRUSES (17-I, pp. 247–249)

Terms

The page-referenced terms are important; they were in boldface type in the chapter. Refer to the instructions given in Chapter 1, p. 1 of this workbook.

microorganism (246)_____

pathogen (246)_____

virus (247)_____

bacteriophage (249)_____

Short Answer

1.   a. State the principal characteristics of viruses (p. 247)._____

_____

    b. Describe the structure of viruses (p. 247)._____

_____

c. Distinguish between the ways viruses replicate themselves (p. 248)._____

_____

2.   a. List five specific viruses that cause human illness (p. 249). _____

_____

    b. Describe how each virus in (a) does its dirty work (p. 249)._____

_____

_____

_____

_____

_____

_____

_____

_____

_____

Fill-in-the-Blanks

A(n) (3)_____ is a noncellular, nonliving infectious agent, each of which consists of a central
(4)_____ _____ core surrounded by a protective (5)_____ _____. (6)_____ contain
the blueprints for making more of themselves but cannot carry on metabolic activities. Chicken pox and
shingles are two infections caused by DNA viruses from the (7)_____ category. Naked strands or circles of RNA that lack a protein coat are called (8)_____. (9)_____ are RNA viruses that infect animal cells; some cause AIDS and follow (10)_____ pathways of replication. (11)_____ are the usual
units of measurement with which to measure viruses, while microbiologists measure bacteria and protistans
in terms of (12)_____. A bacterium 86 micrometers in length is (13)_____ nanometers long.
During a period of (14)_____ , viral genes remain inactive inside the host cell and any of its descendants. Pathogenic protein particles are called (15)_____.

Identification

Identify the virus that causes the following illnesses by writing the name of the virus in the blank preceding the disease.

_____ 16. Common colds

_____ 17. AIDS, ARC

_____ 18. Cold sores, chicken pox

Matching

Match each item below with the correct description.

19. ___ bacteriophage
20. ___ capsid
21. ___ host
22. ___ latent period of temperate pathway
23. ___ lytic pathway
24. ___ microorganism
25. ___ pathogen
26. ___ viroid
27. ___ virus

A. Any organism too small to be seen without a microscope
B. The one infected by a parasite (or virus)
C. A virus that infects a bacterium
D. Damage and destruction to host cells occurs quickly
E. Noncellular infectious agent that must take over a living cell in order to reproduce itself
F. Protein or lipid coat outside the nucleic acid core
G. Viral nucleic acid is integrated into the nucleic acid system of the host cell and replicated during this time
H. Strands or circles of RNA with no protein coat that infect plants
I. Any disease-causing organism or agent

BACTERIA (17-II, pp. 250–255)

Terms

binary fission (251) _____

bacterial conjugation (251) _____

antibiotic (252)_____

archaebacteria (252) _____

eubacteria (252)_____

methanogen (252) _____

halophile (252)_____

thermophile (252) _____

Fill-in-the-Blanks

Photosynthetic autotrophs produce their own organic compounds from simple (1)_____ _____ , using (2)_____ as an energy source. (3) _____ _____ are autotrophs that extract energy from chemical reactions involving inorganic molecules and do *not* use sunlight as an energy source. Anaerobic bacteria can live in canned food, reproduce, and produce deadly toxins. Two examples of pathogenic (disease-causing) bacteria that form endospores harmful to humans are (4)_____ _____ and (5)_____ _____. Eubacterial cell walls are composed of (6)_____ , a substance that never occurs in eukaryotes. Heterocysts are cells in *Anabaena* that carry out (7)_____ _____. When environmental conditions become adverse, many bacteria form (8)_____ , which resist moisture loss, irradiation, disinfectants, and even acids. Most bacteria reproduce by (9)_____ _____ , an asexual form of cell division. Some bacteria can engage in (10)_____—the transfer of a plasmid from a donor to a recipient cell. Bacteria differ from all other kinds of organisms in being (11)_____. Spherical bacteria are (12)_____, rod-shaped bacteria are (13)_____, and helical bacteria are (14)_____. Bacterial behavior depends on (15) _____ _____, which change shape when they absorb or connect with chemical compounds. Some bacteria are photosynthetic autotrophs that use (16) _____ as an energy source. Most of the world's bacteria are (17) [choose one] ( ) producers, ( ) consumers, ( ) decomposers, so we think of them as "good" heterotrophs. Antibiotics such as (18) _____ block protein synthesis in their target cells; other antibiotics such as (19) _____ disrupt the formation of cell walls in their target cells. *Escherichia coli*, which dwell in our gut, synthesize vitamin (20) ____ and substances useful in digesting fats.

Matching

Match each of the items below with a lowercase letter designating its principal bacterial group and an uppercase letter denoting its best descriptor from the right-hand column.

a. Archaebacteria

b. Chemosynthetic eubacteria

c. Heterotrophic eubacteria

d. Photosynthetic eubacteria

e. None of the above

21. ___ , ___ *Anabaena*

22. ___ , ___ *Bacillus, Clostridium*

23. ___ , ___ *Escherichia coli*

24. ___ , ___ *Halobacterium*

25. ___ , ___ *Herpesvirus*

26. ___ , ___ *Lactobacillus*

27. ___ , ___ *Methanobacterium*

28. ___ , ___ *Nitrobacter, Nitrosomonas*

29. ___ , ___ *Rhizobium, Agrobacterium*

30. ___ , ___ *Rhodospirillum*

31. ___ , ___ *Salmonella*

32. ___ , ___ *Spirochaeta, Treponema*

33. ___ , ___ *Staphylococcus, Streptococcus*

34. ___ , ___ *Streptomyces, Actinomyces*

35. ___ , ___ *Thermoplasma, Sulfolobus*

A. Live in anaerobic sediments of lakes and in animal gut; chemosynthetic; used in sewage treatment facilities

B. Purple; generally in anaerobic sediments of lakes or ponds; do not produce oxygen

C. Endospore-forming rods and cocci that live in the soil and in the animal gut; some major pathogens

D. Gram-positive cocci that live in the soil and in the skin and mucous membranes of animals; some major pathogens

E. Gram-positive nonsporulating rods that ferment plant and animal material; some are important in dairy industry; others contaminate milk, cheese

F. In acidic soil, hot springs, hydrothermal vents on seafloor; may use sulfur as a source of electrons for ATP formation

G. Live in extremely salty water; have a unique form of photosynthesis

H. Gram-negative aerobic rods and cocci that live in soil or aquatic habitats or are parasites of animals and/or plants; some fix nitrogen

I. Nitrifying bacteria that live in the soil, fresh water, and marine habitats; play a major chemosynthetic role in the nitrogen cycle

J. Gram-negative anaerobic rod that inhabits the human colon where it produces vitamin K

K. Major gram-negative pathogens of the human gut that cause specific types of food poisoning

L. Mostly in lakes and ponds; cyanobacteria; produce O_2 from water as an electron donor

M. A nonliving infectious particle

N. Major producer of antibiotics; an actinomycete that lives in soil and some aquatic habitats

O. Helically coiled, motile parasites of animals; some are major pathogens

PROTISTANS (17-III, pp. 255–263)

Terms

protistan (255) _____

eukaryote (255) _____

endosymbiosis (256)_____

euglenoid (258) _____

contractile vacuole (258) _____

slime mold (259)_____

chrysophyte (260)_____

plankton (260)_____

dinoflagellate (260)_____

protozoan (260)_____

flagellated protozoan (261) _____

cysts (261) _____

amoeboid protozoan (261) _____

ciliated protozoan (262)_____

sporozoan (262)_____

Fill-in-the-Blanks

Comparisons of nucleotide sequences of the (1)_____ and RNA from different species suggest that
(2)_____ , eubacteria, and the forerunners of eukaryotes diverged from a common ancestor long before
(3)_____ evidence of eukaryotes began accumulating; recent estimates place this divergence as occurring approximately (4)_____ billion years ago. Protistans, plants, fungi, and animals are all
(5)_____ ; they are like each other in having a nucleus and (6)_____-_____ organelles.
Bacteria have no membrane-bound (7)_____ or nucleus. Lynn Margulis and others believe that the
eukaryotic cell arose by the merger of two (8)_____ cells and the maintenance of their (9)_____
arrangement. Although the basic genetic code is identical for all living species (which suggests that the
genetic code was in effect before the first cells on Earth diverged into separate evolutionary lines), the
(10)_____ of several species follow a slightly altered code. Starting about (11)_____ billion years
ago, eukaryotic cells equipped with mitochondria, chloroplasts, or both evolved rapidly into a variety of
multicelled organisms.

Complete the Table

Statement from the Theory of Endosymbiosis	Evidence That Supports This Hypothesis
Mitochondria are descendants of aerobic eubacteria living symbiotically in a host cell.	12. _____ _____ 13. _____ _____ 14. _____ _____
Chloroplasts are descendants of aerobic photosynthetic bacteria that took up residence in predatory host cells.	15. _____ _____ 16. _____ _____ 17. _____ _____

True/False

If the statement is true, write a T in the blank. If the statement is false, make it correct by changing the underlined word(s) and writing the correct word(s) in the answer blank.

_____ 18. At present, there is no <u>direct</u> evidence of how either the single-celled eukaryotes or the multicelled eukaryotes originated.

_____ 19. Among prokaryotes, <u>some photosynthetic bacteria</u> resemble chloroplasts strongly, and aerobic bacteria with "extra" cytochromes are the most likely candidates to be the ancestors of mitochondria.

Fill-in-the-Blanks

The cells of some (20)_____ _____ differentiate and form (21)_____ _____, stalked structures bearing spores at their tips. Slime molds also spend part of their life creeping about like (22)_____ and engulfing food. Euglenoids reproduce by (23)_____ _____, a reproductive mode that is common among all flagellated protistans; the cell grows in circumference while all (24)_____ are being duplicated and then divides along its long axis. (25)_____ undergo explosive population growth and color the seas red or brown, causing a red tide that may kill hundreds or thousands of fish and, occasionally, people. Chrysophytes include 450 species of "yellow-green algae," about 500 species of "golden algae," and more than 5,000 species of golden-brown (26)_____. Diatom cells have external thin, overlapping (27)_____ that fit together like a pill box. Dinoflagellates are photosynthetic members of marine (28)_____ and freshwater ecosystems; some forms are also heterotrophic.

Amoebas move by sending out (29)_____, which surround food and engulf it. *Paramecium* is a ciliate that depends on (30)_____ _____ for eliminating the excess water constantly flowing into the cell. *Paramecium* has a (31)_____, a cavity that opens to the external watery world. Once inside the cavity, food particles become enclosed in (32)_____-_____ _____, where digestion takes place. (33)_____ is a famous sporozoan that causes malaria. When a particular (34)_____ draws blood from an infected individual, (35)_____ of the parasite fuse to form zygotes, which eventually develop within the mosquito.

Labeling

36._____   39._____   42._____ _____

37._____   40._____   43._____

38._____   41._____   44._____ _____

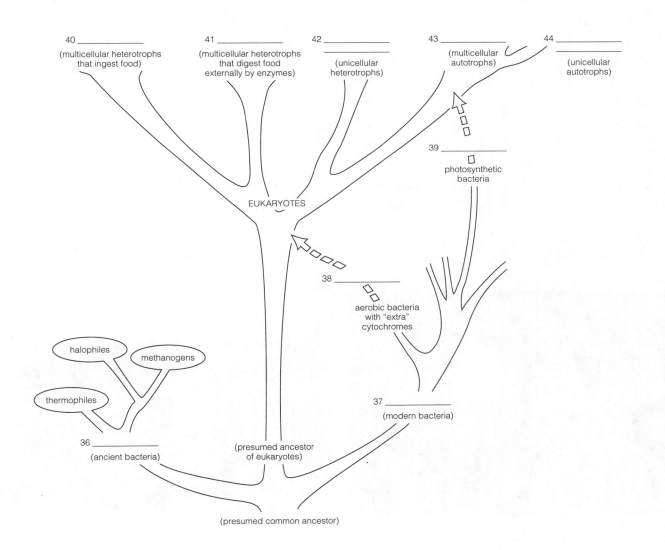

Matching

Put as many letters in each blank as are applicable.

45. _____ diatoms

46. _____ *Entamoeba histolytica*

47. _____ foraminiferans

48. _____ *Ptychodiscus brevis*

49. _____ *Paramecium*

50. _____ *Plasmodium*

51. _____ *Trichomonas vaginalis*

52. _____ *Trypanosoma brucei*

A. Ciliated Protozoans
B. Flagellated Protozoans
C. Amoeboid Protozoans
D. Sporozoans
E. Dinoflagellates
F. Chrysophytes
G. Causes a sexually transmitted disease
H. African sleeping sickness
I. Malaria
J. Traveler's diarrhea
K. Red tide
L. Primary component of many ocean sediments

Matching

Match the pictures on this page and page 197 with the names below.

53. ___
54. ___
55. ___
56. ___
57. ___
58. ___
59. ___
60. ___
61. ___
62. ___

A. *Amoeba proteus*
B. *Bacillus*
C. bacteriophage
D. *Clostridium tetani*
E. Cyanobacterium
F. Diatom
G. foraminiferans
H. *Herpesvirus*
I. HIV
J. *Paramecium*

53

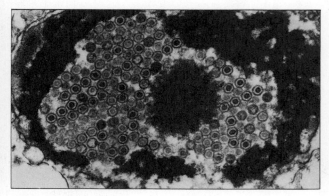

54

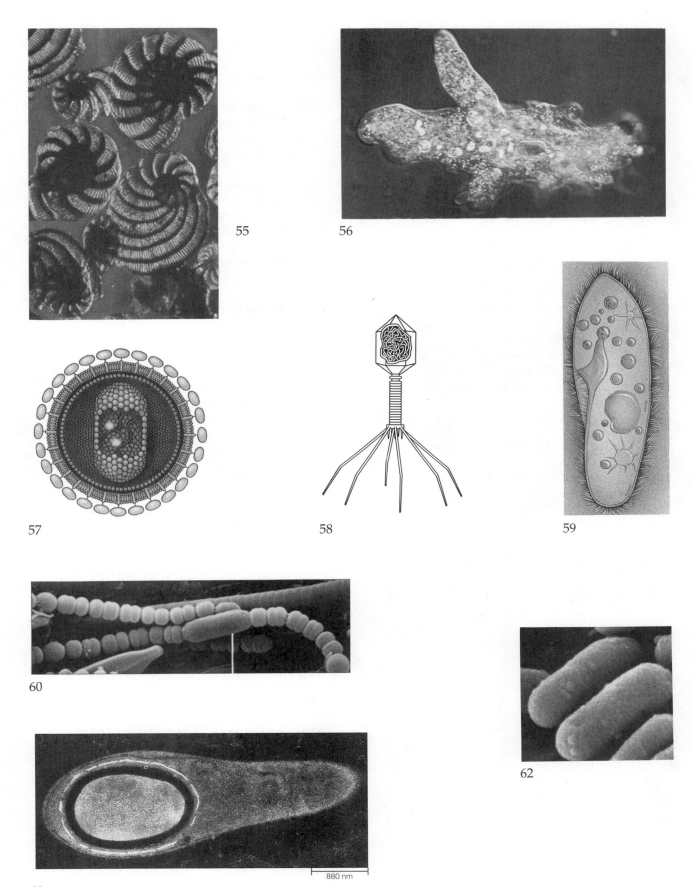

55

56

57

58

59

60

62

880 nm

61

Self-Quiz

___1. Which of the following diseases is not caused by a virus?
 a. smallpox
 b. polio
 c. influenza
 d. syphilis

___2. Bacteriophages are _____.
 a. viruses that parasitize bacteria
 b. bacteria that parasitize viruses
 c. bacteria that phagocytize viruses
 d. composed of a protein core surrounded by a nucleic acid coat

___3. Which of the following specialized structures is not correctly paired with a function?
 a. gullet—ingestion
 b. cilia—food gathering
 c. contractile vacuole—digestion
 d. anal pore—waste elimination

___4. Many biologists believe that chloroplasts are descendants of _____ that were able to live symbiotically within a predatory host cell.
 a. aerobic bacteria with "extra" cytochromes
 b. photosynthetic bacteria
 c. dinoflagellates
 d. bacterial heterotrophs

___5. Population "blooms" of _____ cause "red tides" and extensive fish kills.
 a. *Euglena*
 b. specific dinoflagellates
 c. diatoms
 d. *Plasmodium*

___6. Exposure to free oxygen is lethal for all _____.
 a. obligate anaerobes
 b. bacterial heterotrophs
 c. chemosynthetic autotrophs
 d. facultative anaerobes

___7. Which of the following protists does NOT cause great misery to humans?
 a. *Dictyostelium discoideum*
 b. *Entamoeba histolytica*
 c. *Plasmodium*
 d. *Trypanosoma brucei*

___8. _____ all transform energy into usable forms and have complete genetic systems that maintain and reproduce themselves; hence, they are unquestionably alive.
 a. Bacteria, blue-green algae, and prions
 b. Bacteria and viroids
 c. Bacteria and rickettsias
 d. Bacteria and viruses

___9. When nutrients are scarce, many bacteria _____.
 a. engage in conjugation
 b. switch to photosynthesis
 c. form endospores
 d. become pathogenic

___10. Which of the following play an important role in the cycling of nitrogen-containing substances?
 a. cyanobacteria
 b. prions
 c. viruses
 d. photosynthetic flagellates

Matching

Match all applicable letters with the appropriate terms. A letter may be used more than once, and a blank may contain more than one letter.

11. _____ *Amoeba proteus*

12. _____ *Anabaena*

13. _____ *Clostridium botulinum*

14. _____ diatoms

15. _____ *Dictyostelium*

16. _____ *Escherichia coli*

17. _____ foraminifera

18. _____ *Ptychodiscus brevis* (red tide)

19. _____ *Herpesvirus*

20. _____ HIV

21. _____ *Lactobacillus*

22. _____ *Paramecium*

23. _____ *Plasmodium*

24. _____ *Staphylococcus*

A. Bacteria
B. Protista
C. Virus
D. Slime mold
E. Cyanobacteria
F. Photosynthetic flagellates
G. Dinoflagellates
H. Gram-positive eubacteria
I. Obtain food by using pseudopodia
J. Causes malaria
K. A sporozoan
L. A ciliate
M. Cause cold sores and a type of venereal disease
N. Live in "glass" houses
O. Live in hardened shells that have thousands of tiny holes, through which pseudopodia protrude
P. Associated with AIDS, ARC

Chapter Objectives/Review Questions

This section lists general and detailed chapter objectives that can be used as review questions. You can make maximum use of these items by writing answers on a separate sheet of paper. Fill in answers where blanks are provided. To check for accuracy, compare your answers with information given in the chapter or glossary.

Page	Objectives/Questions
(250–251)	1. Describe the principal body forms of prokaryotic cells (inside and outside).
(250, 251)	2. Explain how, with no nucleus or membrane-bound organelles, prokaryotic cells reproduce themselves and obtain energy to carry on metabolism.
(250)	3. Distinguish chemosynthesis from photosynthesis.
(252–254)	4. State the ways in which archaebacteria differ from eubacteria.
(256)	5. Trace a sequence of events that may have transformed heterotrophic prokaryotes into photosynthetic eukaryotes.
(261)	6. Two flagellated protozoans that cause human misery are _____ and _____.
(261)	7. State the principal characteristics of the amoebas, radiolarians, and foraminiferans. Indicate how they generally move from one place to another and how they obtain food.
(262)	8. List the features common to most ciliated protozoans.
(262)	9. Characterize the sporozoan group, identify the group's most prominent representative, and briefly describe the life cycle of that organism.

Integrating and Applying Key Concepts

The textbook (page 622) identifies natural gas as a nonrenewable fuel resource, yet there is a group of archaebacteria that produce methane, the burning of which can serve as a fuel for heating and/or cooking. Recall or imagine how these bacteria could be incorporated into a system that could serve human societies by generating methane in a cycle that is renewable. Why did your text categorize natural gas as a nonrenewable resource? Is methane a constituent of natural gas? Why or why not?

Critical Thinking Exercise

When bacteria are grown in culture, the number of cells begins at some level and steadily increases. Viruses, on the other hand, are inoculated in a culture at some level, then disappear for a short time, reappear in the culture, and then increase steadily in number. Why was this growth pattern so puzzling to early virologists? With current knowledge, how do you explain it?

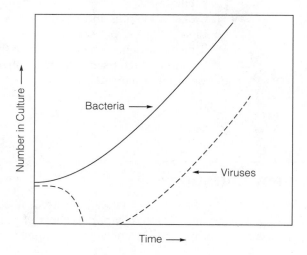

Answers

Answers to Interactive Exercises

VIRUSES (17-I)

1. a. Nonliving, infectious agents, smaller than the smallest cells; require living cells to act as hosts for their replication; not acted upon by antibiotics.
b. The core can be DNA or RNA; the capsid can be protein and/or lipid.
c. Bacteriophage viruses may use the lytic pathway, in which the virus quickly subdues the host cells and replicates itself and descendants are released as the cell undergoes lysis; or they may use a temperate pathway, in which viral genes remain inactive inside the host cell during a period of latency, which may be a long time, before activation and lysis.
2. There can be multiple answers (see Table 17.1 in text).
a. Possible answers include Herpes simplex (a Herpesvirus), Varicella-zoster (a Herpesvirus), Rhinovirus (a Picornavirus), Poliovirus (an enterovirus of the Picorna group), and HIV (a Retrovirus).
b. Herpes simplex: DNA virus. Initial infection is a lytic cycle that causes Herpes (sores) on mucous membranes

of mouth or genitals. Recurrent infections are temperate. Most infected cells are in nerves and skin. No immunity. No cure. Varicella-zoster: DNA virus. Initial infection is a lytic cycle that causes sores on skin. Generally, immunity is conferred by one infection, but in some people subsequent infections follow temperate cycles and cause "shingles." Rhinovirus: RNA virus. Causes the *common cold*. Host cells are generally mucus-producing cells of respiratory tract. Poliovirus: RNA virus causes *polio*. Host cells are in motor nerves that lead to the diaphragm and other important muscles. Destruction of these nerve cells causes paralysis that may be temporary or permanent. Recurrences can occur. Immunize your children! HIV: RNA virus. Host cells are specific white blood cells. Temperate cycle has a latency period that may last longer than a year before host tests positive for HIV. As white blood cells are destroyed, the host's immune system is progressively destroyed (*AIDS*). No cure exists.
3. virus; 4. nucleic acid; 5. protein coat (viral capsid);
6. Viruses; 7. *Herpesvirus* (or *Varicella*); 8. viroids;
9. Retroviruses (HIV); 10. temperate; 11. Nanometers;

12. micrometers; 13. 86,000; 14. latency; 15. prions; 16. Rhinoviruses; 17. Retroviruses; 18. Herpesviruses; 19. C; 20. F; 21. B; 22. G; 23. D; 24. A; 25. I; 26. H; 27. E.

BACTERIA (17-II)

1. inorganic compounds; 2. sunlight; 3. Chemosynthetic bacteria; 4. *Clostridium botulinum* (*Clostridium tetani*); 5. *Clostridium tetani* (*Clostridium botulinum*); 6. peptidoglycan; 7. nitrogen fixation; 8. endospores; 9. binary fission; 10. conjugation; 11. prokaryotic; 12. cocci; 13. bacilli; 14. spirilla; 15. membrane receptors; 16. light; 17. decomposers; 18. streptomycins; 19. penicillins; 20. K; 21. d, L; 22. c, C; 23. c, J; 24. a, G; 25. e, M; 26. c, E; 27. a, A; 28. b, I; 29. c, H; 30. d, B; 31. c, K; 32. c, O; 33. c, D; 34. c, N; 35. a, F.

PROTISTANS (17-III)

1. DNA; 2. archaebacteria; 3. fossil; 4. 3.5; 5. eukaryotes; 6. membrane-bound; 7. organelles; 8. prokaryotic; 9. symbiotic; 10. mitochondria; 11. 1.5; [12–14] Could include any of the following: Mitochondria resemble purple photosynthetic bacteria in size and structure. Mitochondrial DNA is replicated independently of nuclear DNA, which implies that it may once have existed separately. Even today, bacteria live and divide inside a protistan, where they produce ATP and share it with the host cell; [15–17] Could include any of these: Chloroplasts resemble some of the cyanobacteria and prochlorobacteria in size and structure. Chloroplast DNA is replicated independently of nuclear DNA. Chloroplast metabolic pathways and overall nucleic acid sequences resemble those of the prochlorobacteria. Chloroplasts vary in shape and in their light-absorbing pigments, just as different photosynthetic bacteria do; 18. T; 19. T; 20. slime molds; 21. fruiting bodies; 22. animals (amoebae); 23. longitudinal fission; 24. organelles; 25. Dinoflagellates; 26. diatoms; 27. shells; 28. plankton; 29. pseudopodia; 30. contractile vacuoles; 31. gullet; 32. enzyme-filled vesicles; 33. *Plasmodium*; 34. mosquito; 35. gametes; 36. archaebacteria; 37. eubacteria; 38. mitochondria; 39. chloroplasts; 40. animals; 41. fungi; 42. heterotrophic protistans; 43. plants; 44. photosynthetic protistans; 45. F, L, ; 46. C, J; 47. C, L; 48. E, K; 49. A;

50. D, I; 51. B, G ; 52. B, H; 53. F; 54. H; 55. G; 56. A; 57. I; 58. C; 59. J; 60. E; 61. D; 62. B.

Answers to Self-Quiz

1. d; 2. a; 3. c; 4. b; 5. b; 6. a; 7. a; 8. c; 9. c; 10. a; 11. B, I; 12. A, E; 13. A, H; 14. B, N; 15. B, D; 16. A; 17. B, I, O; 18. B, F, G; 19. C, M; 20. C, P; 21. A, H; 22. B, L; 23. B, J, K; 24. A, H.

Answer to Integrating and Applying Key Concepts

The text identifies natural gas as a fossil fuel along with coal and petroleum because most supplies of it are produced as a result of drilling and mining rather than through the 2 billion tons of methane generation by bacteria each year.

$$2\,CH_4\ +\ 4\,O_2\ \rightarrow\ 4\,H_2O\ +\ 2\,CO_2\ +\ \text{heat}$$
(for cooking
and heating)

Plants then use CO_2 in photosynthesis to produce carbon-based organismal flesh. Methanogens generate methane (CH_4) from dead organisms or dung.

Critical Thinking Analysis

Early virologists viewed viruses as living things. As such, the viruses would have to reproduce themselves from existing individuals. What, then, could account for their reappearance in a culture that contained none of them?

We no longer view viruses as living things, and we know that production of viruses is fundamentally different from reproduction of cells. The initial inoculum of viruses first attaches to the outside of the host cells. Then the viruses inject their nucleic-acid component into the host cells and fall apart. At this point, no viruses are present in the culture. Later, new viruses, produced by the host cells under the direction of the injected viral nucleic acid, are continually released and extend the infection to more cells.

18

FUNGI AND PLANTS

Interactive Exercises

PART I. KINGDOM OF FUNGI (18-I, pp. 266–271)

Terms

The page-referenced terms are important; they were in boldface type in the chapter. Refer to the instructions given in Chapter 1, p. 1 of this workbook.

saprobes (266): _____

parasites (266): _____

spore (267): _____

water molds (268):_____

chytrids (268):_____

zygospore-forming fungi (269):_____

sac fungi (269): _____

club fungi (271): _____

Labeling

Identify each life cycle stage below by writing its name in the appropriate blank.

1._____

2._____

3._____ _____

4._____

5._____

Complete the Table

6. Complete the table below by supplying the common name and typical habitats of the major groups of fungi.

Group	Common Name	Typical Habitats
a. Oomycetes		
b. Chytridiomycetes		
c. Zygomycetes		
d. Ascomycetes		
e. Basidiomycetes		
f. Fungi Imperfecti		

Choice

For questions 7–21, choose from the following.

a. water molds   b. chytrids   c. zygospore-forming fungi   d. sac fungi   e. club fungi   f. imperfect fungi

_____ 7. Fungi in which the sexual phase is absent

_____ 8. Possess reproductive structures, ascocarps, that are shaped like globes, flasks, or dishes

_____ 9. Resemble the single-celled organisms that gave rise to fungi

_____ 10. Mycelia appear as cottony growths on goldfish or tropical fish

_____ 11. Sexual reproduction involves a thick-walled resting spore that forms around the zygote

_____ 12. Includes mushrooms, rusts, and smuts

_____ 13. Includes yeasts, truffles, and morels

_____ 14. Includes the commercially useful *Aspergillus* and *Penicillium*

_____ 15. Includes late blight of potato and tomato as well as downy mildew of grapes

_____ 16. Includes dutch elm disease, chestnut blight, apple scab, ergot of rye, and brown rot of stone fruits

_____ 17. Includes black stem wheat rust, corn smut, and poisonous mushrooms

_____ 18. Includes various species causing ringworm, including athlete's foot

_____ 19. Includes black bread mold, a food spoiler

_____ 20. Reproductive structures are club-shaped cells that bear spores on their surfaces

_____ 21. Spores form in pouches within ascocarps

Fill-in-the-Blanks

The numbered items on the illustration below represents missing information; complete the numbered blanks in the narrative below to supply missing information on the illustration.

The mature mushroom is composed of an upper (22) _____ and a supporting stalk. Club-shaped structures develop on the gills; each bears two haploid ($n + n$) nuclei. (23) _____ fusion occurs within the club-shaped structures that yields a (24) _____ stage. (25) _____ occurs within the club-shaped structures and four haploid (26) _____ emerge at the tip of each. The spores are released and each may germinate into a haploid (n) mycelium. When hyphae of two compatible mating strains meet, (27) _____ fusion occurs. Following this, a "dikaryotic" ($n + n$) mycelium gives rise the spore-bearing mushrooms.

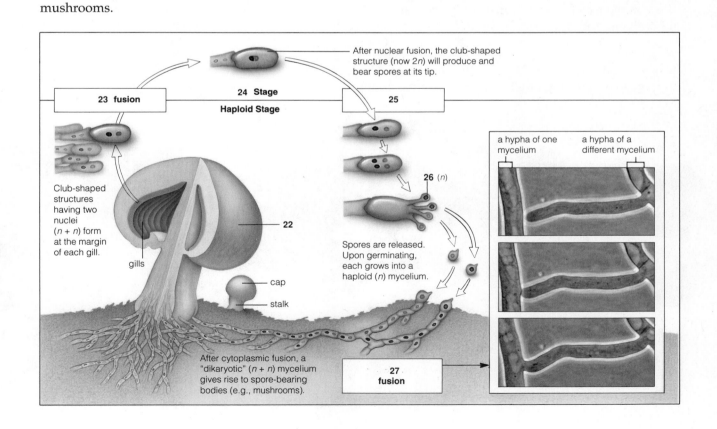

SYMBIOSIS BETWEEN FUNGI AND PLANTS (18-II, pp. 272–273)

Terms

lichens (272):_____

mycorrhizae (273):_____

Fill-in-the-Blanks

The symbiotic associations between fungi and photosynthetic cells are called (1) _____. The fungus involved in the relationship is a sac fungus or a (2) _____ fungus. The photosynthetic cells involved are (3) _____ , green (4) _____ , or both. Lichens grow slowly on bare (5) _____ , tree bark, and fence posts. (6) _____ form slowly through the activities of lichens as they absorb minerals from rock. The death of lichens that live around cities is a signal of (7) _____ deterioration. The rootlets of many vascular plants associate with fungi in an intimate, mutually beneficial way; such associations are called (8) _____. A mycorrhizal fungus absorbs (9) _____ from the host plant, which absorbs (10) _____ ions from the fungus. Without mycorrhizae, many plants do not grow well, for they cannot readily absorb (11) _____ and other important ions. Mycorrhizae are highly susceptible to (12) _____ _____ , and this affects the world's forests.

Label-Match

In the blank beneath each illustration (13–21), identify the organism by common name (or scientific name if a common name is unavailable). Then match each organism with the appropriate item (may be used more than once) from the list (A–F) below by entering the letter in the parentheses.

A. sac fungi
B. algae and fungi
C. zygospore-forming fungi
D. water molds
E. club fungi
F. imperfect fungi

13. _____ ( )

14. _____ ( )

15. _____ ( )

16. _____ (   )

17. _____ (   )

18. _____ (   )

19. _____ (   )

20. _____ (   )

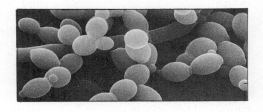

21. _____ (   )

PART II. KINGDOM OF PLANTS (18-III, pp. 274–285)

Terms

vascular plants (274): _____

red alga (274): _____

brown alga (274): _____

green alga (274): _____

bryophytes (274): _____

sporophyte (275): _____

gametophyte (275): _____

pollen grains (275): _____

seed (275): _____

lycophytes (278): _____

horsetails (278): _____

ferns (278): _____

ovules (282): _____

angiosperm (284): _____

monocots (284): _____

dicots (284): _____

Matching

Choose the most appropriate answer to match with each item.

1. _____ xylem and phloem
2. _____ gametophyte
3. _____ vascular plants
4. _____ nonvascular plants
5. _____ seed
6. _____ pollen grains
7. _____ sporophyte

A. Develop into sperm-bearing male gametophytes
B. Diploid, multicelled plant body in which diploid cells undergo meiosis to produce spores
C. Multicelled haploid body in which gametes are produced
D. Contain interconnected tubes distributing water and solutes; well-developed root and shoot systems
E. Package containing an embryo with nourishing and protective tissues
F. Red, brown, green algae and bryophytes; no true roots, stems, and leaves
G. Internal conducting tissues that lend support to root and shoot expansions

Label-Match

Identify each life cycle shown below by entering the correct name in the blank beneath the sketch. Choose the life cycle of algae or the life cycle of vascular plants. Complete the exercise by matching and entering the letter of the correct dominance description in the parentheses following each label.

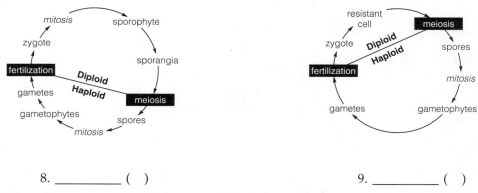

8. _____ (   )   9. _____ (   )

A. Haploid dominance
B. Diploid dominance

Choice

For questions 10–29, choose from the following.

   a. red algae   b. brown algae   c. green algae   d. bryophytes

_____ 10. Most live just offshore or in intertidal zones

_____ 11. Cell walls contain agar; agar is used as a moisture-fixing or setting agent

_____ 12. A group containing mosses, liverworts, and hornworts

_____ 13. Type of algae that bears the greatest resemblance to land plants

_____ 14. Most live in marine habitats, abounding in tropical seas

_____ 15. Xanthophyll pigments help color various species olive-green, golden, or dark-brown

_____ 16. Appear green, red, purple, or greenish-black, depending on pigments other than chlorophylls

_____ 17. May furnish oxygen for space travelers

_____ 18. Grow in freshwater, on open ocean surfaces, on soil, on marine sediments, on rocks, on tree bark and snow

_____ 19. Store starch in chloroplasts, have pigments like land plants, and have cellulose cell walls

_____ 20. The cell walls of some contain algin, a thickening, emulsifying, and suspension agent

_____ 21. Some are found at great depths, at more than 200 meters

_____ 22. Sporophytes remain attached to "free-living" gametophytes

_____ 23. They have hollow, gas-filled bladders that keep blades near the sunlit, surface waters

_____ 24. Eggs and sperm develop in jacketed structures at their shoot tips

_____ 25. Contain phycobolins, which trap blue wavelengths far below the water's surface

_____ 26. Most species have rhizoids that attach gametophytes to soil and absorb water and minerals

_____ 27. A group that comes to mind when eating commercial baked goods and gelatin

_____ 28. Giant kelps are often 50 meters tall

_____ 29. Above-ground parts covered by a cuticle; a "jacket" of cells surrounds gamete-producing parts; embryo sporophyte tissue starts developing inside female gametophyte tissues

Fill-in-the-Blanks (Fig. 18.15, text)

In the life cycle of _Chlamydomonas_, the (30) _____ is the only diploid structure; it develops a thick, resistant wall within which (31) _____ and germination occurs. Haploid cells leaving the ruptured zygote represent (number) (32) _____ mating strains and may develop into (33) _____ or (34) _____ , depending on environmental conditions. Environmental cues for sexual reproduction in _Chlamydomonas_ are low (35) _____ levels and when the (36) _____ is of certain quality and intensity. At that time asexual cells develop into gametes of different mating types that are designated (37) _____ and (38) _____ . When gametes of different mating types meet the first event is (39) _____ fusion and the second event is (40) _____ fusion, which completes formation of a diploid (41) _____ . In the life cycle of _Chlamydomonas_, (42) _____ reproduction is the most common form.

Fill-in-the-Blanks

The numbered items on the illustration below represent missing information; complete the numbered blanks in the narrative below to supply the information missing from the illustration.

The gametophytes are the green "moss plants." Sperms develop in jacketed structures at the shoot tip of the male (43) _____ and eggs develop in jacketed structures at the shoot tip of the female (44) _____. Raindrops transport (45) _____ to the egg-producing structure. (46) _____ occurs within the egg-producing structure. The (47) _____ grows and develops into a mature (48) _____ (with sporangium and stalk) while attached to the gametophyte. (49) _____ occurs within the sporangium of the sporophyte where haploid (50) _____ form, develop, and are released. The released spores grow and develop into male or female (51) _____.

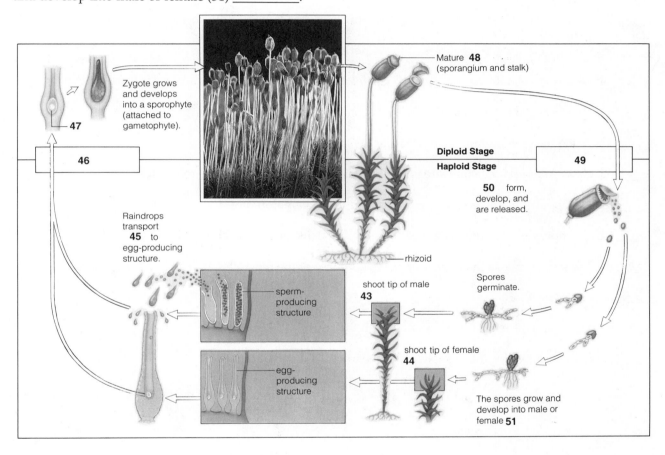

Zygote grows and develops into a sporophyte (attached to gametophyte).

47

46

Mature **48** (sporangium and stalk)

Diploid Stage

Haploid Stage

49

50 form, develop, and are released.

Raindrops transport **45** to egg-producing structure.

sperm-producing structure

shoot tip of male **43**

Spores germinate.

egg-producing structure

shoot tip of female **44**

rhizoid

The spores grow and develop into male or female **51**

Choice

For questions 52–68, choose from the following.

    a. lycophytes   b. horsetails   c. ferns   d. applies to a, b, and c

_____ 52. Only one genus survives, *Equisetum*

_____ 53. Familiar club mosses growing on forest floors

_____ 54. Seedless vascular plants

_____ 55. Rust-colored patches on the lower surface of their fronds

_____ 56. Some tropical species are the size of trees

_____ 57. Plants composing "ancient carbon treasures" of the Carboniferous; became peat and coal

_____ 58. Stems were used by pioneers of the American West to scrub cooking pots

_____ 59. Leaves are usually finely divided, like feathers

_____ 60. The sporophyte has vascular tissues

_____ 61. When the sporangium snaps open, spores catapult through the air

_____ 62. Grow in moist soil along streams and in disturbed habitats, such as roadsides

_____ 63. The sporophyte does not remain attached to the gametophyte

_____ 64. Sporophytes have underground stems and scalelike leaves on aboveground stems

_____ 65. The sporophyte is the larger, longer lived phase of the life cycle

_____ 66. *Lycopodium*

_____ 67. Male gametes must have water to reach the eggs

_____ 68. A spore develops into a small green, heart-shaped gametophyte

Fill-in-the-Blanks

The numbered items on the illustration below represent missing information; complete the numbered blanks in the narrative below to supply the missing information on the illustration.

Fern leaves (fronds) of the sporophyte are usually finely divided, like feathers. The underground stem of the sporophyte is termed a (69) _____. On the undersides of many fern fronds, rust-colored patches of sporangia (the sori) occur. (70) _____ of diploid cells within each sporangium produces haploid (71) _____. The spores are catapulted into the air when each sporangium snaps open. A spore may germinate and grow into a (72) _____ that is small, green, and heart-shaped. Jacketed structures develop on the underside of the mature (73) _____. Each male jacketed structure produces many (74) _____ while each female jacketed structure produces a single (75) _____. These gametes meet in (76) _____. The diploid (77) _____ is first formed inside the female jacketed structure; it divides to form the developing (78) _____ , still attached to the gametophyte.

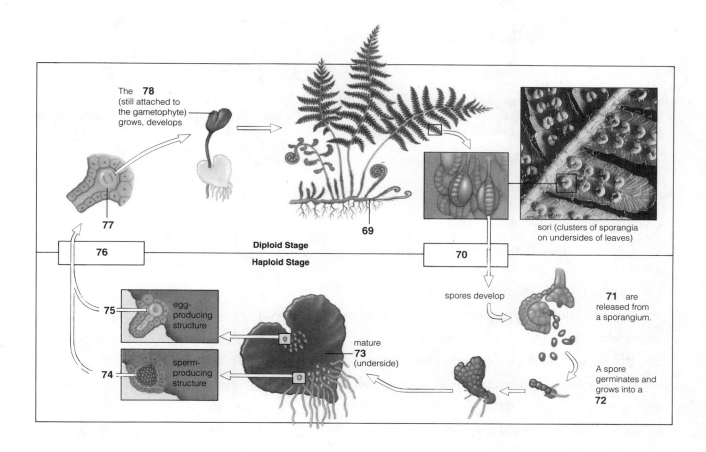

The **78** (still attached to the gametophyte) grows, develops

77

Diploid Stage

Haploid Stage

76

69

70

sori (clusters of sporangia on undersides of leaves)

spores develop

71 are released from a sporangium.

75 egg-producing structure

mature **73** (underside)

A spore germinates and grows into a **72**

74 sperm-producing structure

Choice

For questions 79–90, choose from the following.

   a. cycads   b. ginkgos   c. gnetophytes   d. conifers   e. gymnosperms (includes a, b, c, and d)

_____ 79. Fleshy-coated seeds of female trees produce an awful stench when stepped on

_____ 80. Includes pines, spruces, firs, hemlocks, junipers, cypresses, and redwoods

_____ 81. In parts of Asia, people eat the seeds and a starchy flour is made from the trunks

_____ 82. Only a single species survives

_____ 83. Includes the *Welwitschia* of hot deserts of south and west Africa

_____ 84. Have massive, cone-shaped structures that bear either pollen or ovules

_____ 85. Seeds are mature ovules

_____ 86. The male trees are now planted in cities because of their attractive, fan-shaped leaves and their resistance to insects, disease, and air pollutants

_____ 87. Their seeds are not covered; they are perched on scales of cones, which are attached to the sporophyte

_____ 88. The plant is mostly a deep root; the exposed part is a woody disk-shaped stem with cones and two strap-shaped leaves that split into a scraggly pile

_____ 89. Most species are evergreen trees and shrubs with needlelike or scalelike leaves

_____ 90. Includes conifers, cycads, ginkgos, and gnetophytes

Fill-in-the-Blanks

The numbered items in the illustration on the next page represent missing information; complete the numbered blanks in the narrative below to supply the missing information on the illustration.

The familiar pine tree, a conifer, represents the mature (91) _____. Pine trees produce two kinds of spores in two kinds of cones. Pollen grains are produced in male (92) _____. Ovules are produced in young female (93) _____. Inside each (94) _____, (95) _____ occurs to produce haploid megaspores; one develops into a many-celled female (96) _____ that contains haploid (97)_____. Diploid cells within pollen sacs of male cones undergo (98) _____ to produce haploid microspores. Microspores develop into (99) _____ grains. (100) _____ occurs when spring air currents deposit pollen grains near ovules of female cones. A pollen (101) _____ representing a male gametophyte grows toward the female gametophyte. (102) _____ nuclei form within the pollen tube as it grows toward the egg. (103) _____ follows and the ovule becomes a seed that is composed of an outer seed coat, the (104) _____ diploid sporophyte plant, and nutritive tissue.

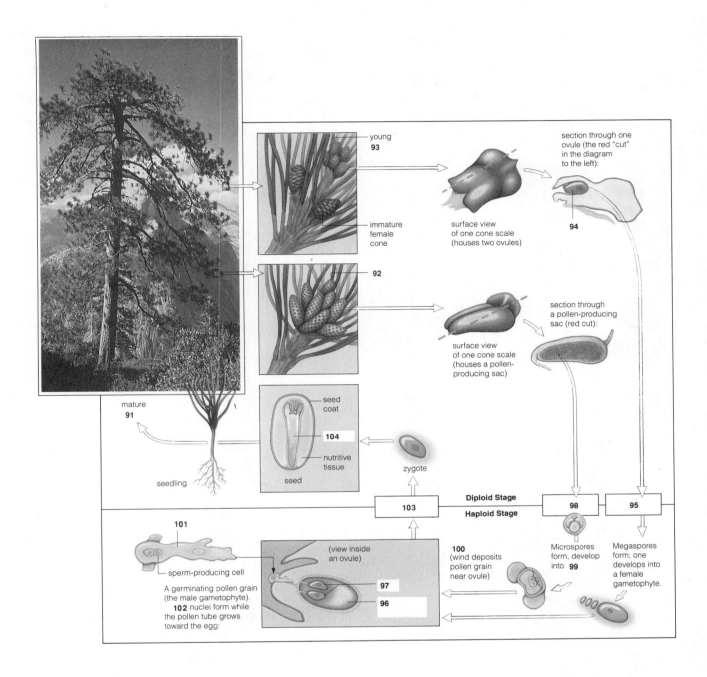

young
93

immature
female
cone

92

section through one
ovule (the red "cut"
in the diagram
to the left):

surface view
of one cone scale
(houses two ovules)

94

section through
a pollen-producing
sac (red cut):

surface view
of one cone scale
(houses a pollen-
producing sac)

mature
91

seed
coat

104

nutritive
tissue

zygote

seed

seedling

Diploid Stage

103

98

95

Haploid Stage

101

sperm-producing cell

A germinating pollen grain
(the male gametophyte).
102 nuclei form while
the pollen tube grows
toward the egg:

(view inside
an ovule)

100
(wind deposits
pollen grain
near ovule)

97

96

Microspores
form, develop
into **99**

Megaspores
form; one
develops into
a female
gametophyte.

Matching

105. _____ monocot examples
106. _____ flower
107. _____ endosperm
108. _____ dicot examples
109. _____ seeds

A. Nutritive seed tissue
B. Palms, lilies, orchids, wheat, corn, rice, rye, sugarcane, and barley
C. Unique angiosperm reproductive structure
D. Packaged in fruits
E. Most shrubs and trees, most nonwoody plants, cacti, and water lilies

Complete the Table

110. Complete the table below to compare the plant groups studied in this chapter.

Plant Group	Dominant Generation	Vascular Tissue	Seeds
a. Algae			
b. Bryophytes			
c. Lycophytes			
d. Horsetails			
e. Ferns			
f. Gymnosperms			
g. Angiosperms			

Self-Quiz

___ 1. Most true fungi send out cellular filaments called _____.
   a. mycelia
   b. hyphae
   c. mycorrhizae
   d. asci

___ 2. Fungi _____.
   a. are producers
   b. are generally saprobic
   c. usually have life cycles in which the diploid phase dominates
   d. include *Fucus* and liverworts

For questions 3–5, choose from the following.
   a. club fungi
   b. imperfect fungi
   c. sac fungi
   d. water molds
   e. zygosporangium-forming fungi

___ 3. The group that includes *Saprolegnia*, a cottony growth often parasitic on goldfish or tropical fish, is _____.

___ 4. The group that includes shelf fungi, which decompose dead and dying trees, and mycorrhizal symbionts that help trees extract mineral ions from the soil, is

   _____.

___ 5. The group that includes the commercial mushroom *Agaricus brunnescens*, as well as the fly agaric mushroom, *Amanita muscaria*, is _____.

___ 6. Plants possessing xylem and phloem are called _____ plants.
   a. gametophyte
   b. nonvascular
   c. vascular
   d. seedless

___ 7. Red, brown, and green "algae" are found in the kingdom _____.
a. Plantae
b. Monera
c. Protista
d. all of the above

___ 8. Because of pigmentation, cellulose walls, and starch storage similarities, the _____ algae are thought to be ancestral to more complex plants.
a. red
b. brown
c. blue-green
d. green

___ 9. Bryophytes _____.
a. have vascular systems that enable them to live on land
b. include lycopods, horsetails, and ferns
c. include mosses, liverworts, and hornworts
d. have true roots but not stems

___ 10. In horsetails, lycopods, and ferns, _____.
a. spores give rise to gametophytes
b. the main plant body is a gametophyte

c. the sporophyte bears sperm- and egg-producing structures
d. all of the above

___ 11. _____ are seed plants.
a. Cycads and ginkgos
b. Conifers
c. Angiosperms
d. all of the above

___ 12. In complex land plants, the diploid stage is resistant to adverse environmental conditions such as dwindling water supplies and cold weather. The diploid stage progresses through this sequence: _____.
a. gametophyte → male and female gametes
b. spores → sporophyte
c. zygote → sporophyte
d. zygote → gametophyte

_____ 13. Monocots and dicots are groups of _____.
a. gymnosperms
b. club mosses
c. angiosperms
d. horsetails

Chapter Objectives/Review Questions

This section lists general and detailed chapter objectives that can be used as review questions. You can make maximum use of these items by writing answers on a separate sheet of paper. Fill in answers where blanks are provided. To check for accuracy, compare your answers with information given in the chapter or glossary.

Page	Objectives/Questions
(266)	1. Do fungi function as autotrophs or heterotrophs?
(266–267)	2. Distinguish between parasitic and saprobic fungi. Mention one way in which parasitic fungi harm humans and one way in which saprobic fungi benefit humans.
(267)	3. Describe the general structure of fungi and its relation to their method of obtaining nutrients.
(267)	4. Describe the type of environment in which fungi are most likely to be found.
(267)	5. Generally describe the reproductive modes of fungi.
(267)	6. List the five principal groups of fungi, and give one example of an organism in each group.
(268)	7. Give two examples of parasitic fungi that have played havoc with the production of crop plants.
(271)	8. Explain why the imperfect fungi are viewed as "imperfect."
(272–273)	9. Symbiotic associations between a fungus and photosynthetic cells are called _____ ; symbiotic associations between vascular plant rootlets and fungi are known as _____.
(266–267)	10. State the fundamental contribution of fungi to ecosystems.
(275)	11. The dominant diploid phase of vascular plants is called the _____.
(274–275)	12. _____ plants include red, brown, and green algae and bryophytes.

(274–275) 13. Be able to state the differences in sporophyte/gametophyte emphasis when comparing algae, bryophyte, and vascular plant life cycles.

(274) 14. Describe the evolution of root and shoot systems.

(274) 15. State the advantages of strong, thick-walled cells for plants living on land.

(274) 16. Land plants have a waxy _____ covering their young stems and leaves to reduce water loss.

(275) 17. A plant spore is a reproductive cell that can develop into a _____ phase, which produces gametes.

(278) 18. What are the three adaptive bryophyte features that allowed ancient plants to make the transition to land?

(282, 284) 19. _____ are the plants with naked seeds; _____ have protected ovules.

(282–283) 20. Be able to generally describe cycads, ginkgos, gnetophytes, and conifers.

(283) 21. The arrival of a pollen grain on female reproductive parts is called _____.

(283) 22. A pine _____ includes the embryo, female gametophyte, and outer coats.

(284) 23. The _____ are the most successful plant division.

(284) 24. The two classes of angiosperms are the _____ and the _____.

(275) 25. The evolution of pollen grains freed gymnosperms and angiosperms from dependence on free water for _____.

Integrating and Applying Key Concepts

Suppose humans acquired a few well-placed fungal genes that caused them to reproduce in the manner of a "typical" fungus (Figure 18.4, text). Try to imagine the behavioral changes that humans would likely undergo. Would their food supplies necessarily be different? Table manners? Stages of their life cycle? Courtship patterns? Habitat? Would the natural limits to population increase be the same? Would their body structure change? Would there necessarily have to be separate sexes? Compose a descriptive science-fiction tale about two mutants who find each other and set up "housekeeping" together.

Critical Thinking Exercise

1. The text argues that the appearance of strong-walled cells in plants conferred an advantage because it allowed plants to become taller and to compete better for light. A similar argument has been applied to a case involving two species of duckweed, a very small green plant that floats in or on water, forming a thin layer of plants. Both species could be grown successfully in aquaria, but when both species were placed in a single tank, *Lemna polyrhiza* inevitably died out, and *L. gibba* survived. The reason is thought to be that *L. gibba* has tiny air sacs that cause it to float higher in the water and absorb all the light before it penetrates to the competing species. Which of the following observations would add the most strength to this interpretation?

   a. *L. polyrhiza* does not survive in water removed from a tank where *L. gibba* is growing.

   b. Both species survive in water to which nitrogen and phosphorus fertilizer has been added.

   c. *L. polyrhiza* does survive in the same tank with *L. gibba* if *L. polyrhiza* is introduced first and allowed to grow for a while before the other species is added.

   d. *L. gibba* dies out and *L. polyrhiza* survives in a tank illuminated from the bottom only.

   e. Both species die when the tank is illuminated with green light only.

Answers

Answers to Interactive Exercises

PART I. KINGDOM OF FUNGI (18-I)

1. zygote; 2. spores; 3. absorptive body; 4. spores;
5. gametes; 6. a. water molds, aquatic, some parasitic;
b. chytrids, aquatic, some parasitic; c. zygospore-forming fungi, soil, decaying plant parts, mostly saprobic, a few parasitic; d. sac fungi, soil, decaying plant parts; e. club fungi, soil, decaying plant parts; f. imperfect fungi, diverse (for example, soil, human body); 7. f; 8. d; 9. b; 10. a; 11. c; 12. e; 13. d; 14. f; 15. a; 16. d; 17. e; 18. f; 19. c; 20. e; 21. d; 22. cap; 23. Nuclear; 24. diploid; 25. Meiosis; 26. spores; 27. cytoplasmic.

SYMBIOSIS BETWEEN FUNGI AND PLANTS (18-II)

1. lichens; 2. club; 3. cyanobacteria; 4. algae; 5. rocks;
6. Soils; 7. environmental; 8. mycorrhizae; 9. carbohydrates; 10. mineral; 11. phosphorus; 12. air pollution;
13. Downy mildew of grapes (*Phytophthora infestans*) (D);
14. lichen (B); 15. cup fungus (A); 16. shelf fungus (E);
17. *Pilobolus* (C); 18. *Penicillium* (F); 19. lichen (B); 20. big laughing mushroom (E); 21. yeast (*Candida albicans*) (A).

PART II. KINGDOM OF PLANTS (18-III)

1. G; 2. C; 3. D; 4. F; 5. E; 6. A; 7. B; 8. vascular plants (B);
9. some algae (A); 10. b; 11. a; 12. d; 13. c; 14. a; 15. b;
16. a; 17. c; 18. c; 19. c; 20. b; 21. a; 22. d; 23. b; 24. d; 25. a;
26. d; 27. a; 28. b; 29. d; 30. zygote; 31. meiosis; 32. two;
33. gametes (spores); 34. spores (gametes); 35. nitrogen;
36. light; 37. plus (minus); 38. minus (plus); 39. cytoplasmic; 40. nuclear; 41. zygote; 42. asexual; 43. gametophyte;
44. gametophyte; 45. sperm; 46. Fertilization; 47. zygote;
48. sporophyte; 49. Meiosis; 50. spores; 51. gametophytes;
52. b; 53. a; 54. d; 55. c; 56. c; 57. d; 58. b; 59. c; 60. d; 61. c;
62. b; 63. d; 64. b; 65. d; 66. a; 67. d; 68. c; 69. rhizome;
70. Meiosis; 71. spores; 72. gametophyte; 73. gametophyte; 74. sperms; 75. egg; 76. fertilization; 77. zygote;
78. embryo; 79. b; 80. d; 81. a; 82. b; 83. c; 84. a; 85. e;
86. b; 87. e; 88. c; 89. d; 90. e; 91. sporophyte; 92. cones;
93. cones; 94. ovule; 95. Meiosis; 96. gametophyte; 97. eggs; 98. meiosis; 99. pollen; 100. Pollination; 101. tube;
102. Sperm; 103. Fertilization; 104. embryo; 105. B; 106. C;
107. A; 108. E; 109. D; 110. a. gametophyte, no, no; b. gametophyte, none or simple vascular tissue, no; c. sporophyte, yes, no; d. sporophyte, yes, no; e. sporophyte, yes, no; f. sporophyte, yes, yes; g. sporophyte, yes, yes.

Answers to Self-Quiz

1. b; 2. b; 3. d; 4. a; 5. a; 6. c; 7. a; 8. d; 9. c; 10. a; 11. d;
12. c; 13. c.

Critical Thinking Analysis

To answer this question, it is necessary to think of alternative hypotheses. How else could one plant species inhibit the growth of another? What other resources might two plant species compete for?

a. This observation would be consistent with the hypothesis that one species secretes into the water a substance that is toxic to the other species. If the basis for differential survival is competition for light, then either species should survive alone in a medium conditioned by growth of the other species as long as illumination is satisfactory.

b. Plants might compete for nitrogen and phosphorus nutrients, and one species might take them up faster than the other species. If this is the reason why one species dies out, survival of that species is predicted if the nutrients are supplied in high enough amounts that the plants no longer compete. This observation would support an alternative hypothesis and indicate that competition for light does not account for the differential survival.

c. Sometimes organisms compete for space, and one survives better because it grows faster. If the slower-growing species is started first and allowed to become established, it can survive in the same area as the other species. Once again, this observation is predicted by an alternative hypothesis and not by the one being evaluated.

d. The hypothesis of competition for light assumes that light comes from above. If the hypothesis is valid and the direction of light is reversed, the survival pattern should also be reversed. The plants that float lower would receive the light first and would shade out the plants above them.

e. This would be predicted no matter what the basis for differential survival was. All green plants depend on light, and since they all use chlorophyll, which does not absorb green light, none of them grow very well without red and/or blue light.

19

ANIMALS

Interactive Exercises

OVERVIEW OF THE ANIMAL KINGDOM/PART I. THE INVERTEBRATES

(19-I, pp. 288–293)

Sponges
Cnidarians

Terms

The page-referenced terms are important; they were in boldface type in the chapter. Refer to the instructions given in Chapter 1, p. 1 of this workbook.

animal (288)_____

radial symmetry (288)_____

bilateral symmetry (288)_____

cephalization (289) _____

gut (289) _____

coelom (289)_____

vertebrate (290)_____

invertebrate (290)_____

sponges (290)_____

larva, -ae (290)_____

cnidarians (292)_____

epithelium, -ia (292)_____

nerve cells (292)_____

Complete the Table

1. Complete the table below by filling in the appropriate phylum or representative group name.

Phylum	Some Representatives	Number of Known Species
a.	Sponges	8,000
b.	i. _____ , jellyfishes, corals, sea anemones	11,000
c.	Turbellarians, flukes, tapeworms	15,000
d.	Pinworms, hookworms	20,000
Rotifera	Species with crown of cilia	1,800
e.	j. _____ , slugs, clams, squids, octopuses	110,000
f.	k. _____ , leeches, polychaetes	15,000
g.	Crabs, lobsters, spiders, insects	1,000,000+
h.	Sea stars, sea urchins, sea cucumbers	6,000
Chordata	Invertebrate chordates: Tunicates, lancelets	2,100
Chordata	l. _____	
Chordata	Fishes	21,000
Chordata	Amphibians	3,900
Chordata	Reptiles	7,000
Chordata	Birds	8,600
Chordata	Mammals	4,500

Fill-in-the-Blanks

(2)_____ form the most primitive major group of multicellular animals. Sticky (3) _____ _____ extract microscopic organisms from the water that flows in through pores in the body wall. Sponges lack (4)_____ cells, muscles, and a gut. Cnidarians are (5)_____ symmetrical and have stinging cells called (6)_____ , which aid in defense and food capture. Both the polyp and medusa forms have a (7) _____ gut, and (8) _____ _____ that form a "nerve net."

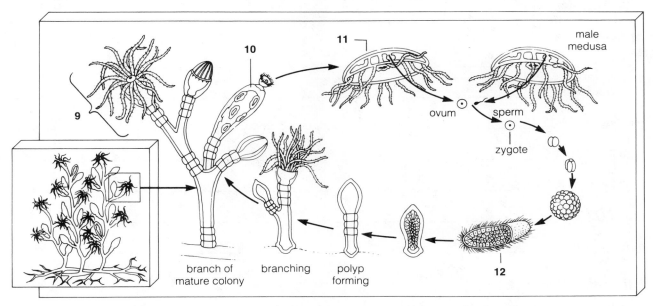

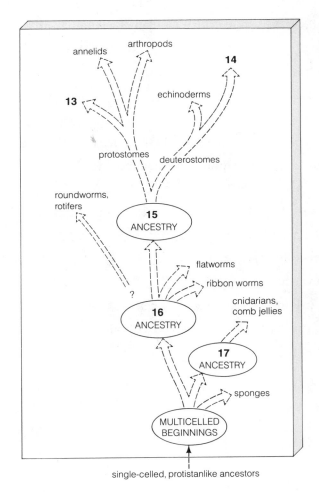

Labeling

Identify each indicated part of the illustration.

9. _____ _____

10. _____ _____

11. _____ _____

12. _____

Identify the groups in the family tree shown at the left by writing the group name or descriptor in the appropriate blank.

13. _____

14. _____

15. _____

16. _____

17. _____

PART I. THE INVERTEBRATES (cont.) (19-II, pp. 294–299)

Flatworms
Roundworms
Focus on Health: **A Rogues' Gallery of Parasitic Worms**
Rotifers
Two Main Evolutionary Roads
Mollusks

Terms

flatworms (294) _____

pharynx (294) _____

mesoderm (294) _____

parasite (294) _____

roundworms (295) _____

rotifer (297) _____

mollusk (298) _____

Fill-in-the-Blanks

A shift from radial to bilateral symmetry could have led to (1)_____ _____ of the sort seen in many flatworms; for example, turbellarians have units called (2)_____ that regulate the volume and salt concentrations of their body fluid. Flatworms have no (3)_____ or _____ systems, and their (4)_____ system is saclike. (5)_____ have complete digestive tracts, a false coelom, and a bilateral, cylindrical body. Between the gut and body wall of a nematode is a (6)_____ _____, which contains paired organs and serves as both a(n) (7) _____ system and a hydrostatic skeleton. (8)_____ are tiny abundant aquatic pseudocoelomates that have crowns of cilia that attract single-celled organisms to them as prey. (9)_____ include echinoderms and chordates; in this group, the first opening to the gut becomes the (10)_____, and the second one to appear becomes the (11)_____. The situation is reversed in the (12)_____, which includes annelids (such as (13)_____), arthropods (such as (14)_____ and crabs), and (15)_____ (such as abalones, limpets, squids, and chambered nautiluses). The most highly evolved invertebrates are generally considered to be the (16)_____, which include squids and octopuses. Like vertebrates, these animals have acute (17)_____ and refined (18)_____ control, which is well integrated with the activities of the nervous system. In less highly evolved mollusks, a structure known as the (19)_____ secretes one or more pieces of calcareous shell that protect these soft-bodied animals from predation.

Labeling

Identify the parts of the animal shown in the drawings at the right.

20. _____ _____

21. _____

22. _____

23. _____ _____

24. _____

25. _____

Answer exercises 26–29 for the animal illustrated at the right.

26. What is the common name of the animal?

   _____

27. Is the animal parasitic? _____

28. Is the animal hermaphroditic? _____

29. Does the animal have a coelom? _____

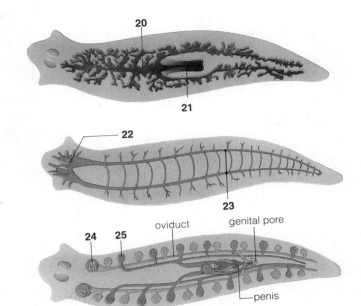

20

21

22

23

oviduct   genital pore

24   **25**

penis

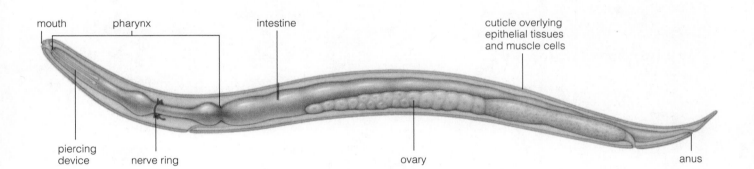

mouth   pharynx   intestine   cuticle overlying
                                                                                epithelial tissues
                                                                                and muscle cells

piercing
device   nerve ring   ovary   anus

Answer exercises 30–32 for the drawing of the dissected animal above.

30. What is the common name of the animal? _____

31. Is the animal hermaphroditic? _____

32. Does the animal have any kind of coelom? _____

Labeling

Identify each numbered part in the drawings by writing its name in the appropriate blank.

33. _____

34. _____

35. _____

36. _____ _____

37. _____

38. _____

39. _____

40. _____

Matching

Identify the animals pictured on this page.

41. _____ Animal A   I. Bivalve

42. _____ Animal B   II. Cephalopod

43. _____ Animal C   III. Gastropod

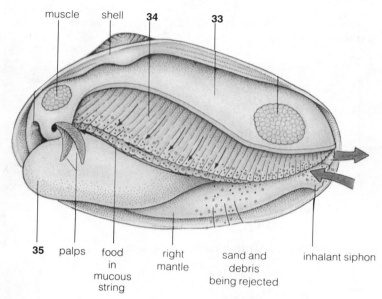

muscle   shell   **34**   **33**

35   palps   food in mucous string   right mantle   sand and debris being rejected   inhalant siphon

Animal A

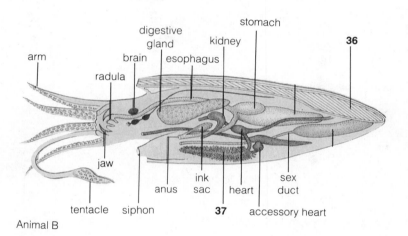

digestive gland   stomach   kidney   **36**

arm   brain   esophagus

radula

jaw

tentacle   siphon   anus   ink sac   heart   **37**   sex duct   accessory heart

Animal B

Animal C (two views)

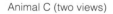

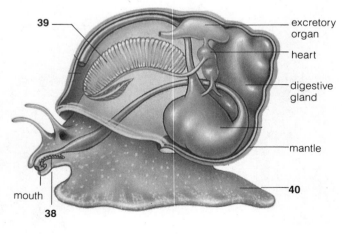

39   excretory organ

heart

digestive gland

mantle

mouth   **40**

38

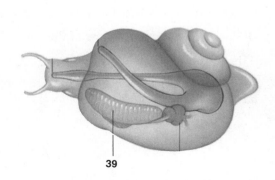

39

PART I. THE INVERTEBRATES (cont.) (19-III, pp. 300–309)

Annelids
Insects and Other Arthropods
Echinoderms

Terms

annelid (300) _____

nerve cord (300) _____

arthropod (302)_____

molting (302)_____

metamorphosis (302)_____

echinoderm (309)_____

Fill-in-the-Blanks

(1)_____ include truly segmented worms such as earthworms, (2)_____, and leeches; they differ from flatworms in having (first) a complete digestive system with a mouth and (3)_____ and (second) a (4)_____ , a fluid-filled space between the gut and body wall. In a circulatory system, blood in (5)_____ _____ provides a means for transporting materials between internal and external environments. (6)_____ , which is the tendency to develop a repeating series of body parts, is well developed in annelids. In each segment, there are swollen regions of the (7)_____ _____ that control local activity and a pair of (8)_____ that act as kidneys, as well as bristles embedded in the body wall. Earthworms (9) _____ soil and help to make nutrients available to plants; polychaetes live in (10) _____ habitats, where they are important food for fishes and birds.

Arthropods develop a lightweight armor called the (11) _____ ; it protects its owner from predators and supports a body deprived of water's buoyancy. In land-dwelling arthropods, it protects against (12) _____ loss. The name arthropod means (13) _____ _____. Aquatic arthropods such as lobsters extract oxygen from water with (14) _____, but most land-dwelling insects use (15) _____ systems.

Arthropods grow by (16) _____ and many undergo a radical transformation called (17) _____ as they change from the feeding larval stage to the reproductive (18) _____. Only two prominent groups of (19)_____ have survived to the present—the (20)_____ and the chordates. In echinoderms, (21)_____ symmetry has been overlaid on an earlier bilateral heritage; most echinoderms still go through a free-swimming (22)_____ symmetrical larval stage. Echinoderm locomotion is based on constant circulation of seawater through a (23)_____-_____ system of canals and (24)_____ _____.

Short Answer

25. Name this animal.

   _____

26. Name this animal's phylum.

   _____

27. Name two distinguishing characteristics of this group. _____

   _____

28. Protostome ☐ or deuterostome ☐ ?

29. Symmetry of adult: ☐ radial   ☐ bilateral

30. Does this animal have a true coelom?
   ☐ yes   ☐ no

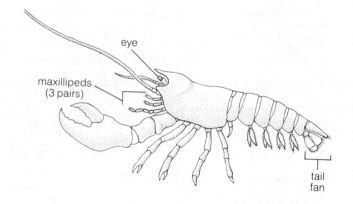

eye

maxillipeds
(3 pairs)

tail
fan

31. Name this animal.

   _____

32. Name this animal's phylum.

   _____

33. Name two distinguishing characteristics of this group._____

   _____

34. Protostome ☐ or deuterostome ☐ ?

35. Is this animal segmented? ☐ yes   ☐ no

36. Symmetry of adult: ☐ radial   ☐ bilateral

37. Does this animal have a true coelom?
   ☐ yes   ☐ no

38. Name this animal.

_____

39. Name this animal's phylum.

_____

40. Name two distinguishing characteristics
   of this group. _____

_____

41. Protostome ☐ or deuterostome ☐ ?

42. Is this animal segmented? ☐ yes ☐ no

43. Symmetry of adult:
   ☐ radial ☐ bilateral

44. Does this animal have a true coelom?
   ☐ yes ☐ no

"hearts"

blood
vessels

PART II. VERTEBRATES AND THEIR KIN (19-IV, pp. 310–316)

The Chordate Heritage
Invertebrate Chordates
Evolutionary Trends Among the Vertebrates
Fishes

Terms

chordate (310)_____

sea squirt (310)_____

lancelet (311)_____

vertebra (312)_____

jaws (312) _____

fins (312)_____

gills (313) _____

lungs (313) _____

fishes (314) _____

Fill-in-the-Blanks

Four major features distinguish chordates from all other animals: a hollow dorsal (1)_____ _____ , a (2)_____ with slits in its wall, a (3)_____ , and a tail that extends past the anus at least during part of its life. In some chordates, the (4)_____ chordates, the notochord is *not* divided into a skeletal column of separate, hard segments; in others, the (5)_____ , it is. Invertebrate chordates living today are represented by tunicates and (6)_____ , which obtain their food by (7)_____ - _____ ; they draw in plankton-laden water through the mouth and pass it over sheets of mucus, which trap the particulate food before the water exits through the (8)_____ _____ in the pharynx. The ancestors of the vertebrate line may have been mutated forms of their closest relatives, the (9)_____ . (10)_____ may have formed from notochordal segments (in the mutated forms) that became hardened by accumulating calcium salts. The evolution of (11)_____ intensified the competition for prey and the competition to avoid being preyed upon; animals in which mutations expanded the nerve cord into a (12)_____ that enabled the animal to compete effectively survived more frequently than their duller-witted fellows and passed along their genes into the next generations. Fins became (13)_____ ; the paired fins in some fishes became (14)_____ and equipped with skeletal supports; these forms set the stage for the development of legs, arms and wings in later groups.

Labeling

Name the structures numbered in the illustrations at the right.

15._____ _____

16._____ _____ _____

17._____ _____ _____

18._____ _____ _____ _____

19._____

20._____

Exercises 21–22 also refer to the illustrations at the right.

21. Name the creature in illustration A.

_____

22. Name the creature in illustration B.

_____

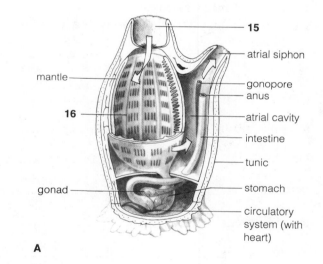

15
atrial siphon
mantle
gonopore
anus
16
atrial cavity
intestine
tunic
gonad
stomach
circulatory system (with heart)

A

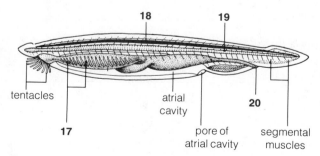

18   **19**

tentacles   atrial cavity   **20**

17   pore of atrial cavity   segmental muscles

B

Analysis and Short Answer

Although Figure 19.14 is *not* in Chapter 19 of the text, the text *does* mention features that distinguish the amphibian-lungfish lineage from the lineage that leads to sturgeons and other bony fishes. At least one of those features may have developed near ⑤ in this evolutionary tree.

Exercises 23–33 refer to the evolutionary diagram illustrated below.

23. What single feature do the lampreys, hagfishes, and extinct ostracoderms have in common that is different from the placoderms? ____________________

24. How did ostracoderms feed? ____________________

25. A mutation in ostracoderm stock led to the development of what characteristic in all organisms that descended from ①? ____________________

26. Mutation at ② led to the development of an endoskeleton made of what? ____________________

27. Mutations at ③ led to an endoskeleton of what? ____________________

28. Mutations at ④ led to which spectacularly diverse fishes that have delicate fins originating from the dermis? ____________________

29. Mutations at ⑤ led to which fishes whose fins incorporate fleshy extensions from the body?

____

30. Which branch, ④ or ⑤, gave rise to the amphibians? ____________________

31. Which branch gave rise to the modern bony fishes? ____________________

32. In which period was this (#29) thought to have occurred?

____

33. Approximately how many million years ago did the fork in the evolutionary path that led to the amphibians occur?

____

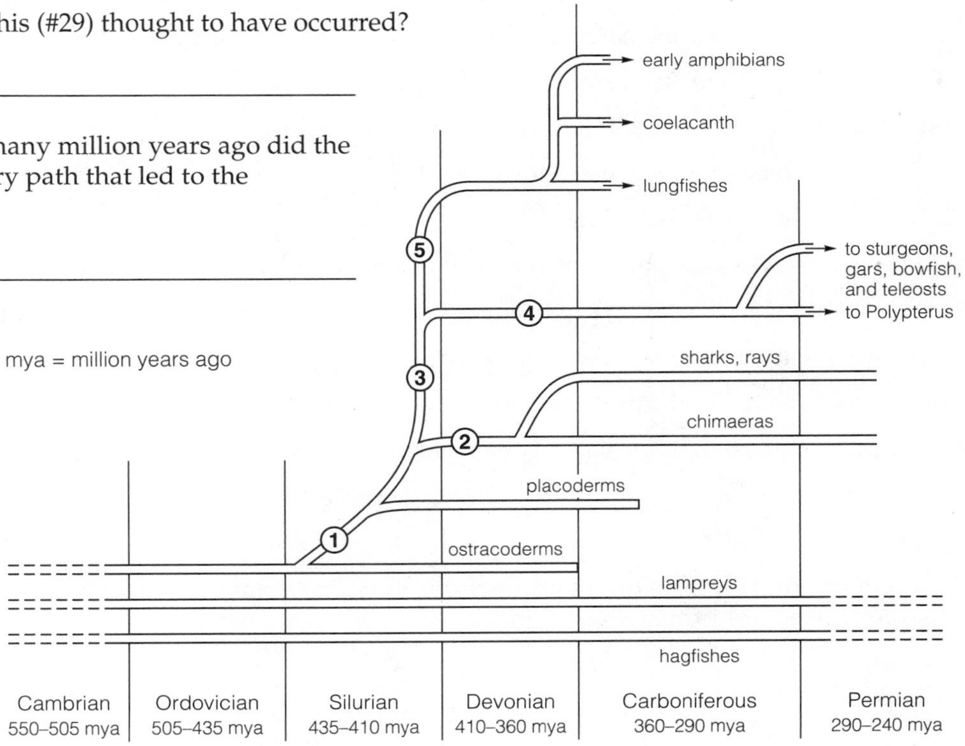

early amphibians
coelacanth
lungfishes
to sturgeons, gars, bowfish, and teleosts
to Polypterus
sharks, rays
chimaeras
placoderms
ostracoderms
lampreys
hagfishes
mya = million years ago
1
2
3
4
5
Cambrian
550–505 mya
Ordovician
505–435 mya
Silurian
435–410 mya
Devonian
410–360 mya
Carboniferous
360–290 mya
Permian
290–240 mya

PART II. VERTEBRATES AND THEIR KIN (cont.) (19-V, pp. 316–323)

Amphibians
Reptiles
Birds
Mammals

Terms

amphibian (316) _____

reptile (318)_____

amniote egg (318)_____

bird (320)_____

mammal (322)_____

Fill-in-the-Blanks

Natural selection acting on lobe-finned fishes during the Devonian period favored the evolution of ever more efficient (1)_____ used in gas exchange and stronger (2)_____ used in locomotion. Without the buoyancy of water, an animal traveling over land must support its own weight against the pull of gravity. The (3)_____ of early amphibians underwent dramatic modifications that involved evaluating incoming signals related to vision, hearing, and (4)_____. Although fish have (5)_____-chambered hearts, amphibians have (6)_____-chambered hearts (see Fig. 19.35). Early in amphibian evolution, mutations may have created the third chamber, which added a second (7)_____ in addition to the already existing atrium and ventricle. The Carboniferous period brought humid, forested swamps with an abundance of aquatic invertebrates some of which were (8)_____—ideal prey for amphibians.

There are three groups of existing amphibians: (9)_____ , frogs and toads, and caecilians. Amphibians require free-standing (10)_____ or at least a moist habitat to (11)_____. Amphibian skin generally lacks scales but contains many glands, some of which produce (12)_____. In the late Carboniferous, (13)_____ began a major adaptive radiation into the lush habitats on land, and only amphibians that mutated and developed certain (14) _____ features were able to follow them and exploit an abundant food supply. Several features helped: Modification of (15)_____ bones favored swiftness, modification of teeth and jaws enabled them to feed efficiently on a variety of prey items, and the development of a (16)_____ _____ egg protected the embryo inside from drying out, even in dry habitats.

(Consult Fig. 19.42 of the main text, the time line below, and the figure at the Analysis and Short Answer of page 234 of this Study Guide.)

Today's reptiles include (17)_____ , crocodilians, snakes, and (18)_____ . All rely on (19)_____ fertilization, and most lay leathery eggs. Although amphibians originated during Devonian times, ancestral "stem" reptiles appeared during the (20)_____ period, about 340 million years ago. Reptilian groups living today that have existed on Earth longest are the (21)_____ ; their ancestral path diverged from that of the "stem" reptiles during the (22)_____ period. Crocodilian ancestors appeared in the early (23)_____ period, about 220 million years ago. Snake and lizard stocks diverged from the tuatara line during the late (24)_____ period, about 140 million years ago. (25)_____ are more closely related to extinct dinosaurs and crocodiles than to any other existing vertebrates; they, too, have a (26)_____-chambered heart. Mammals have descended from therapsids, which in turn are descended from the (27)_____ group of reptiles, which diverged earlier from the stem reptile group during the (28)_____ period, approximately 320 million years ago.

In blanks 29–35, arrange the following groups in sequence according to their earliest appearance in the fossil record:

A. Birds   B. Crocodilians   C. Dinosaurs
D. Early ancestors of mammals (= synapsid reptiles)
E. Early ancestors of turtles (= anapsid reptiles)
F. Lizards and snakes   G. "Stem" reptiles.

Earliest   Latest

29. _____   30. _____   31. _____   32. _____   33. _____   34. _____   35. _____

36. _____   37. _____   38. _____   39. _____   40. _____   41. _____   42. _____

In blanks 36–42, select from the choices below the geologic period in which each group first appeared and write it in the correct space:

Paleozoic Era		Mesozoic Era		
A. Carboniferous	B. Permian	C. Triassic	D. Jurassic	E. Cretaceous
360–290 mya	290–240 mya	240–205 mya	205–138 mya	138–65 mya

Analysis and Short Answer

43. Although the figure below is not in Chapter 19 of the text, the text *does* mention the features that distinguish, say, the mammals from all of the remaining groups on the right-hand side of the diagram; therefore, at some time during the evolution of mammals, mutations that produced those features appeared. Consult the evolutionary diagram below and imagine what sort of mutation(s) occurred at the numbered places.
    a. What may have occurred at ①?
    b. What may have occurred at ②?
    c. What may have occurred at ③?
    d. What may have occurred at ④?
    e. What may have occurred at ⑤?

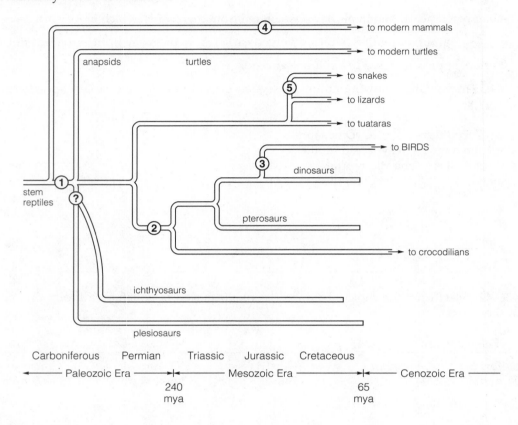

Labeling

Label the structures pictured at the right.

44._____

45._____

46._____ _____

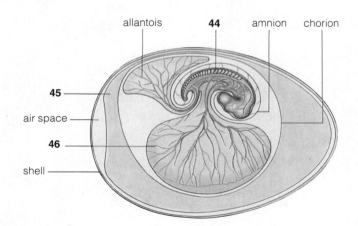

Fill-in-the-Blanks

Birds descended from (47)_____ that ran around on two legs some 160 million years ago. All birds have (48)_____ that insulate and help get the bird aloft. Generally, birds have a greatly enlarged (49)_____ to which flight muscles are attached. Bird bones contain (50)_____ _____ , and air flows through sacs, not into and out of them, for gas exchange in the lungs. Birds also lay (51)_____ _____ eggs, have complex courtship behaviors, and generally nurture their offspring. Birds are able to regulate their body (52)_____ , which is generally higher than that of mammals. Existing birds do not have socketed (53)_____ in their beaks.

There are three groups of existing mammals: those that lay eggs (examples are the (54)_____ and the spiny anteater), those that are (55)_____ (examples are the opossum and the kangaroo), and those that are (56)_____ mammals (there are more than 4,500 species of these). Mammals regulate their body temperature, have a (57)_____-chambered heart, and show a high degree of parental nurture. Most mammals have (58)_____ as a means of insulation, and mammalian mothers generally suckle their young with milk.

Self-Quiz

___1. Which of the following is *not* true of sponges? They have no _____.
a. distinct cell types
b. nerve cells
c. muscles
d. gut

___2. Which of the following is *not* a protostome?
a. earthworm
b. crayfish or lobster
c. sea star
d. squid

___3. Bilateral symmetry is characteristic of _____.
a. cnidarians
b. sponges
c. jellyfish
d. flatworms

___4. Flukes and tapeworms are parasitic _____.
a. leeches
b. flatworms
c. nematodes
d. annelids

___5. Insects include _____.
a. spiders, mites, and ticks
b. centipedes and millipedes
c. termites, aphids, and beetles
d. all of the above

___6. Creeping behavior and a mouth located toward the "head" end of the body may have led, in some evolutionary lines, to _____.
a. development of a circulatory system with blood
b. sexual reproduction
c. feeding on nutrients suspended in the water (filter feeding)
d. concentration of sense organs in the head region

___7. Which of the following is associated with the shift from radial to bilateral body form?
a. a circulatory system
b. a one-way gut
c. paired organs
d. the development of a water-vascular system

___8. The _____ body plan is characterized by simple gas-exchange mechanisms, two-way traffic through a relatively unspecialized gut, and a thin body with all cells fairly close to the gut.
a. annelid
b. nematode
c. echinoderm
d. flatworm

___9. The _____ have a tough cuticle, longitudinal muscles, and a complete digestive system, and they can live under anaerobic conditions.
a. nematodes
b. cnidarians
c. flatworms
d. echinoderms

___10. _____ insulates various internal organs from the stresses of body-wall movement and bathes them in a liquid through which nutrients and waste products can diffuse.
a. A coelom
b. Mesoderm
c. A mantle
d. A water-vascular system

___11. The annelid _____ may resemble the ancestral structure from which the vertebrate kidney evolved.
a. trachea
b. nephridium
c. mantle
d. parapodia

___12. In true fishes, the gills serve primarily _____ function.
a. a gas-exchange
b. a feeding
c. a water-elimination
d. both a feeding and a gas-exchange

___13. Filter-feeding chordates rely on _____, which have cilia that create water currents and mucous sheets that capture nutrients suspended in the water.
a. notochords
b. differentially permeable membranes
c. filiform tongues
d. gill slits

___14. The feeding behavior of true fishes selected for highly developed _____.
a. parapodia
b. notochords
c. sense organs
d. gill slits

___15. The heart in amphibians _____.
a. pumps blood more rapidly than the heart of fish
b. is efficient enough for amphibians but would not be efficient for birds and mammals
c. has three chambers (ventricle and two atria)
d. all of the above

Matching

Match each phylum below with the corresponding characteristics (a–j) and representatives (A–O). A phylum may match with more than one letter from the group of representatives. **If no representative is listed, let the blank remain blank.**

16. ___ , ___ Annelids

17. ___ , ___ Arthropods

18. ___ , ___ Chordates

19. ___ , ___ Cnidarians

20. ___ , ___ Echinoderms

21. ___ , ___ Molluscs

22. ___ , ___ Roundworms

23. ___ , ___ Flatworms

24. ___ , ___ Sponges

25. ___ , ___ Rotifers

a. choanocytes (= collar cells) + spicules
b. jointed legs + an exoskeleton
c. gill slits in pharynx + dorsal, tubular nerve cord + notochord
d. false coelom + crown of cilia + soft body
e. soft body + mantle; may or may not have radula or shell
f. bilateral symmetry + saclike gut
g. radial symmetry + saclike gut; stinging cells
h. body compartmentalized into repetitive segments; coelom containing nephridia (= primitive kidneys)
i. tube feet + calcium carbonate structures in skin
j. complete gut + bilateral symmetry + cuticle; includes many parasitic species, some of which are harmful to humans

A. Dinosaurs
B. Corals and *Hydra*
C. Salamanders and toads
D. Whales and opossums
E. Tapeworms and *Planaria*
F. Insects
G. Jellyfish and sea anemones
H. Sand dollars and starfishes

I. Earthworms and leeches
J. Lobsters, shrimp, and crayfish
K. Organisms with spicules and collar cells
L. Scorpions and millipedes
M. Octopuses and oysters
N. Flukes
O. Hookworm, trichina worm

Matching

Match the following groups and classes with the corresponding characteristics (a–i) and representatives (A–I).

26. ___ , ___ Amphibians

27. ___ , ___ Birds

28. ___ , ___ Bony fishes

29. ___ , ___ Cartilaginous fishes

30. ___ , ___ Invertebrate chordates

31. ___ , ___ Jawless fishes

32. ___ , ___ Mammals

33. ___ , ___ Reptiles

a. hair + vertebrae
b. feathers + hollow bones
c. jawless + cartilaginous skeleton (in existing species)
d. two pairs of limbs (usually) + glandular skin + "jelly"-covered eggs
e. amniote eggs + scaly skin + bony skeleton
f. sessile adult
g. jaws + cartilaginous skeleton + vertebrae
h. invertebrates; notochord stretches from head to tail
i. bony skeleton + skin covered with scales and mucus

A. lancelet
B. loons, penguins, and eagles
C. tunicates, sea squirts
D. sharks and manta rays
E. lampreys and hagfishes (and ostracoderms)

F. true eels and sea horses
G. lizards and turtles
H. caecilians and salamanders
I. platypuses and opossums

Chapter Objectives/Review Questions

This section lists general and detailed chapter objectives that can be used as review questions. You can make maximum use of these items by writing answers on a separate sheet of paper. Fill in answers where blanks are provided. To check for accuracy, compare your answers with information given in the chapter or glossary.

Page	Objectives/Questions
(290)	1. Be able to reproduce from memory a phylogenetic tree that expresses the relationships between the major groups of animals.
(292)	2. Describe the two cnidarian body types.
(288)	3. Explain how radial symmetry might be more advantageous to floating or sedentary animals than bilateral symmetry.
(292–293)	4. Tell how cnidarians obtain and digest food and tell what they do with food they cannot digest.
(300)	5. Name the three groups of annelids and give a specific example from each group.
(302)	6. Explain how the development of a thickened cuticle and a hardened exoskeleton affected the ways that arthropods lived.
(309)	7. Describe how locomotion occurs in echinoderms.
(310)	8. List three characteristics found only in chordates.
(312–313)	9. State what sort of changes occurred in the primitive chordate body plan that could have promoted the emergence of vertebrates.
(316)	10. Describe the changes that enabled aquatic fishes to give rise to land dwellers.
(313, 316–318, 321–322)	11. State what kind of heart each of the four groups of four-limbed vertebrates has and list the principal skin structures that each produces.

Integrating and Applying Key Concepts

a. Birds and mammals both have four-chambered hearts, high metabolic rates, and regulate their body temperatures efficiently. Both groups evolved from reptiles, so one would think that those same traits would have developed in ancestral reptiles. Data suggest that most reptiles have a heart intermediate between three and four chambers, lower metabolic rates, and body temperatures that are not well regulated and tend to rise and fall in accord with the environmental temperature. If the three traits mentioned in the first sentence had developed in reptilian groups, how might their lives have been different?

b. Most highly evolved animals have a complete gut, a closed blood-vascular system, both central and peripheral nervous systems, and are dioecious. Why do you suppose having two sexes in separate individuals is considered to be more highly evolved than the monoecious condition utilized by earthworms? Wouldn't it be more efficient if all individuals in a population could produce both kinds of gametes? Cross-fertilization would then result in both individuals being able to produce offspring.

Critical Thinking Exercise

Suppose you found a fossil mid-Silurian placoderm with bony fins like a coelacanth. How would you have to modify the lineages shown below?

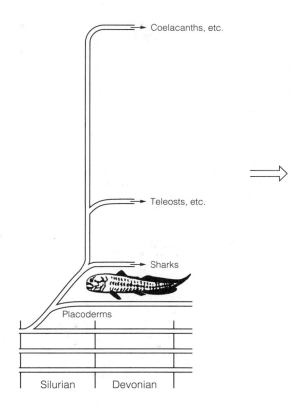

Answers

Answers to Interactive Exercises

OVERVIEW OF THE ANIMAL KINGDOM/ SPONGES, CNIDARIANS

1. a. Porifera; b. Cnidaria; c. Platyhelminthes; d. Nematoda; e. Mollusca; f. Annelida; g. Arthropoda; h. Echinodermata; i. *Hydra* (*Obelia,* Portuguese man-of-war); j. Snails (nudibranchs, oysters); k. earthworms (oligochaetes); l. Vertebrates; 2. Sponges; 3. collar cells; 4. nerve; 5. radially; 6. nematocysts; 7. saclike; 8. nerve cells; 9. feeding polyp; 10. reproductive polyp; 11. female medusa; 12. planula; 13. mollusks; 14. chordates; 15. coelomate; 16. bilateral; 17. radial.

FLATWORMS, ROUNDWORMS, ROTIFERS, MOLLUSKS

1. organ systems; 2. protonephridia; 3. respiratory, circulatory; 4. digestive; 5. Roundworms; 6. false coelom; 7. circulatory; 8. Rotifers; 9. Deuterostomes; 10. anus; 11. mouth; 12. protostomes; 13. earthworms; 14. insects (spiders); 15. mollusks; 16. cephalopods; 17. vision; 18. muscular; 19. mantle; 20. branched gut; 21. pharynx;

22. brain; 23. nerve cord; 24. ovary; 25. testis; 26. planarian; 27. no; 28. yes; 29. no; 30. roundworm; 31. no; 32. yes, a false coelom; 33. mantle; 34. gill; 35. foot; 36. internal shell; 37. gill; 38. radula; 39. gill; 40. foot; 41. I; 42. II; 43. III.

ANNELIDS, INSECTS AND OTHER ARTHROPODS, ECHINODERMS

1. Annelids; 2. polychaetes; 3. anus; 4. coelom; 5. blood vessels; 6. Segmentation; 7. nerve cord; 8. nephridia; 9. aerate; 10. marine (salt water); 11. exoskeleton; 12. water; 13. jointed feet; 14. gills; 15. tracheal; 16. molting; 17. metamorphosis; 18. adult; 19. deuterostomes; 20. echinoderms; 21. radial; 22. bilaterally; 23. water-vascular; 24. tube feet; 25. Sea urchin; 26. Echinodermata; 27. water vascular system + spiny skin; 28. deuterostome; 29. radial; 30. yes; 31. lobster; 32. Arthropoda; 33. chitinous exoskeleton + jointed appendages; 34. protostome; 35. yes; 36. bilateral; 37. yes; 38. earthworm; 39. Annelida; 40. body segmented + coelom with nephridia; 41. protostome; 42. yes; 43. bilateral; 44. yes.

PART II/THE CHORDATE HERITAGE, INVERTE-BRATE CHORDATES, EVOLUTIONARY TRENDS AMONG THE VERTEBRATES, FISHES

1. nerve cord; 2. pharynx; 3. notochord; 4. invertebrate; 5. vertebrates; 6. lancelets; 7. filter-feeding; 8. gill slits; 9. lancelets; 10. vertebrae; 11. jaws; 12. brain; 13. paired; 14. fleshy; 15. oral opening (oral siphon); 16. pharynx with slits; 17. pharyngeal gill slits; 18. dorsal, tubular nerve cord; 19. notochord; 20. anus; 21. sea squirt tunicate; 22. lancelet; 23. all jawless; 24. filter-feeders; 25. jaws; 26. cartilage; 27. bone; 28. ray-finned fishes; 29. lobe-finned fishes; 30. branch ⑤; 31. branch ③; 32. Devonian; 33. about 375 million years ago.

AMPHIBIANS, REPTILES, BIRDS, AND MAMMALS

1. lungs; 2. fins; 3. brain; 4. balance; 5. two; 6. three; 7. atrium; 8. insects; 9. salamanders; 10. water; 11. reproduce; 12. toxins; 13. insects; 14. reptilian; 15. limb; 16. shelled amniote; 17. turtles (lizards); 18. lizards (turtles); 19. internal; 20. Carboniferous; 21. turtles; 22. Carboniferous; 23. Triassic; 24. Jurassic; 25. Birds; 26. four; 27. synapsid; 28. Carboniferous; 29. G; 30. D; 31. E; 32. B; 33. C; 34. A; 35. F; 36. A; 37. A; 38. A; 39. C; 40. C; 41. D; 42. D; 43. a. dry, scaly skin; b. four-chambered heart; c. feather development; d. hair or mammary gland development; e. loss of limbs; 44. embryo (notochord); 45. albumin; 46. yolk sac; 47. reptiles; 48. feathers; 49. sternum (breastbone); 50. air cavities; 51. shelled amniote; 52. temperature; 53. teeth; 54. platypus; 55. pouched (marsupials); 56. placental; 57. four; 58. hair.

Answers to Self-Quiz

1. a; 2. c; 3. d; 4. b; 5. c; 6. d; 7. c; 8. d; 9. a; 10. a; 11. b; 12. a; 13. d; 14. c; 15. d; 16. h, I; 17. b, F, J, L; 18. c, A, C, D; 19. g, B, G; 20. i, H; 21. e, M; 22. k, O; 23. f, E, N; 24. a, K; 25. d, no representative listed (blank); 26. d, H; 27. b, B; 28. i, F; 29. g, D; 30. f, h, A, C; 31. c, E; 32. a, I; 33. e, G.

Critical Thinking Analysis

This finding would force a revision of the divergence of fish lineages occurring at the end of the Silurian. This fossil would show that lobe-fin bone patterns were already present in ancestral placoderm populations. It would indicate that the lobe-fin lineage diverged from the other lineages before, not after, the sharks diverged from the teleosts. This interpretation would also mean that the development of the jaw had to occur independently in both lineages.

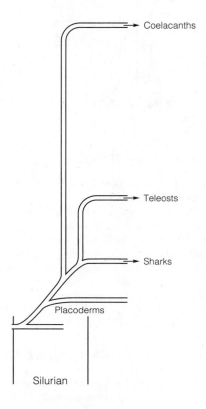

20

PLANT TISSUES

Interactive Exercises

THE PLANT BODY: AN OVERVIEW (20-I, pp. 327–330)

Terms

The page-referenced terms are important; they were in boldface type in the chapter. Refer to the instructions given in Chapter 1, p. 1 of this workbook.

shoots (327): _____

roots (327): _____

ground tissue system (328): _____

parenchyma (328): _____

collenchyma (328): _____

sclerenchyma (328): _____

lignin (329): _____

vascular tissues (329): _____

xylem (329): _____

phloem (329): _____

epidermis (329): _____

periderm (329):_____

apical meristem (328):_____

lateral meristems (328):_____

monocots (330): _____

dicots (330): _____

Complete the Table

1. Three basic kinds of tissues compose the plant body. Complete the table below to indicate their major functions.

Tissue	Function(s)
a. Ground	
b. Vascular	
c. Dermal	

2. Three kinds of ground tissue compose most of the bulk of the plant body. Complete the table below, which summarizes information about ground tissues.

Ground Tissue	Secondary Wall Present	Function(s)
a. Parenchyma		
b. Collenchyma		
c. Sclerenchyma		

Label-Match

Identify each of the ground tissues below by entering the correct name in the blank adjacent to the sketch. Complete the exercise by matching and entering the letter of the correct description in the parentheses following each label.

3. _____ ( )

4. _____ ( )

5. _____ ( )

A. Usually thick, lignified secondary walls; fibers and sclereids
B. Living, thin-walled cells with ample air spaces between them
C. Living cells; primary walls thickened with cellulose and pectin; cell corners appear thickened

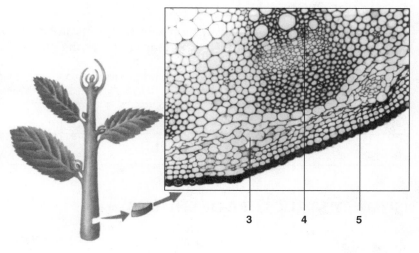

3   4   5

Complete the Table

6. Two kinds of vascular tissues occur in strands (vascular bundles) within the plant body. Complete the table below, which summarizes information about these vascular tissues.

Vascular Tissue	Main Cell Types	Cells Alive	Function(s)
a. Xylem			
b. Phloem			

7. Two kinds of dermal tissues cover the plant body. Complete the table below, which summarizes information about the dermal tissues.

Dermal Tissue	Primary/Secondary Plant Body	Function(s)
a. Epidermis		
b. Periderm		

Dichotomous Choice

Circle one of two possible answers given between parentheses in each statement.

8. Tissues originating through the meristem activity at root and shoot tips are called (primary/secondary) growth.
9. Dome-shaped tips of roots and shoots each consist of a mass of undifferentiated cells, the (apical/lateral) meristem.
10. Descendants of dividing cells in the (apical/lateral) meristem develop into specialized tissues of the *lengthening* root or stem.
11. Plants with a woody body show (primary/secondary) growth at regions other than root and shoot tips.
12. Secondary growth originates at self-perpetuating tissue masses called (apical/lateral) meristems; this increases the diameter of older roots and stems.

Complete the Table

13. There are two classes of flowering plants, monocots and dicots. Complete the table below, which summarizes information about the two groups of flowering plants.

Class	Number of Embryo Cotyledons	Number of Floral Parts	Leaf Venation	Pollen Grains	Vascular Bundles
a. Monocots					
b. Dicots					

PRIMARY STRUCTURE OF SHOOTS (20-II, pp. 330–335)

Terms

vascular bundles (330): _____

leaves (332): _____

bud (334): _____

mesophyll (333): _____

veins (333): _____

Matching

Choose the most appropriate answer to match with each term.

1. _____ terminal bud
2. _____ dicot stems
3. _____ leaves
4. _____ internode
5. _____ monocot stems
6. _____ lateral bud
7. _____ compound leaf
8. _____ node
9. _____ deciduous
10. _____ bud

A. Located along the side of a branch in the upper angles where leaves attach to stems
B. Stem location where one or more leaves are attached
C. Portion of a stem between two successive nodes
D. An undeveloped shoot of mostly meristematic tissue, often covered and protected by modified leaves
E. Leaf composed of many leaflets
F. Most have a ring of vascular bundles that divides ground tissue into an outer cortex and an inner pith
G. A plant that drops its leaves as winter approaches
H. Located at the end of a branch
I. Most have vascular bundles distributed throughout the ground tissue
J. Metabolic food factories equipped with photosynthetic cells; form as bulges from meristems

Label-Match

Identify each indicated part of the accompanying illustration (Fig. 20.11, text). Complete the exercise by matching and entering the letter of the proper function description in the parentheses following each label.

11. _____ _____ ( )

12. _____ _____ ( )

13. _____ ( )

14. _____ _____ ( )

15. _____ ( )

A. Lowermost cuticle-covered cell layer
B. Loosely packed photosynthetic parenchyma cells just above the lower epidermal layer
C. Allows movement of oxygen and water vapor out of leaves and allows carbon dioxide to enter
D. Photosynthetic parenchyma cells just beneath the upper epidermis
E. Move water and solutes to photosynthetic cells and carry products away from them

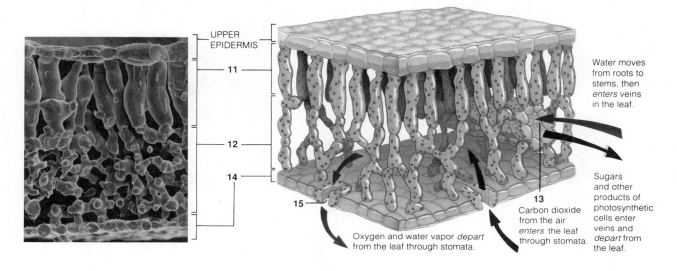

UPPER EPIDERMIS

11

12

14

15

13

Water moves from roots to stems, then *enters* veins in the leaf.

Sugars and other products of photosynthetic cells enter veins and *depart* from the leaf.

Oxygen and water vapor *depart* from the leaf through stomata.

Carbon dioxide from the air *enters* the leaf through stomata.

PRIMARY STRUCTURE OF ROOTS (20-III, pp. 335–337)

Terms

lateral roots (335):_____

taproot system (335):_____

fibrous root system (335):_____

root hairs (337):_____

vascular cylinder (337):_____

Label-Match

Identify each indicated part of the accompanying illustration. Complete the exercise by matching the letter of the proper description in the parentheses following each label. Some choices are used more than once.

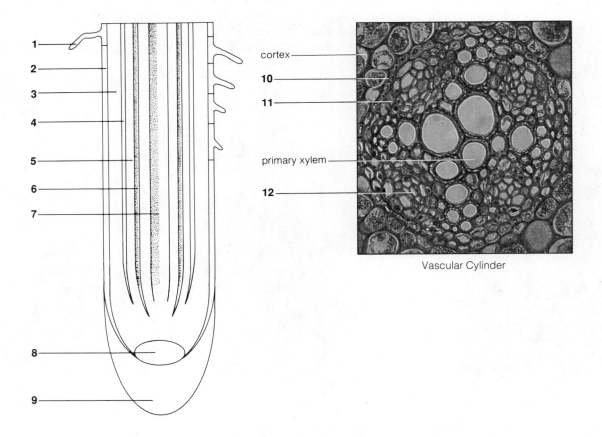

Vascular Cylinder

1._____ _____ ( )

2._____ ( )

3._____ ( )

4._____ ( )

5._____ ( )

6._____ _____ ( )

7._____ _____ ( )

8._____ _____ _____ ( )

9._____ _____ ( )

10._____ ( )

11._____ ( )

12._____ _____ ( )

A. Dome-shaped cell mass produced by the apical meristem
B. Part of the vascular column; gives rise to lateral roots
C. Part of the vascular column; transports photosynthetic products
D. Ground tissue region
E. The absorptive interface with the root's environment
F. Part of the vascular column; transports water and minerals
G. The region of dividing cells
H. Innermost part of the root cortex; helps control water and mineral movement into the vascular column
I. Greatly increases the surface available for taking up water and solutes

Fill-in-the-Blanks

Carrots and dandelions are examples of plants whose primary root and its lateral branchings represent a (13)_____ system. In monocots such as grasses, the primary root is short-lived; in its place, numerous (14)_____ roots arise from the stem of the young plant. Such roots and their branches are somewhat alike in length and diameter and form a (15)_____ root system. Cytoplasm of all adjacent living plant cells is interconnected at cell junctions, the (16)_____. Water entering the root moves from cell to cell until it reaches the (17) _____, the innermost part of the root cortex. Abutting (18) _____ of the endodermal cells are cemented together and help control the movement of water and dissolved minerals into the vascular cylinder. The (19) _____ is located just inside the endodermis; cell divisions in this portion of the vascular cylinder give rise to (20) _____ roots, which grow out through the cortex and epidermis.

WOODY PLANTS (20-IV, pp. 338–339)

Terms

annuals (338):_____

biennials (338): _____

perennials (338): _____

lateral meristems (338): _____

Matching

Choose the most appropriate answer to match with each term.

1. _____ "tree rings"
2. _____ vascular cambium
3. _____ early wood
4. _____ biennial
5. _____ girdling
6. _____ perennial
7. _____ cork cambium
8. _____ late wood
9. _____ annual
10. _____ bark

A. These cells produce periderm—a corky replacement for lost epidermis
B. Plant life cycle completed in one growing season
C. All living and nonliving tissues between the vascular cambium and the stem or root surface
D. As the growing season progresses, the xylem cell diameters become smaller and the walls become thinner
E. Vegetative growth and seed formation continue year after year; some have secondary tissues, others do not
F. Stripping off a band of phloem all the way around a tree's circumference; interrupts phloem transport
G. A plant life cycle completed in two growing seasons
H. Alternating bands representing annual growth layers
I. A cylinder of cells in older stems and roots that give rise to secondary xylem and phloem; xylem forms on the inner face, phloem on the outer
J. The first xylem cells produced at the start of the growing season; tend to have large diameters and thin walls

Label-Match

Identify each indicated part of the accompanying illustration. Complete the exercise by matching and entering the letter of the proper description in the parentheses following each label.

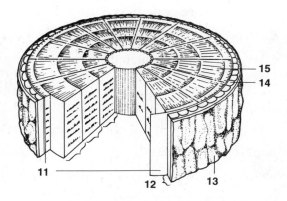

11. _____ _____ ( )

12. _____ ( )

13. _____ ( )

14. _____ ( )

15. _____ ( )

A. Corky replacement for epidermis
B. Vascular tissue that conducts water and dissolved minerals absorbed from soil; gives mechanical support to the plant
C. Has meristematic cells that give rise to secondary xylem and phloem tissues
D. The vascular tissue that transports sugars and other solutes through the plant body
E. All living and nonliving tissues between the vascular cambium and the stem or root surface

Self-Quiz

___ 1. _____ develops into the plant's surface layers.
   a. Ground tissue
   b. Dermal tissue
   c. Vascular tissue
   d. Pericycle

___ 2. Which of the following is *not* considered a ground cell type?
   a. epidermis
   b. parenchyma
   c. collenchyma
   d. sclerenchyma

___ 3. The _____ produces secondary xylem growth.
   a. apical meristem
   b. lateral meristem
   c. cork cambium
   d. endodermis

___ 4. The _____ is a leaflike structure that is part of the embryo; monocot embryos have one, dicot embryos have two.
   a. shoot tip
   b. root tip
   c. cotyledon
   d. apical meristem

___ 5. Leaves are differentiated and buds develop at specific points along the stem called

   _____.
   a. nodes
   b. internodes
   c. vascular bundles
   d. cotyledons

___ 6. Which of the following structures is *not* considered to be meristematic?
   a. vascular cambium
   b. lateral meristem
   c. cork cambium
   d. endodermis

___ 7. Which of the following statements about monocots is *false*?
   a. They are usually herbaceous.
   b. They develop one cotyledon in their seeds.
   c. Their vascular bundles are scattered throughout the ground tissue of their stems.
   d. They have a single central vascular cylinder in their stems.

___ 8. New plants grow and older plant parts lengthen through cell divisions at _____ meristems present at root and shoot tips; older roots and stems of woody plants increase in diameter through cell divisions at _____ meristems.
a. lateral; lateral
b. lateral; apical
c. apical; apical
d. apical; lateral

___ 9. Vascular bundles called _____ form a network through a leaf blade.
a. xylem
b. phloem
c. veins
d. stomata

___ 10. A primary root and its lateral branchings represent a _____ system.
a. lateral root
b. adventitious root

c. taproot
d. branch root

___ 11. Plants whose vegetative growth and seed formation continue year after year are _____ plants.
a. annual
b. perennial
c. biennial
d. herbaceous

___ 12. The _____ layer of a root divides to produce lateral roots.
a. endodermis
b. pericycle
c. xylem
d. cortex

Chapter Objectives/Review Questions

Page	Objectives/Questions
(327)	1. _____ are flowering plants.
(327)	2. Define shoot system and root system.
(328)	3. Name and define the three tissue systems that extend throughout the plant body.
(328–329)	4. _____ with thin primary walls forms the most abundant ground tissue; _____ is a ground tissue that strengthens the plant body; sclereids and fibers are _____ cells with lignified secondary walls that give mechanical support and protection to mature plant parts.
(329)	5. Distinguish cuticle, epidermis, and periderm from one another in structure, plant location, and function.
(328)	6. _____ growth is initiated at root and shoot tips in dome-shaped cell masses called _____ meristems.
(328)	7. Secondary growth originates at self-perpetuating tissue masses called _____ meristems.
(330)	8. Describe a vascular bundle.
(330)	9. Most dicot stems have a ring of vascular bundles that divides ground tissue into two zones, an outer _____ and an inner _____.
(334)	10. Briefly describe how leaves develop and how they are arranged on stems.
(337)	11. Root _____ are epidermal cell extensions that increase absorptive surface area.
(337)	12. Define the vascular cylinder.
(338)	13. Distinguish between herbaceous plants and woody plants; give examples of each.
(328)	14. Describe the activity of the vascular cambium and the cork cambium.

Integrating and Applying Key Concepts

Try to imagine the specific behavioral restrictions that might be imposed if the human body resembled the plant body in having (1) open growth with apical meristematic regions, (2) stomata in the epidermis, (3) cells with chloroplasts, (4) excess carbohydrates stored primarily as starch rather than as fat, and (5) dependence on the soil as a source of water and inorganic compounds.

Critical Thinking Exercise

1. You drive a nail into a tree trunk. Twenty years later the nail is still at the same height above the ground, although the tree has grown many meters taller. Which of the following is the best explanation?
    a. Tree growth is mainly secondary growth of vascular cambium.
    b. The functional vascular tissue in a tree is a thin layer near the surface.
    c. Tree elongation is by growth of the apical meristems.
    d. Tree elongation is by expansion of the junction zone between the root and the shoot.
    e. Vascular bundles elongate as the shoot elongates.

Answers

Answers to Interactive Exercises

THE PLANT BODY: AN OVERVIEW (20-I)
1. a. photosynthesis, storage, secretion, strength, support, and protection; b. conducts substances through roots, stems, and leaves; c. covers the primary and secondary plant body to provide protection and restrict water loss; 2. a. no, various—storage, photosynthesis, secretion, divides to heal wounds and regenerate missing parts; b. no, strengthens and supports plant parts; c. yes, supports and protects mature plant parts; 3. parenchyma (B); 4. sclerenchyma (A); 5. collenchyma (C); 6. a. vessel members and tracheids, no, conducts water and dissolved minerals absorbed from soil; mechanical support; b. sieve tube members and companion cells, yes, transports sugar and other solutes; 7. a. primary, cutin in the cuticle layer over epidermal cells restricts water loss and resists microbial attack; openings (stomates) allow water vapor and gases to enter and leave the plant; b. secondary, replaces epidermis to cover stems and roots; 8. primary; 9. apical; 10. apical; 11. secondary; 12. lateral; 13. a. one, in threes or multiples thereof, usually parallel, one pore or furrow, distributed throughout ground stem tissue; b. two, in fours or fives or multiples thereof, usually netlike, three pores or pores with furrows, positioned in a ring in the stem.

PRIMARY STRUCTURE OF SHOOTS (20-II)
1. H; 2. F; 3. J; 4. C; 5. I; 6. A; 7. E; 8. B; 9. G; 10. D; 11. palisade mesophyll (D); 12. spongy mesophyll (B); 13. vein (E); 14. lower epidermis (A); 15. stoma (C).

PRIMARY STRUCTURE OF ROOTS (20-III)
1. root hair (I); 2. epidermis (E); 3. cortex (D); 4. endodermis (H); 5. pericycle (B); 6. primary phloem (C); 7. primary xylem (F); 8. root apical meristem (G); 9. root cap (A); 10. endodermis (H); 11. pericycle (B); 12. primary phloem (C); 13. taproot; 14. adventitious; 15. fibrous; 16. plasmodesmata; 17. endodermis; 18. walls; 19. pericycle; 20. lateral.

WOODY PLANTS (20-IV)
1. H; 2. I; 3. J; 4. G; 5. F; 6. E; 7. A; 8. D; 9. B; 10. C; 11. vascular cambium (C); 12. bark (E); 13. periderm (A); 14. phloem (D); 15. xylem (B).

Answers to Self-Quiz

1. b; 2. a; 3. b; 4. c; 5. a; 6. d; 7. d; 8. d; 9. c; 10. c; 11. b; 12. b.

Critical Thinking Analysis

a. Secondary growth is important in trees, but it only enlarges the diameter of the shoot system parts. If this were the only form of growth, it would explain why the nail didn't rise but not why the tree became taller.

b. This is true, but it does not explain the growth pattern. Puncturing this layer with a nail would not significantly reduce the flow of water and solutes up and down the trunk of a tree and thus would not affect growth.

c. Apical meristems add cells and length to the tips of the shoot (and the root) system. Parts closer to the center of the plant remain in the same place. This explains both observations.

d. If this happened, tree growth would be by addition of height below the nail and would move the segment containing the nail to a higher position. It would not explain the observation, and it does not occur.

e. They must do so, but they, like the other tissue types, grow by addition of cells at their tips.

21

PLANT NUTRITION AND TRANSPORT

NUTRITIONAL REQUIREMENTS

UPTAKE OF WATER AND NUTRIENTS
    Specialized Absorptive Structures
    Controlling Nutrient Uptake

WATER TRANSPORT AND CONSERVATION
    Transpiration

Control of Water Loss

TRANSPORT OF ORGANIC SUBSTANCES
    Storage and Transport Forms of Organic
        Compounds
    Translocation
    Pressure Flow Theory

Interactive Exercises

NUTRITIONAL REQUIREMENTS (21-I, p. 343)

Terms

The page-referenced terms are important; they were in boldface type in the chapter. Refer to the instructions given in Chapter 1, p. 1 of this workbook.

mineral ions (343):_____

Fill-in-the-Blanks

The three essential elements that plants use as their main metabolic building blocks are oxygen, carbon, and

(1)_____. The thirteen essential elements available to plants as dissolved salts are known as

(2)_____ _____. Of these, six dissolved salts are present in easily detectable concentrations in

plant tissues and are known as (3)_____. The remainder of the dissolved salts occur in very small

amounts in plant tissues and are known as (4)_____.

Complete the Table

5. Thirteen essential elements are available to plants as mineral ions. Complete the table below, which summarizes information about these important plant nutrients. Refer to Table 21.1 in the text.

Mineral Element	Macronutrient or Micronutrient	Known Functions
a. iron		roles in chlorophyll synthesis, electron transport
b.	macronutrient	activation of enzymes, role in maintaining water- solute balance
c. magnesium		
d. chlorine		role in root, shoot growth; role in photolysis
e.		role in chlorophyll synthesis; coenzyme activity
f. molybdenum	micronutrient	
g.	macronutrient	component of proteins, nucleic acids, coenzymes, chlorophyll
h. sulfur		component of most proteins, two vitamins
i.	micronutrient	roles in flowering, germination, fruiting, cell division, N metabolism
j. zinc		
k. calcium		roles in cementing cell walls, regulation of many cell functions
l. copper		component of several enzymes
m. phosphorus	macronutrient	

UPTAKE OF WATER AND NUTRIENTS (21-II, pp. 344–345)

Terms

symbiosis (344): _____

root nodules (344): _____

mycorrhizae (344): _____

root hairs (344): _____

vascular cylinder (345): _____

endodermis (345): _____

Casparian strip (345): _____

exodermis (345): _____

Dichotomous Choice

Circle one of two possible answers given between parentheses in each statement.

1. (Symbiosis/Parasitism) refers to permanent and intimate interactions between species that are permanent and in which benefits flow both ways.
2. (Gaseous nitrogen/Nitrogen "fixed" by bacteria) represents the chemical form of nitrogen plants can use in their metabolism.
3. Nitrogen-fixing bacteria reside in localized swellings on legume plants known as (root hairs/root nodules).
4. Mycorrhizae represent symbiotic relationships in which fungi and the roots they coat both benefit; the roots receive (sugars and nitrogen-containing compounds/scarce minerals).
5. (Root nodules/Root hairs) greatly increase a plant's surface area capacity for absorbing water and nutrients from the soil.

Sequence

Arrange in correct chronological sequence the path that nutrients take from the soil to cells in living plant tissues. Write the letter of the first step next to 6, the letter of the second step next to 7, and so on.

6. _____	A.	Cortex cells lacking Casparian strips
7. _____	B.	Endodermis cells with Casparian strips
	C.	Root hairs
8. _____	D.	ATP energy (from photosynthesis and/or respiration) actively transporting nutrients at membrane proteins
9. _____	E.	Exodermis cells with Casparian strips
10. _____	F.	Vascular cylinder
11. _____		

Label-Match

Identify each indicated part of the illustration below. Choose from the following: water movement, cytoplasm, vascular cylinder, endodermal cell wall, exodermis, endodermis, and Casparian strip. Complete the exercise by matching from the list below and entering the correct letter in the parentheses following each label.

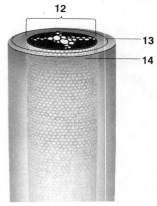

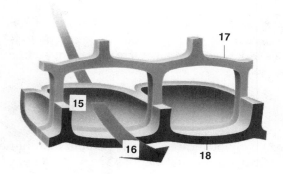

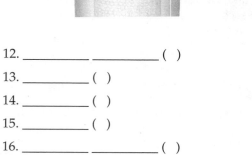

12. _____ _____ ( )

13. _____ ( )

14. _____ ( )

15. _____ ( )

16. _____ _____ ( )

17. _____ _____ ( )

18. _____ _____ _____ ( )

A. Cellular area through which water and dissolved nutrients must move due to Casparian strips
B. A layer of cortex cells just inside the epidermis; also equipped with Casparian strips
C. Waxy band acting as an impermeable barrier between the walls of abutting endodermal cells; forces water and dissolved nutrients through the cytoplasm of endodermal cells
D. Specific location of the waxy strips known as Casparian strips
E. Substance whose diffusion occurs through the cytoplasm of endodermal cells due to Casparian strips
F. Sheetlike layer of single cortex cells wrapped around the vascular cylinder
G. Tissues include the xylem, phloem, and pericycle

WATER TRANSPORT AND CONSERVATION (21-III, pp. 346-349)

Terms

transpiration (346): _____

cohesion-tension theory of water transport (346): _____

cuticle (348): _____

stomata (348): _____

guard cells (348): _____

Fill-in-the-Blanks

The cohesion-tension theory explains (1) _____ transport to the tops of plants. The water travels inside tubelike strands of (2) _____ , which are formed by hollow, dead, cells called tracheids and vessel members. The process begins with the drying power of air, which causes (3) _____ , the evaporation of water from plant parts exposed to air. (4) _____ of hydrogen-bonded water molecules in the xylem of roots, stems, and leaves provides a continuous column of water. This places the xylem water in a state of (5) _____ that extends from veins in leaves, down through the stems, to roots. As water continues to escape from plant surfaces, water enters the roots to replace that which was lost.

True/False

If the statement is true, write a T in the blank. If the statement is false, make it correct by changing the underlined word(s) and writing the correct word(s) in the answer blank.

_____ 6. Of the water moving into a leaf, 2 percent or more is lost by transpiration.

_____ 7. When evaporation exceeds water uptake by roots, plant tissues wilt and water-dependent activities are seriously disrupted.

_____ 8. All plant epidermal cell walls have a waxy covering, the cuticle, which reduces rate of water loss and limits inward diffusion of carbon dioxide.

_____ 9. The cuticle-covered epidermis is peppered with tiny openings called guard cells through which water leaves the plant and carbon dioxide enters.

_____ 10. When a pair of guard cells swells with water, the stoma between them closes.

_____ 11. In most plants, stomata remain open during the daylight photosynthetic period; water is lost but they gain carbon dioxide.

_____ 12. Stomata stay closed at night in most plants.

_____ 13. Photosynthesis starts when the sun comes up; as the morning progresses, carbon dioxide levels increase in cells—including guard cells.

_____ 14. A drop in carbon dioxide level within guard cells triggers an inward active transport of potassium ions; water follows by osmosis and the fluid pressure closes the stoma.

_____ 15. When the sun goes down and photosynthesis stops, carbon dioxide levels rise; stomata close when potassium, then water, moves out of the guard cells.

TRANSPORT OF ORGANIC SUBSTANCES (21-IV, pp. 350–351)

Terms

phloem (350):_____

sieve-tube members (350):_____

companion cells (350): _____

pressure flow theory (351): _____

Fill-in-the-Blanks

Sucrose and other organic compounds resulting from (1) _____ are used throughout the plant. Most plant cells store their carbohydrates as (2) _____. Quantities of (3) _____ become stored in some fruits; (4) _____ store proteins and fats. (5) _____ molecules are too large to cross cell membranes and too insoluble to be transported to other regions of the plant body. (6) _____ are largely insoluble in water and cannot be transported from storage sites. (7) _____ proteins do not lend themselves to transport. Specific chemical reactions such as (8) _____ convert storage forms of organic compounds to their subunits, which are transportable forms. For example, the hydrolysis of starch liberates glucose units, which combine with fructose to form (9) _____, the main transport form for sugars. CAM plants such as cacti and other succulents open stomata during the (10) [choose one] ( ) day, ( ) night when they fix carbon dioxide by way of a special C4 metabolic pathway. Such plants use carbon dioxide the next (11) [choose one] ( ) day, ( ) night when stomata are closed.

Matching

Match the appropriate letter with its numbered partner.

12. ___ translocation
13. ___ sieve tube members
14. ___ companion cells
15. ___ aphids
16. ___ source
17. ___ sink
18. ___ pressure flow theory

A. Any region where organic compounds are being loaded into the sieve tube system
B. Nonconducting cells adjacent to sieve tube members that supply energy to load sucrose at the source
C. Any region of the plant where organic compounds are being unloaded from the sieve tube system and used or stored
D. Process occurring in phloem that distributes sucrose and other organic compounds through the plant
E. States that pressure builds up at the source end of a sieve tube system and pushes solutes toward a sink, where they are removed
F. Passive conduits for translocation within vascular bundles; water and organic compounds flow rapidly through large pores on their end walls
G. Insects used to verify that in most plant species sucrose is the main carbohydrate translocated under pressure

Self-Quiz

___ 1. The _____ theory of water transport states that hydrogen bonding allows water molecules to maintain a continuous fluid column as water is pulled from roots to leaves.
a. pressure flow
b. evaporation
c. cohesion-tension
d. abscission

___ 2. The three elements that are present in carbohydrates, lipids, proteins, and nucleic acids are _____.
a. oxygen, carbon, and nitrogen

b. oxygen, hydrogen, and nitrogen
c. oxygen, carbon, and hydrogen
d. carbon, nitrogen, and hydrogen

___ 3. Macronutrients are the six mineral ions that _____.
a. play vital roles in photosynthesis and other metabolic events
b. occur in only small traces in plant tissues
c. become heavily incorporated in plant tissues
d. can function only without the presence of micronutrients
e. both a and c

___ 4. Without _____ , plants would rapidly wilt and die during hot, dry spells.
   a. a cuticle
   b. a mycorrhiza
   c. phloem
   d. cotyledons

___ 5. Gaseous nitrogen is converted to a plant-usable form by _____ .
   a. root nodules
   b. mycorrhizae
   c. nitrogen-fixing bacteria
   d. Venus flytraps

___ 6. _____ prevent(s) water from moving past the abutting walls of the root endo-dermal cells.
   a. Cytoplasm
   b. Plasma membranes
   c. Casparian strips
   d. Osmosis

___ 7. Most of the water moving into a leaf is lost through _____ .
   a. osmotic gradients being established
   b. transpiration
   c. pressure-flow forces
   d. translocation

___ 8. Stomata remain _____ during day-light, when photosynthesis occurs, but remain _____ during the night when carbon dioxide accumulates through aero-bic respiration.
   a. open; open
   b. closed; open
   c. closed; closed
   d. open; closed

___ 9. By control of _____ levels inside the guard cells of stomata, the activity of sto-mata is controlled when leaves are losing more water than roots can absorb.
   a. oxygen
   b. potassium
   c. carbon dioxide
   d. ATP

___ 10. Leaves represent _____ regions; growing leaves, stems, fruits, seeds, and roots represent _____ regions.
   a. source; source
   b. sink; source
   c. source; sink
   d. sink; sink

Chapter Objectives/Review Questions

This section lists general and detailed chapter objectives that can be used as review questions. You can make maximum use of these items by writing answers on a separate sheet of paper. Fill in answers where blanks are provided. To check for accuracy, compare your answers with information given in the chapter or glossary.

Page	Objectives/Questions
(343)	1. Plants generally require _____ (number) essential elements.
(343)	2. Distinguish between macronutrients and micronutrients in relation to their role in plant nutrition.
(345)	3. Differentiate between the endodermis and the exodermis of the root cortex.
(345)	4. Due to the presence of the _____ strip in the walls of endodermal cells, water can move into the vascular cylinder only by crossing the plasma membrane and diffusing through the cytoplasm.
(346)	5. The _____-_____ theory explains how water moves upward in an unbroken col-umn through xylem to the tops of tall trees.
(346)	6. Transpiration occurs mostly at _____ , tiny epidermal passageways of leaves and stems.
(348)	7. Explain the mechanism by which stomata open during daylight and close during the night.
(350)	8. Carbohydrates are stored as _____ in most plants.
(350)	9. _____ is the main form in which sugars are transported through most plants.
(350)	10. Describe the role of sieve tube members and companion cells in translocation.

(351)   11. A _____ is any region of the plant where organic compounds are loaded into the sieve tube system; a _____ is any region where organic compounds are unloaded from the sieve tube system and used or stored.

(351)   12. According to the _____ _____ theory, pressure builds up at the source end of a sieve tube system and pushes solutes toward a sink, where they are removed.

(351)   13. Companion cells supply the _____ that loads sucrose at the source.

(351)   14. Describe the gradients responsible for continuous flow of organic compounds through phloem.

Integrating and Applying Key Concepts

How do you think maple syrup is made from maple trees? Which specific systems of the plant are involved, and why are maple trees tapped only at certain times of the year?

Critical Thinking Exercise

1. The Casparian strip is clearly visible in the microscope and forms a continuous band among the cells of the endodermis. Which of the following observations provides the strongest support for the hypothesis that the Casparian strip is impermeable to water and solutes?
    a. Fluid in the xylem vessels flows upward with water and ions.
    b. The composition of the intercellular fluid in the cortex is different from the composition of the water in the vascular column.
    c. Radioactive water supplied to the outside of the roots can later be detected in the intercellular fluid of the cortex, the cytoplasm of the endodermis cells, and the xylem fluid.
    d. Photosynthesis takes place in the leaves, not in the roots.
    e. Glucose, containing radioactive carbon provided to the plant as carbon dioxide gas around the leaves, can be found in the cortex cells of the roots.

Answers

Answers to Interactive Exercises

NUTRITIONAL REQUIREMENTS (21-I)
1. hydrogen; 2. mineral ions; 3. macronutrients; 4. micronutrients; 5. a. micronutrient; b. potassium; c. macronutrient, component of chlorophyll; activation of enzymes; d. micronutrient; e. manganese, micronutrient; f. component of enzyme used in nitrogen metabolism; g. nitrogen; h. macronutrient; i. boron; j. micronutrient, role in formation of auxin, chloroplasts, starch; enzyme component; k. macronutrient; l. micronutrient; m. component of nucleic acids, phospholipids, ATP.

UPTAKE OF WATER AND NUTRIENTS (21-II)
1. Symbiosis; 2. Nitrogen "fixed" by bacteria; 3. root nodules; 4. scarce minerals; 5. Root hairs; 6. C; 7. E; 8. A; 9. B; 10. F; 11. D; 12. vascular cylinder (G); 13. exodermis (B); 14. endodermis (F); 15. cytoplasm (A); 16. water movement (E); 17. Casparian strip (C); 18. endodermal cell wall (D).

WATER TRANSPORT AND CONSERVATION (21-III)
1. water; 2. xylem; 3. transpiration; 4. Cohesion; 5. tension; 6. F, 90 percent; 7. T; 8. T; 9. F, stomata; 10. F, opens; 11. T; 12. T; 13. F, decrease; 14. F, opens; 15. T.

TRANSPORT OF ORGANIC SUBSTANCES (21-IV)
1. photosynthesis; 2. starch; 3. fats; 4. seeds; 5. Starch; 6. Fats; 7. Storage; 8. hydrolysis; 9. sucrose; 10. night; 11. day; 12. D; 13. F; 14. B; 15. G; 16. A; 17. C; 18. E.

Answers to Self-Quiz

1. c; 2. c; 3. c; 4. a; 5. c; 6. c; 7. b; 8. d; 9. b; 10. c.

Critical Thinking Analysis

a. This observation shows what happens to water and ions after they pass the endodermis. It neither supports nor refutes a hypothesis about the route of substances moving between the root cortex and the xylem.

b. Fluid moves freely among the cells of the cortex. If the Casparian strip was permeable, water and ions would diffuse through it, and the solutions on both sides would have the same composition. The observation that the solutions are different means that a permeability barrier exists between them. In the root, that barrier has two components—the membranes of the endodermis cells and the Casparian strip among the cells.

c. This observation would directly support the idea that water is absorbed from the roots into the vascular column and distributed to cells. It does not indicate whether the water moved through the endodermal cells, among them and then into them from the inside of the vascular column, or both.

d. This observation would lead us to expect products of photosynthesis to move downward in the vascular column to the roots. It would not lead to any conclusion as to whether the molecules moved through or among the endodermal cells.

e. This observation would be direct evidence of distribution of photosynthetic products downward within the plant, but it, too, would not support any conclusion regarding the pathway of movement.

22

PLANT REPRODUCTION AND DEVELOPMENT

Interactive Exercises

REPRODUCTIVE MODES (22-I, pp. 354–355)

Terms

The page-referenced terms are important; they were in boldface type in the chapter. Refer to the instructions given in Chapter 1, p. 1 of this workbook.

sporophyte (354):_____

flowers (354): _____

gametophytes (354): _____

Label-Match

Study the generalized life cycle for flowering plants as shown in Figure 22.2 in the text. Identify each indicated part of the accompanying illustration. Complete the exercise by matching and entering the letter of the proper description in the parentheses following each label.

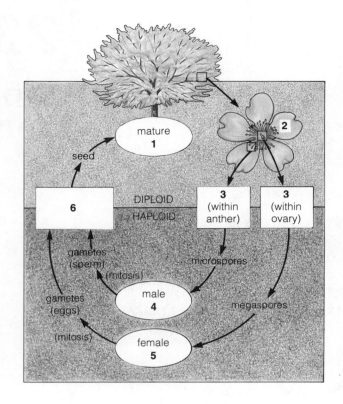

1. _____ ( )

2. _____ ( )

3. _____ ( )

4. _____ _____ ( )

5. _____ _____ ( )

6. _____ ( )

A. An event that produces a young sporophyte
B. A reproductive shoot produced by the sporophyte
C. The "plant"; a vegetative body that develops from a zygote
D. Cellular division event occurring within flowers to produce spores
E. Produces haploid eggs by mitosis
F. Produces haploid sperm by mitosis

GAMETE FORMATION IN FLOWERS (22-II, pp. 355–356)

Terms

stamens (355): _____

carpels (355): _____

ovary (355): _____

pollen grain (355): _____

microspores (356): _____

ovule (356): _____

megaspores (356): _____

endosperm (356): _____

Labeling

Identify each indicated part of the accompanying illustration.

1. _____

2. _____

3. _____

4. _____

5. _____

6. _____

7. _____

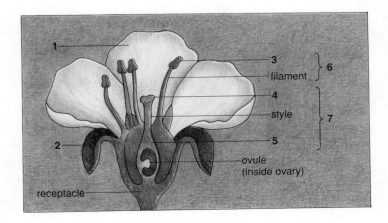

Matching

Choose the one most appropriate answer for each

8. ___ sepals

9. ___ petals

10. ___ stamens

11. ___ ovule

12. ___ pollen sacs

13. ___ carpels

14. ___ ovaries

15. ___ perfect flowers

16. ___ imperfect flowers

A. Have both male and female parts
B. Four chambers within an anther where pollen grains develop
C. Collectively, the flower's "corolla"
D. Have male or female parts, but not both
E. Female reproductive parts; includes stigma, style, and ovary
F. Structure that matures to become a seed
G. Chambers in the carpels where egg formation, fertilization, and seed development occur
H. Outermost leaflike whorl of floral organs
I. Male reproductive parts; a stalk capped by an anther

Fill-in-the-Blanks

The numbered items on the adjacent illustration represent missing information; complete the numbered blanks of the narrative below to supply the missing information on the illustration.

Within each (17) _____ , mitotic divisions produce four masses of spore-forming cells, each mass forming within a (18) _____ _____.

Each one of these diploid cells is known as a (19) _____ _____ cell and undergoes (20) _____ to produce four haploid (21) _____. Mitosis within each haploid microspore results in a two-celled haploid body, the immature male gametophyte. One of these cells will give rise to a (22) _____ _____ ; the other cell will develop into a (23) _____-_____ cell. Mature microspores are eventually released from the pollen sacs of the anther as (24) _____. Pollination occurs and after the pollen lands on a (25) _____ of a carpel, the pollen tube develops from one of the cells in the pollen grain; the other cell within the pollen grain divides to form two sperm cells. As the pollen tube grows through the carpel tissues, it contains the two sperm cells and a tube nucleus. The pollen tube with its included two sperm cells and the tube nucleus is known as the mature (26) _____ _____.

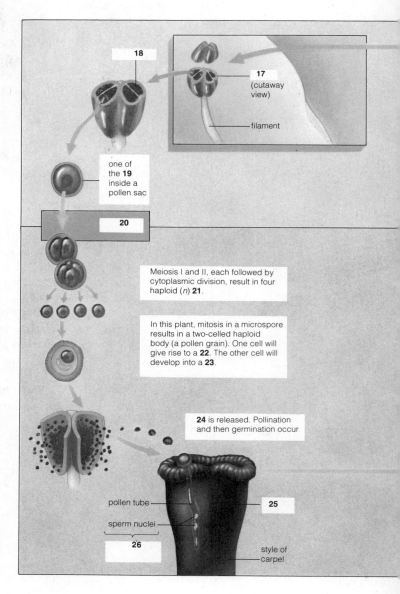

18

17 (cutaway view)

filament

one of the **19** inside a pollen sac

20

Meiosis I and II, each followed by cytoplasmic division, result in four haploid (*n*) **21**.

In this plant, mitosis in a microspore results in a two-celled haploid body (a pollen grain). One cell will give rise to a **22**. The other cell will develop into a **23**.

24 is released. Pollination and then germination occur.

pollen tube

sperm nuclei

26

25

style of carpel

Fill-in-the-Blanks

The numbered items on the illustration on the next page represent missing information; complete the numbered blanks in the narrative below to supply the missing information on the illustration.

In the carpel of the flower, one or more dome-shaped, diploid tissue masses develop on the inner wall of the ovary. Each mass is the beginning of an (27) _____. A tissue forms inside a domed mass as it grows, and one or two protective layers called (28) _____ form around it. Inside each mass, a cell (the megaspore mother cell) divides by (29) _____ to form four haploid spores, the (30) _____. Commonly,

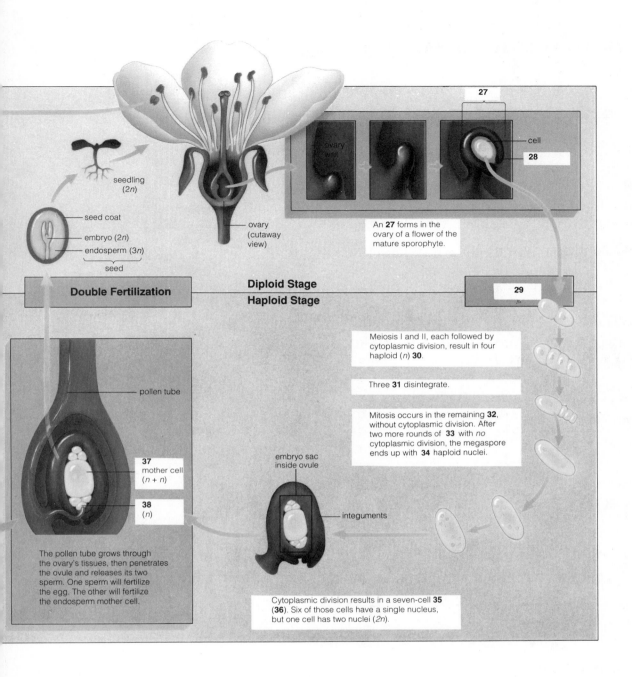

seedling
(2n)

seed coat

embryo (2n)

endosperm (3n)

seed

Double Fertilization

Diploid Stage

Haploid Stage

ovary
(cutaway
view)

ovary
wall

cell

28

27

An **27** forms in the
ovary of a flower of the
mature sporophyte.

29

Meiosis I and II, each followed by
cytoplasmic division, result in four
haploid (n) **30**.

Three **31** disintegrate.

Mitosis occurs in the remaining **32**,
without cytoplasmic division. After
two more rounds of **33** with *no*
cytoplasmic division, the megaspore
ends up with **34** haploid nuclei.

pollen tube

37
mother cell
(n + n)

38
(n)

embryo sac
inside ovule

integuments

The pollen tube grows through
the ovary's tissues, then penetrates
the ovule and releases its two
sperm. One sperm will fertilize
the egg. The other will fertilize
the endosperm mother cell.

Cytoplasmic division results in a seven-cell **35**
(**36**). Six of those cells have a single nucleus,
but one cell has two nuclei (2n).

all but one (31) _____ disintegrates. The remaining (32) _____ undergoes (33) _____ three

times without cytoplasmic division. At first, this structure is a cell with (34) _____ haploid nuclei.

Cytoplasmic division results in a seven-cell (35) _____ _____ , which represents the mature

(36) _____ _____. One of these cells, the (37) _____ mother cell, has two nuclei and will help

form the 3n endosperm, a nutritive tissue for the forthcoming embryo. Another haploid cell within the

embryo sac is the (38) _____ , which will combine with one sperm to form the diploid embryo.

FROM POLLINATION TO FERTILIZATION (22-III, pp. 356–359)

Terms

pollination (356): _____

double fertilization (357): _____

Short Answer

1. What guides the growth of the pollen tube down through the female floral tissues toward the chamber holding the egg? _____

2. Describe the site of double fertilization in flowering plants. _____

Complete the Table

3. Complete the table below to summarize the double fertilization occurring only in flowering plant life cycles. Refer to Figure 22.5 in the text.

Double Fertilization Products	Origin	Produces	Function
a. zygote (2n) nucleus			
b. endosperm (3n) nucleus			

EARLY DEVELOPMENT (22-IV, pp. 360–361)

Terms

cotyledons (360): _____

seed (360): _____

fruit (360): _____

Complete the Table

1. Complete the following table, which summarizes concepts associated with seeds and fruits.

Structure	Origin	Function(s)
a. Cotyledons		
b. Seed		
c. Seed Coat		
d. Fruit		

Labeling

Identify each indicated part of the accompanying illustration.

2._____   8._____ _____

3._____   9._____

4._____   10._____

5._____   11._____ _____

6._____   12._____

7._____   13._____

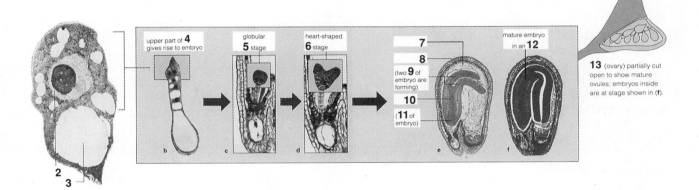

Choice

For questions 14–20, choose from the following.

a. simple, wall dry, splits at maturity   b. simple, wall dry; intact at maturity
c. simple, fleshy; often leathery   d. aggregate
e. multiple   f. accessory, simple
g. accessory, aggregate

___ 14. apple, pear

___ 15. pea, magnolia, mustard

___ 16. strawberry

___ 17. sunflower, wheat, rice, maple

___ 18. pineapple, fig, mulberry

___ 19. grape, banana, lemon, cherry

___ 20. blackberry, raspberry

PATTERNS OF GROWTH AND DEVELOPMENT (22-V, pp. 362–368)

Terms

germination (363):_____

hormone (364):_____

auxins (364):_____

gibberellins (364): _____

cytokinins (364): _____

abscisic acid (364): _____

ethylene (364): _____

phototropism (366):_____

gravitropism (366): _____

thigmotropism (366): _____

photoperiodism (367): _____

senescence (368): _____

dormancy (368): _____

Fill-in-the-Blanks

Before or after seed dispersal, the growth of the (1) _____ idles. After the embryo absorbs water, resumes growth, and breaks through the seed coat, (2) _____ has occurred. This process depends on water, (3) _____ , temperature, light and day length, and other environmental factors. Usually germination coincides with the return of spring (4) _____. As more and more water molecules move inside it, the seed swells and its (5) _____ ruptures. More oxygen now reaches the embryo, and (6) _____ respiration moves into high gear. Cells of the embryo rapidly grow and divide, become specialized, and so produce the (7) _____. The seedling develops into a mature (8) _____ , which forms flower, fruits, and seeds. There are two basic patterns of growth and development among flowering plants, that of a (9) _____ and that of a (10) _____. Basic growth patterns are (11) _____ and dictated by the plant's genes. The growth patterns can be adjusted in response to the (12) _____. Enzymes and other proteins within the cells carry out the growth responses and (13) _____ stimulate them to act.

Complete the Table

14. Complete the following table, which summarizes the five types of plant hormones and their actions.

Hormone	Action
a. Auxins	
b. Gibberellins	
c. Cytokinins	
d. Abscisic acid	
e. Ethylene	

Choice

For questions 15–22, choose from the following.

    a. auxins   b. gibberellins   c. cytokinins   d. abscisic acid   e. ethylene

___ 15. Chinese burning incense to hurry fruit ripening

___ 16. Natural and synthetic versions are used to prolong the shelf life of cut flowers, lettuces, mushrooms, and other vegetables

___ 17. Orchardists spraying trees with IAA

___ 18. Proper doses produce radishes as big as beach balls and cabbage shoots as tall as you are

___ 19. Applied to nursery stock about to be shipped as dormant plants are not damaged as much

___ 20. Synthetic versions are used as herbicides

___ 21. Used to ripen green fruit after shipment

___ 22. Cause celery to grow longer and crisper; keep navel orange skins pliable; give seedless grapes market appeal

Matching

Choose the one most appropriate answer for each.

23. ___ 2,4-D
24. ___ hormone
25. ___ apical dominance
26. ___ herbicide
27. ___ abscission
28. ___ IAA
29. ___ 2,4,5-T
30. ___ target cell

A. Agent Orange; used to defoliate during the Vietnam conflict
B. Dropping of flowers, leaves, fruits, or other plant parts
C. A cell with receptors for a given signaling molecule
D. Hormonal effect that inhibits lateral bud growth, which promotes stem elongation
E. Synthetic auxin used as an herbicide
F. A signaling molecule released from one cell that changes the activity of target cells
G. The most important naturally occurring auxin
H. Any compound used to selectively kill plants

Complete the Table

31. Complete the following table with information about plant growth responses called tropisms.

Tropism	Definition
a. Phototropism	
b. Gravitropism	
c. Thigmotropism	

Choice

For questions 32–36, choose from the following.

a. phototropism   b. gravitropism   c. thigmotropism

___ 32. More intense sunlight on one side of a plant—stems curve toward the light

___ 33. Vines climbing around a fence post as they grow upward

___ 34. A root turned on its side will curve downward

___ 35. A potted seedling turned on its side—the growing stem curves upward

___ 36. Leaves turn until their flat surfaces face light

Matching

Choose the one most appropriate answer for each.

37. _____ Circadian rhythms

38. _____ Photoperiodism

39. _____ Senescence

40. _____ Dormancy

A. Any biological response to a change in the relative length of daylight and darkness in a cycle of twenty-four hours

B. When a plant stops growing under conditions that seem quite suitable for growth

C. Plant activities that occur regularly in cycles of about twenty-four hours, even when outside conditions are held constant

D. The sum total of processes leading to the death of plant parts or the whole plant

Choice

For questions 41–45, choose from the following.

a. circadian rhythms   b. photoperiodism   c. senescence   d. dormancy

___ 41. A Delicious apple tree grown in Utah requires 1,230 hours near 43°F (6°C) before it will break dormancy and begin a new season of growth

___ 42. Dropping of leaves, flowers, and fruits

___ 43. "Short-day," "day-neutral," and "long-day" plants

___ 44. Gardeners routinely remove flower buds from many plants to maintain vegetative growth

___ 45. A plant that positions its leaves horizontally during the day but folds them closer to the stem at night

Fill-in-the-Blanks

(46) _____ is defined as any biological response to a change in the relative length of daylight and darkness in a twenty-four hour cycle. A blue-green pigment molecule, (47) _____, serves as a switching mechanism in the photoperiodism response. Phytochrome is converted to an active form (Pfr) at sunrise, when (48) _____ wavelengths dominate the sky. Phytochrome reverts to an active form (Pr) at sunset, at night, or in the shade, where (49) _____ - _____ wavelengths predominate. Pfr may control which types of (50) _____ are being produced in particular cells for different growth responses. Photoperiodism is especially apparent in the (51) _____ process, which is often keyed to day-length changes throughout the year. "Long-day plants" flower in spring when day-length becomes (52) [choose one] ( ) shorter, ( ) longer than some critical value. "Short-day plants" flower in late summer or early autumn when daylength becomes (53) [choose one] ( ) shorter, ( ) longer than some critical value. "Day-neutral plants" flower whenever they become (54) _____ enough to do so without regard to daylength.

Labeling

Identify each indicated part of the accompanying illustration.

55. _____

56. _____

57. _____

58. _____

59. _____

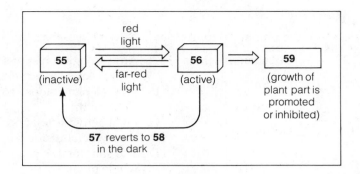

Self-Quiz

___1. A stamen is _____.
  a. composed of a stigma
  b. the mature male gametophyte
  c. the site where microspores are produced
  d. part of the vegetative phase of an angiosperm

___2. The phase in the life cycle of plants that gives rise to spores is known as the

  _____.
  a. gametophyte
  b. embryo
  c. sporophyte
  d. seed

___3. Fertilization and seed maturation occurs in

  _____.
  a. the style of the carpel
  b. the ovary of the carpel
  c. pollen sacs of anthers
  d. microspores

___4. An immature fruit is a(n) _____ and an immature seed is a(n) _____.
  a. ovary; megaspore
  b. ovary; ovule
  c. megaspore; ovule
  d. ovule; ovary

___5. In flowering plants, one sperm nucleus fuses with that of an egg, and a zygote forms that develops into an embryo. Another sperm fuses with _____.
  a. a primary endosperm cell to produce three cells, each with one nucleus
  b. a primary endosperm cell to produce one cell with one triploid nucleus

  c. both nuclei of the endosperm mother cell, forming a primary endosperm cell with a single triploid nucleus
  d. one of the smaller megaspores to produce what will eventually become the seed coat

___6. Simple, aggregate, multiple, and accessory refer to types of _____.
  a. carpels
  b. seeds
  c. fruits
  d. ovaries

___7. Promoting fruit ripening and abscission of leaves, flowers, and fruits is a function ascribed to _____.
  a. gibberellins
  b. ethylene
  c. abscisic acid
  d. auxins

___8. Auxins _____.
  a. cause flowering
  b. promote stomatal closure
  c. promote cell division
  d. promote cell elongation in coleoptiles and stems

___9. _____ is demonstrated by a germinating seed whose first root always curves down while the stem always curves up.
  a. Phototropism
  b. Photoperiodism
  c. Gravitropism
  d. Thigmotropism

___10. Plants whose leaves are open during the day but fold at night are exhibiting

_____.
a. a growth movement
b. a circadian rhythm
c. a biological clock
d. both b and c

___11. Phytochrome is converted to an active form, _____, at sunrise and reverts to an inactive form, _____, at sunset, at night, or in the shade.
a. Pr; Pfr
b. Pfr; Pfr

c. Pr; Pr
d. Pfr; Pr

___12. When a perennial or biennial plant stops growing under conditions suitable for growth, it has entered a state of

_____.
a. senescence
b. vernalization
c. dormancy
d. abscission

Chapter Objectives/Review Questions

This section lists general and detailed chapter objectives that can be used as review questions. You can make maximum use of these items by writing answers on a separate sheet of paper. Fill in answers where blanks are provided. To check for accuracy, compare your answers with information given in the chapter or glossary.

Page	*Objectives/Questions*
(354)	1. _____ reproduction requires formation of gametes, followed by fertilization.
(354)	2. Describe the relationship of a flower to a sporophyte.
(354)	3. Distinguish between sporophytes and gametophytes.
(354–355)	4. _____ have several means of reproducing themselves asexually.
(356)	5. Relate the sequence of events that give rise to microspores and megaspores.
(356)	6. Describe the development of the male gametophyte.
(355)	7. What represents the male gametophyte and female gametophyte in flowering plants?
(357)	8. Describe the double fertilization that occurs uniquely in the flowering-plant life cycle.
(360)	9. From the plant's perspective, what is the function of fruits?
(363)	10. The plant embryo absorbs water, resumes growth, and breaks the seed coat at _____.
(363)	11. What portion of the embryo emerges from the seed first?
(364)	12. Plant growth and development are under _____ controls.
(364–365)	13. Describe the general role of plant hormones and target cells.
(365)	14. Define herbicide and give examples.
(365)	15. _____ dominance is the hormonal effect that inhibits lateral bud growth while plant resources are diverted to stem elongation.
(366)	16. Plants have internal time-measuring mechanisms called biological _____.
(366)	17. What are circadian rhythms? Give an example.
(367)	18. Phytochrome is converted to an active form, _____, at sunrise, when red wavelengths dominate the sky. It reverts to an inactive form, _____, at sunset, at night, or even in shade, where far-red wavelengths predominate.
(367)	19. List the plant activities influenced by Pfr.
(367)	20. _____ serves as a switching mechanism in the biological clock governing photoperiodism.
(367)	21. Describe the photoperiodic responses of "long-day," "short-day," and "day-neutral" plants.
(368)	22. List environmental cues that send a plant into dormancy.

Integrating and Applying Key Concepts

An oak tree has grown up in the middle of a forest. A lumber company has just cut down all the surrounding trees except for a narrow strip of woods that includes the oak. How is the oak likely to respond as it adjusts to its changed environment? To what new stresses will it be exposed? Which hormones will most probably be involved in the adjustment?

Critical Thinking Exercise

1. It has long been thought that the growth of the pollen tube into the style is stimulated and directed by a chemical substance released from the ovule. This substance is thought to produce a chemotropic response in the growing pollen tube. Experiments with extracts of ovules indicated that the chemotropic substance was small, stable to heating, and readily soluble in water. These observations indicated that the substance is an inorganic ion. In one type of experiment on chemotropic responses of pollen tubes, pollen grains are placed on the surface of an agar gel. Calcium ions are placed in a shallow well nearby in the gel. After several hours, the pollen tubes are examined. The data are given in the table below.

Calcium in Well (micrograms)	Pollen Tubes Growing Toward Well
0.0	21
0.8	46
1.6	70
2.4	90
4.0	110
6.0	198
8.0	226
9.6	260
20.0	269

The authors concluded from these observations that calcium does exert a chemotropic effect on pollen tubes. Which of the following additional observations would be most important in evaluating the conclusion?

a. The concentration of calcium in the agar at the pollen tubes
b. The length of the pollen tubes
c. The results achieved with a longer incubation time
d. The number of pollen tubes growing away from the wells
e. The incubation temperature

Answers

Answers to Interactive Exercises

REPRODUCTIVE MODES (22-I)

1. sporophyte (C); 2. flower (B); 3. meiosis (D); 4. male gametophyte (F); 5. female gametophyte (E); 6. fertilization (A).

GAMETE FORMATION IN FLOWERS (22-II)

1. petal; 2. sepal; 3. anther; 4. stigma; 5. ovary; 6. stamen; 7. carpel; 8. H; 9. C; 10. I; 11. F; 12. B; 13. E; 14. G; 15. A; 16. D; 17. anther; 18. pollen sac; 19. microspore mother; 20. meiosis; 21. microspores; 22. pollen tube; 23. sperm-producing; 24. pollen; 25. stigma; 26. male gametophyte; 27. ovule; 28. integuments; 29. meiosis; 30. megaspores; 31. megaspore; 32. megaspore; 33. mitosis; 34. eight; 35. embryo sac; 36. female gametophyte; 37. endosperm; 38. egg.

FROM POLLINATION TO FERTILIZATION (22-III)

1. Chemical and molecular cues guide the pollen tube's growth down through the female floral tissues. 2. Double fertilization occurs in flowering plants alone. The embryo sac is the site of double fertilization. 3. a. fusion of one egg nucleus (n) and one sperm nucleus (n), plant embryo ($2n$), eventually develops into a new sporophyte plant; b. fusion of one sperm nucleus (n) with the endosperm mother cell ($2n$), endosperm tissues ($3n$), nourishes the embryo within the seed.

EARLY DEVELOPMENT (22-IV)

1. a. "seed leaves" that develop as part of the embryo, plants with large cotyledons absorb endosperm and store food; thin cotyledons may produce enzymes for transferring food from the embryo to the germinating seedling; b. ovule within the ovary, contains a $2n$ embryo sporophyte, $3n$ endosperm, these wrapped within a seed coat; c. integuments of ovule, protection for embryo with its endosperm (stored food); d. usually a mature, ripened ovary, seed protection and dispersal in specific environments; 2. nucleus; 3. vacuole; 4. zygote; 5. embryo; 6. embryo; 7. endosperm; 8. seed coat; 9. cotyledons; 10. embryo; 11. root tip; 12. ovule; 13. fruit; 14. f; 15. a; 16. g; 17. b; 18. e; 19. c; 20. d.

PATTERNS OF GROWTH AND DEVELOPMENT (22-V)

1. embryo; 2. germination; 3. oxygen; 4. rains; 5. coat; 6. aerobic; 7. seedling; 8. sporophyte; 9. dicot (monocot); 10. monocot (dicot); 11. heritable; 12. environment; 13. hormones; 14. a. make stems lengthen and may influence growth responses to light and gravity; b. make stems grow longer; they induce dormant seeds, buds, and perhaps flowers to grow; c. promote cell division and leaf expansion, and retard aging; d. inhibits cell growth, promotes bud dormancy, and keeps seeds from germinating prematurely; causes stomata to close to conserve water in dry spells; e. promotes fruit ripening and the dropping of flowers, fruits, and leaves from trees; 15. e; 16. c; 17. a; 18. b; 19. d; 20. a; 21. e; 22. b; 23. E; 24. F; 25. D; 26. H; 27. B; 28. G; 29. A; 30. C; 31. a. adjustment of a plant's direction and rate of growth in response to light; b. a growth response to the earth's gravitational force; c. unequal plant growth after contacting solid objects in the surroundings; 32. a; 33. c; 34. b; 35. b; 36. a; 37. C; 38. A; 39. D; 40. B; 41. d; 42. c; 43. b; 44. c; 45. a; 46. Photoperiodism; 47. phytochrome; 48. red; 49. far-red; 50. enzymes; 51. flowering; 52. longer; 53. shorter; 54. mature; 55. Pr; 56. Pfr; 57. Pfr; 58. Pr; 59. response.

Answers to Self-Quiz

1. c; 2. c; 3. b; 4. b; 5. c; 6. c; 7. b; 8. d; 9. c; 10. d; 11. d; 12. c.

Critical Thinking Analysis

a. In order to conclude that calcium mediates the response, it is important to show that the response is concentration-dependent. However, it is reasonable to assume that calcium diffuses through the agar gel and that the concentration at some distance from the well is dependent on the concentration in the well. This additional observation would be a reasonable check on the validity of the assumption, but it is not necessary in order to interpret the data.

b. If calcium stimulates growth of pollen tubes, they should be longer given higher concentrations of calcium. This observation would be useful to confirm the interpretation, but the data on numbers alone can evaluate the hypothesis.

c. As long as the response is clear-cut at the time used, longer times might only confound the results. The test of chemotropism depends on applying the calcium from one side only and showing that the pollen tubes grow nonrandomly to that side. After a longer incubation, the calcium would diffuse beyond the pollen grains and destroy the gradient.

d. The first well, containing no calcium, shows that calcium is not absolutely essential for pollen-tube growth in one direction. The conclusion now depends on showing that more tubes grow toward calcium than away from it. If equal numbers of tubes grow in both directions, the conclusions would simply be that calcium stimulates growth, not that it is chemotropic.

e. Knowing the effects of temperature on the response and the rate of growth might be valuable. Assuming there is a temperature effect, you would need to know the temperature in order to repeat the experiment, but the evaluation of the hypothesis on the basis of the present data does not require knowledge of the temperature.

23

TISSUES, ORGAN SYSTEMS, AND HOMEOSTASIS

ANIMAL STRUCTURE AND FUNCTION: AN
OVERVIEW
    How Tissues Form
    Epithelial Tissue
    Connective Tissue
    Muscle Tissue

Nervous Tissue
MAJOR ORGAN SYSTEMS
HOMEOSTASIS AND SYSTEMS CONTROL
    The Internal Environment
    Mechanisms of Homeostasis

Interactive Exercises

ANIMAL STRUCTURE AND FUNCTION: AN OVERVIEW (23-I, pp. 373–378)

Terms

The page-referenced terms are important; they were in boldface type in the chapter. Refer to the instructions given in Chapter 1, p. 1 of this workbook.

tissue (373) _____

organ (373) _____

organ system (373) _____

ectoderm (373) _____

mesoderm (373) _____

endoderm (373) _____

epithelium, -lia (374) _____

gland (375) _____

connective tissue proper (376) _____

cartilage (376) _____

bone (376) _____

adipose tissue (376)_____

blood (376)_____

muscle tissue (377)_____

nervous tissue (378)_____

True/False

If the statement is true, write a T in the blank. If the statement is false, make it correct by changing the underlined word(s) and writing the correct word(s) in the answer blank.

_____ 1. Physiology is the study of the way body parts are arranged in an organism.

_____ 2. Groups of like cells that work together to perform a task are known as an organ.

_____ 3. Most animals are constructed of only four types of tissue: epithelial, connective, nervous, and muscle.

_____ 4. Mammalian skin contains squamous epithelium and other tissues.

_____ 5. The more tight junctions there are in a tissue, the more permeable the tissue will be.

_____ 6. Endocrine glands secrete their products through ducts that empty onto an epithelial surface.

_____ 7. Endocrine-cell products include digestive enzymes, saliva, and mucus.

_____ 8. Muscle bundles are identical to individual skeletal muscle cells.

_____ 9. Both skeletal and cardiac muscle tissues are striated.

_____ 10. Cardiac muscle cells are fused, end-to-end, but each cell contracts independently of other cardiac muscle cells.

_____ 11. Smooth muscle tissue is involuntary and not striated.

_____ 12. Smooth muscle tissue is located in the walls of the intestine.

_____ 13. Neurons conduct messages to other neurons or to muscles or glands.

Fill-in-the-Blanks

Groups of like cells that work together to perform a task are known as a(n) (14)_____. Groups of *different* types of tissues that interact to carry out a task are known as a(n) (15)_____. Each cell engages in basic (16)_____ activities that assure its own survival. The combined contributions of cells, tissues, organs, and organ systems help maintain a stable (17)_____ _____ that is required for individual cell survival.

In (18)_____ _____ (immature reproductive cells that later develop into (19)_____), a special form of cell division known as (20)_____ occurs. In all other body tissues that consist of (21)_____ cells, the usual form of cell division, (22)_____, occurs. The life of almost any animal begins with two gametes merging to form a fertilized egg, which undergoes reorganization and then divides by mitosis to form first undifferentiated cells, then three types of (23) "_____" _____ (groups of similar cells that perform similar activities) in the early embryo.

Complete the Table

Supply the name of the "primary" tissue of the embryo that does the job indicated by becoming specialized in particular ways.

(24)	forms internal skeleton, muscle, circulatory, reproductive, and urinary systems
(25)	forms inner lining of gut and linings of major organs formed from the embryonic gut
(26)	forms outer layer of skin and the tissues of the nervous system

While a specific kind of tissue (for example, simple squamous epithelium) is composed of cells that look very similar and do similar jobs, a specific (27)_____ (for example, a kidney) is composed of different tissues that cooperate to do a specific job (in this case, remove waste products from blood, produce urine, and maintain the composition of body fluids). Kidneys, a urinary bladder, and various tubes are grouped together into a(n) (28)_____ _____ , which adds the functions of urine storage and elimination to the jobs that the kidneys do. A multicellular (29)_____ is most often composed of organ systems that cooperate to keep activities running smoothly in a coordinated fashion; this maintenance of stable operating conditions in the internal environment is known as (30)_____.

All the diverse body parts found in different animals can be assembled from a few (31)_____ types through variations in the way they are combined and arranged. Somatic cells compose the physical structure of the animal body; they become differentiated into four main types of specialized tissue—(32)_____ , (33)_____ , muscle, and nervous.

Cells of connective tissues are scattered throughout an extensive extracellular (34)_____ _____. (35)_____ connective tissue contains a weblike scattering of strong, flexible protein fibers and a few highly elastic protein fibers and serves as a packing material that holds in place blood vessels, nerves, and internal organs. (36)_____ and (37)_____ are examples of dense, regular connective tissue that help connect elements of the skeletal and muscular systems.

Label-Match

Identify each of the illustrations below by labeling it with one of the following: connective, epithelial, muscle, nervous, germ cells and/or gametes. Complete the exercise by matching and entering *all* appropriate letters/numbers from the groups below in the parentheses following each label.

38. _____ (   )

39. _____ (   )

40. _____ (   )

41. _____ (   )

42. _____ (   )

43. _____ (   )

44. _____ (   )

45. _____ (   )

46. _____ (   )

47. _____ (   )

48. _____ (   )

49. _____ (   )

50. _____ (   )

38.

39.

40.

41.

42.

43.

44. Haversian canal

cells in spaces

45.

46.

lymphocyte

platelets

neutrophils

erythrocytes

47.

48.

49.

50.

A. Adipose
B. Bone
C. Cardiac
D. Dense, regular
E. Loose
F. Simple columnar
G. Simple cuboidal
H. Simple squamous
I. Smooth
J. Skeletal

1. Absorption
2. Maintain diploid number of chromosomes in sexually reproducing populations
3. Communication by means of electrochemical signals
4. Energy reserve
5. Contraction for voluntary movements
6. Diffusion
7. Padding
8. Contract to propel substances along internal passageways; not striated
9. Attaches muscle to bone and bone to bone
10. In vertebrates, provides the strongest internal framework of the organism
11. Elasticity
12. Secretion
13. Pumps circulatory fluid; striated
14. Insulation
15. Transport of nutrients and waste products to and from body cells

MAJOR ORGAN SYSTEMS (23-II, pp. 378–379)

Terms

circulatory system (378–379) _____

digestive system (378–379) _____

endocrine system (378–379) _____

integumentary system (378–379) _____

lymphatic system (378–379) _____

muscular system (378–379) _____

nervous system (378–379) _____

reproductive system (378–379) _____

respiratory system (378–379) _____

skeletal system (378–379) _____

urinary system (378–379) _____

Labeling

Supply the name of the correct organ system.

1. _____ Picks up nutrients absorbed from gut and transports them to cells throughout body

2. _____ Helps cells use nutrients by supplying them with oxygen and relieving them of CO_2 wastes

3. _____ Helps maintain the volume and composition of body fluids that bathe the body's cells

4. _____ Provides basic framework for the animal and supports other organs of the body

5. _____ Uses chemical messengers to control and guide body functions

6. _____ Protects the body from viruses, bacteria, and other foreign agents

7. _____ Produces younger, temporarily smaller versions of the animal

8. _____ Breaks down larger food molecules into smaller nutrient molecules that can be absorbed by body fluids and transported to body cells.

9. _____ Consists of contractile parts that move the body through the environment and propel substances about in the animal

10. _____ Serves as an electrochemical communications system in the animal's body

11. _____ In the meerkat, served as a heat catcher in the morning and protective insulation at night

Matching

Match each system shown on the next page with the most appropriate function.

12. ___ Male: production and transfer of sperm to the female. Female: production of eggs; provision of a protected nutritive environment for developing embryo and fetus. Both systems have hormonal influences on other organ systems.
13. ___ Ingestion of food, water; preparation of food molecules for absorption; elimination of food residues from the body.
14. ___ Movement of internal body parts; movement of whole body; maintenance of posture; heat production.
15. ___ Detection of external and internal stimuli; control and coordination of responses to stimuli; integration of activities of all organ systems.
16. ___ Protection from injury and dehydration; body temperature control; excretion of some wastes; reception of external stimuli; defense against microbes.
17. ___ Provisioning of cells with oxygen; removal of carbon dioxide wastes produced by cells; pH regulation.
18. ___ Support, protection of body parts; sites for muscle attachment, blood cell production, and calcium and phosphate storage.
19. ___ Hormonal control of body functioning; works with nervous system in integrative tasks.
20. ___ Maintenance of the volume and composition of extracellular fluid.
21. ___ Rapid internal transport of many materials to and from cells; helps stabilize internal temperature and pH.
22. ___ Return of some extracellular fluid to blood; roles in immunity (defense against specific invaders of the body).

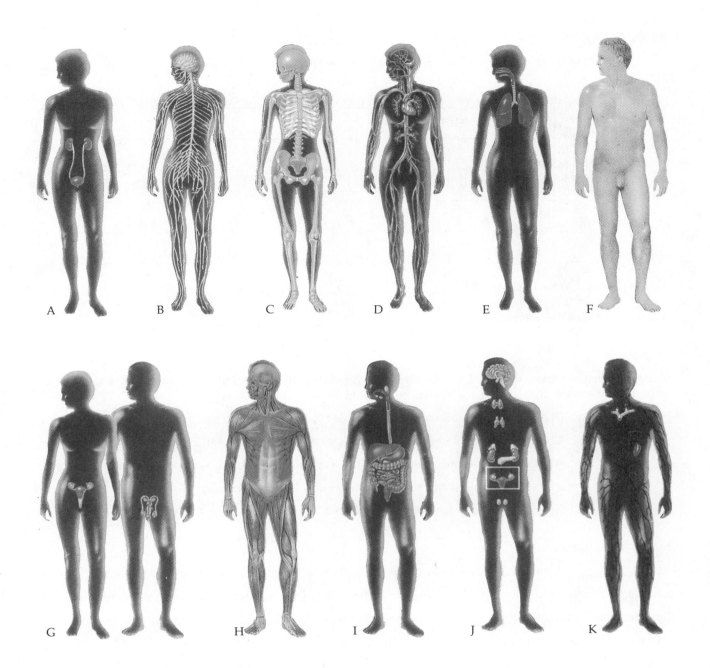

Fill-in-the-Blanks

There are four major body cavities in humans. Lungs are located in the (23)_____ cavity, the brain is in the (24)_____ cavity, and, in the female, ovaries and the urinary bladder are in the (25)_____ cavity. The urinary system of an animal is responsible for the disposal of (26)_____ wastes; fecal material is not included in this category. The endocrine system is generally responsible for internal (27)_____ control; together with the (28)_____ system, it integrates physiological processes.

HOMEOSTASIS AND SYSTEMS CONTROL (23-III, pp. 380–381)

Terms

extracellular fluid (380) _____

homeostasis (380) _____

sensory receptor (380) _____

stimulus (380) _____

integrator (380) _____

effector (380) _____

negative feedback mechanism (380) _____

positive feedback mechanism (381) _____

True/False

If the statement is true, write a T in the blank. If the statement if false, make it correct by changing the underlined word(s) and writing the correct word(s) in the answer blank.

_____ 1. The process of childbirth is an example of a <u>negative</u> feedback mechanism.

_____ 2. The human body's <u>effectors</u> are, for the most part, muscles and glands.

_____ 3. An integrator is constructed in such a way that it is informed of specific energy changes in the environment and relays messages about them to a <u>receptor</u>.

_____ 4. <u>Interstitial fluid</u> is a synonym for plasma.

Fill-in-the-Blanks

In a (5)_____ feedback mechanism, a chain of events is set in motion that intensifies the original condition before returning to set point; sexual arousal and childbirth are two examples. Generally, physiological controls work by means of (6)_____ feedback, in which an activity changes some condition in the internal environment, and the feedback causes the changed condition to be reversed back to "normal"; the maintenance of body temperature close to a "set point" is an example. Your brain is a(n) (7)_____ , a control point where different bits of information are pulled together in the selection of a response. Muscles and (8)_____ are examples of effectors. Internal temperature control of the human body is achieved by (9)_____ in the skin and elsewhere sensing a temperature change at the body's surface and relaying the neural information to an integrator. In this example, the (10)_____ in the brain (see Fig. 23.12) compares neural input against a set point. This part of the brain then sends output signals to (11)_____. Examples of these are the (12)_____ glands, which excrete water from the skin.

Self-Quiz

___1. Which of the following is *not* included in connective tissues?
a. bone
b. blood
c. cartilage
d. skeletal muscle

___2. A surrounding material within which something originates, develops, or is contained is known as a _____.
a. lamella
b. ground substance
c. plasma
d. lymph

___3. Blood is considered to be a(n) _____ tissue.
a. epithelial
b. muscular
c. connective
d. none of these

___4. _____ are abundant in tissues of the heart and liver where they promote diffusion of ions and small molecules from cell to cell.
a. Adhesion junctions
b. Filter junctions
c. Gap junctions
d. Tight junctions

___5. Muscle that is not striped and is involuntary is _____.
a. cardiac
b. skeletal
c. striated
d. smooth

___6. Chemical and structural bridges link groups or layers of like cells, uniting them in structure and function as a cohesive

_____.
a. organ
b. organ system

c. tissue
d. cuticle

___7. A fish embryo was accidentally stabbed by a graduate student in developmental biology. Later, the embryo developed into a creature that could not move and had no supportive or circulatory systems. Which embryonic tissue had suffered the damage?
a. ectoderm
b. endoderm
c. mesoderm
d. protoderm

___8. A tissue whose cells are striated and fused at the ends by cell junctions so that the cells contract as a unit is called _____ tissue.
a. smooth muscle
b. dense fibrous connective
c. supportive connective
d. cardiac muscle

___9. The secretion of tears, milk, sweat, and oil are functions of _____ tissues.
a. epithelial
b. loose connective
c. lymphoid
d. nervous

___10. Memory, decision making, and issuing commands to effectors are functions of _____ tissue.
a. connective
b. epithelial
c. muscle
d. nervous

___11. An animal that feels heated from the sun moves to an environment that tends to cool its body. This is an example of _____.
a. intensifying an original condition
b. positive feedback mechanism
c. positive phototropic response
d. negative feedback mechanism

Chapter Objectives/Review Questions

This section lists general and specific chapter objectives that can be used as review questions. You can make maximum use of these items by writing answers on a separate sheet of paper. Fill in answers where blanks are provided. To check for accuracy, compare your answers with information given in the chapter or glossary.

Page   *Objectives/Questions*

(373)   1. Cells are the basic units of life; in a multicellular animal, like cells are grouped into a _____.

(374–378)   2. Know the characteristics of the various types of tissues. Know the types of cells that compose each tissue type and be able to cite some examples of organs that contain significant amounts of each tissue type.

(374)   3. _____ tissues cover the body surface of all animals and line internal organs from gut cavities to vertebrate lungs; this tissue always has one _____ surface; the opposite surface adheres to a _____ _____.

(374–375)   4. Explain the nature of three different cell-to-cell junctions and state the types of tissues in which these junctions occur.

(374–375)   5. List the functions carried out by epithelial tissue and state the general location of each type.

(375)   6. Explain the meaning of the term *gland*, cite three examples of glands, and state the extracellular products secreted by each.

(376)   7. Describe the basic features of connective tissue, and explain how they enable connective tissue to carry out its various tasks.

(376)   8. Connective tissue cells and fibers are surrounded by a _____ _____.

(377)   9. List three functions of blood.

(377)   10. Distinguish among skeletal, cardiac, and smooth muscle tissues in terms of location, structure, and function.

(377)   11. Muscle tissues contain specialized cells that can _____.

(378)   12. Neurons are organized as lines of _____.

(379)   13. Explain how, if each cell can perform all its basic activities, organ systems contribute to cell survival.

(380–381) 14. Describe the relationships among receptors, integrators, and effectors in a negative feedback system; draw a diagram to illustrate the concept.

Interpreting and Applying Key Concepts

Explain why, of all places in the body, marrow is located on the interior of long bones. Explain why your bones are remodeled after you reach maturity. Why does your body not keep the same mature skeleton throughout life?

Critical Thinking Exercise

Suppose a person consumes the same amount of glucose every day and maintains a constant blood glucose concentration. If that person doubles the daily glucose intake, several predictions might be made about the effect on blood glucose concentration, including the following:

a. Blood glucose would constantly rise.
b. Blood glucose would rise to a new constant level about twice the former level.
c. Blood glucose would rise slightly to a new constant level.
d. Blood glucose would rise slightly, then return to the former level.
e. Blood glucose would be unchanged.

For each prediction given above, state whether it assumes that blood glucose is part of a negative feedback loop, a positive feedback loop, or is not subject to control by a feedback mechanism.

Answers

Answers to Interactive Exercises

ANIMAL STRUCTURE AND FUNCTION: AN OVERVIEW (23-I)

1. F, Anatomy; 2. F, unlike tissues; 3. T; 4. T; 5. F, impermeable; 6. F, Exocrine; 7. F, Exocrine; 8. F, groups of; 9. T; 10. F, synchronously with; 11. T; 12. T; 13. T; 14. tissue; 15. organ; 16. metabolic; 17. internal environment; 18. germ cells; 19. gametes; 20. meiosis; 21. somatic; 22. mitosis; 23. "primary" tissues; 24. mesoderm; 25. endoderm; 26. ectoderm; 27. organ; 28. organ system (urinary system, excretory system); 29. organism; 30. homeostasis; 31. tissue; 32. epithelial; 33. connective; 34. ground substance; 35. Loose; 36. Tendons (Ligaments); 37. ligaments (tendons); 38. connective, D, 9, 11; 39. epithelial, G, 1, 6, 12; 40. muscle, I, 8, (11); 41. muscle, J, 5, (11); 42. connective, E, 7, 11, (14); 43. gametes, 2; 44. connective, B, 10; 45. epithelial, H, 1, 6, 12; 46. connective, 1, 6, 15; 47. nervous, 3; 48. muscle, C, 13; 49. epithelial, F, 1, 6, 12; 50. connective, A, 4, 14.

MAJOR ORGAN SYSTEMS (23-II)

1. circulatory; 2. respiratory; 3. urinary (= excretory); 4. skeletal; 5. endocrine; 6. immune; 7. reproductive; 8. digestive; 9. muscular; 10. nervous; 11. integumentary; 12. G; 13. I; 14. H; 15. B; 16. F; 17. E; 18. C; 19. J; 20. A; 21. D; 22. K; 23. thoracic; 24. cranial; 25. pelvic; 26. fluid; 27. chemical/hormonal; 28. nervous.

HOMEOSTASIS AND SYSTEMS CONTROL (23-III)

1. F, positive; 2. T; 3. F, an effector; 4. F, *Plasma* is the fluid portion of blood located *in* blood vessels. *Interstitial fluid* is located outside of blood vessels, lymphatic vessels and body cells, but within the body proper; 5. positive; 6. negative; 7. integrator; 8. glands; 9. receptors; 10. hypothalamus; 11. effectors; 12. sweat.

Answers to Self-Quiz

1. d; 2. b; 3. c; 4. c; 5. d; 6. c; 7. c; 8. d; 9. a; 10. d; 11. d.

Critical Thinking Analysis

a. Although the prediction is of a constant rise, it does not assume a positive feedback mechanism. If blood glucose concentration is constant, the rate of removal of glucose is equal to the rate of consumption of glucose. If consumption increases and removal mechanisms are not adjusted, blood glucose will increase at a constant rate and will not reach a steady state. This prediction assumes there is no feedback control of blood glucose that can adjust removal mechanisms in response to changes in intake.

b. This prediction does assume that the removal mechanisms increase but does not assume a homeostatic feedback loop. Removal of glucose is accomplished by enzymes and membrane transport carriers. As the intake of glucose increases, the blood concentration increases, and the enzymes and carriers work at higher rate because their substrate concentration is increased. The system establishes a new, higher steady state without any receptors, integrators, or information transmission.

c. This prediction assumes a negative feedback control mechanism. A slight increase in blood glucose triggers an increase in the removal mechanisms that balances the increased rate of intake and maintains a constant value. The response is maintained by a constant, slightly higher value of the variable. In effect, the set point has been slightly raised.

d. This prediction also assumes a negative feedback control mechanism. A slight increase in blood glucose triggers the response, but enough extra enzymes are produced not only to balance the increased glucose intake but also to reduce the blood glucose level to the original set point. The enzyme level must then be reduced to the balance level. This is a slightly more elaborate version of (c).

e. This prediction assumes not a negative feedback system but a system that can anticipate a stimulus and prepare for it. A negative feedback system is activated only after the control variable changes, and the system then reverses the change. Here, no change at all is predicted. This can happen only if another receptor, perhaps for glucose concentration in the stomach or the mouth, activates glucose removal mechanisms before the glucose is absorbed into the blood. Unless this kind of mechanism is precisely timed, it would result in decreases in blood glucose below the set point.

Authors' note: Some of the Critical Thinking sections that follow include descriptions of real experimental procedures that manipulate organisms in various ways to test hypotheses and isolate variables. These graphic descriptions are included to inform students about part of the "real world" of science and, at the same time, lead them to question if and when ethics and morals should be applied to the treatment of organisms used in experimentation. Do any of the other organisms on Earth have the right not to be surgically altered, not to be infected with deadly viruses, not to be dosed with carcinogenic or mutagenic chemicals? Is any experimental procedure in the pursuit of knowledge that might help humans legitimate? We hope you will discuss these issues.

24

PROTECTION, SUPPORT, AND MOVEMENT

Interactive Exercises

INTEGUMENTARY SYSTEM (24-I, pp. 384–385)

Terms

The page-referenced terms are important; they were in boldface type in the chapter. Refer to the instructions given in Chapter 1, p. 1 of this workbook.

integument (384) _____

epidermis (385) _____

dermis (385) _____

keratin (385) _____

melanin (385) _____

Fill-in-the-Blanks

Human skin is an organ system that consists of two layers: the outermost (1)_____ , which contains mostly dead squamous epithelial cells, and the (2)_____ , which contains hair follicles, nerves, tiny muscles associated with the hairs, and various types of glands embedded in dense connective tissue. The (3)_____ layer, with its loose connective tissue and store of fat in (4)_____ tissue, lies beneath the skin.

Labeling

Label the numbered parts of the illustration at the right.

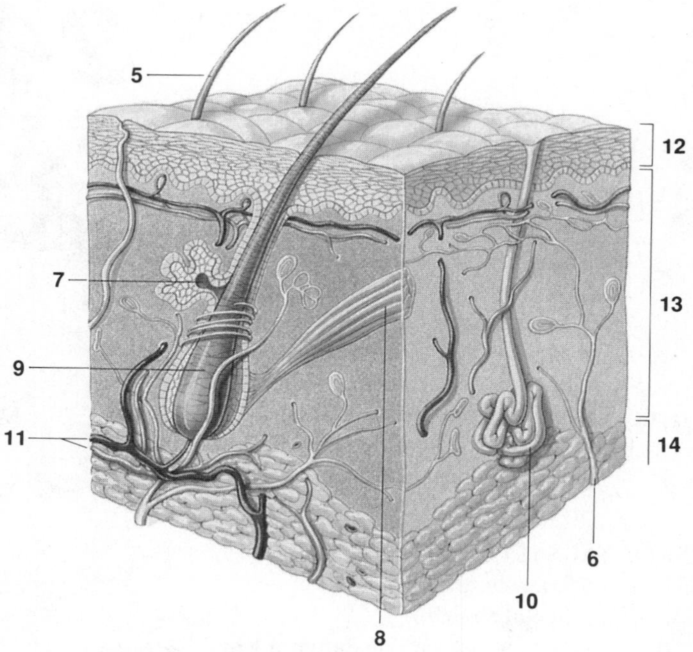

5. _____

6. _____ _____ _____

7. _____ _____

8. _____ _____

9. _____ _____

10. _____ _____

11. _____ _____

12. _____

13. _____

14. _____

Choice

Match the following proteins (or protein derivatives) with the particular ability that each substance lends to the skin. For questions 15–21, choose from these letters:

(a) collagen   (b) elastin   (c) keratin   (d) melanin   (e) hemoglobin

___ 15. Fibers that run through the dermis and lend a flexible, but substantive structure to it

___ 16. Protects against loss of moisture

___ 17. Protects against ultraviolet radiation and sunburn

___ 18. Helps skin to be stretchable, yet return to its previous shape

___ 19. Beaks, hooves, hair, fingernails and claws contain a lot of this

___ 20. Helps ward off bacterial attack by making the skin surface rather impermeable

___ 21. Located in red blood cells; binds with O_2

SKELETAL SYSTEM (24-II, pp. 386–389)

Terms

bones (386) _____

red marrow (386) _____

yellow marrow (386) _____

tendon (389) _____

ligament (389) _____

joint (389) _____

Fill-in-the-Blanks

Bones support and anchor (1)_____ and soft organs, such as eyes. (2)_____ systems are found in the long bones of mammals and contain living bone cells that receive their nutrients from the blood. (3)_____ _____ is a major site of blood cell formation. Bone tissue serves as a "bank" for (4)_____ , and other mineral ions; depending on metabolic needs, the body deposits ions into and withdraws ions from this "bank."

Bones develop from (5)_____ secreting material inside the model's shaft and on the surface of the cartilage model. Bone can also give ions back to interstitial fluid as osteoclasts dissolve out component minerals and remodel bone in response to (6)_____ _____ _____ and calcium deficiencies in the diet. Extreme decreases in bone density result in (7)_____ , particularly among older women.

The (8)_____ skeleton includes the skull, vertebral column, ribs, and breastbone; the (9)_____ skeleton includes the pectoral and pelvic girdles and the forelimbs and hindlimbs, when they exist in vertebrates.

Some joints are (10) _____ _____ and are lubricated by a fluid secreted into the capsule of dense connective tissue that surrounds the bones of the joint. Bones are often tipped with (11)_____ ; as a person ages, the cartilage at (12)_____ _____ joints may simply wear away, a condition called (13)_____ . By contrast, in (14)_____ _____ , the synovial membrane becomes inflamed, cartilage degenerates, and bone becomes deposited in the joint.

Labeling

Identify each indicated part of the accompanying illustrations.

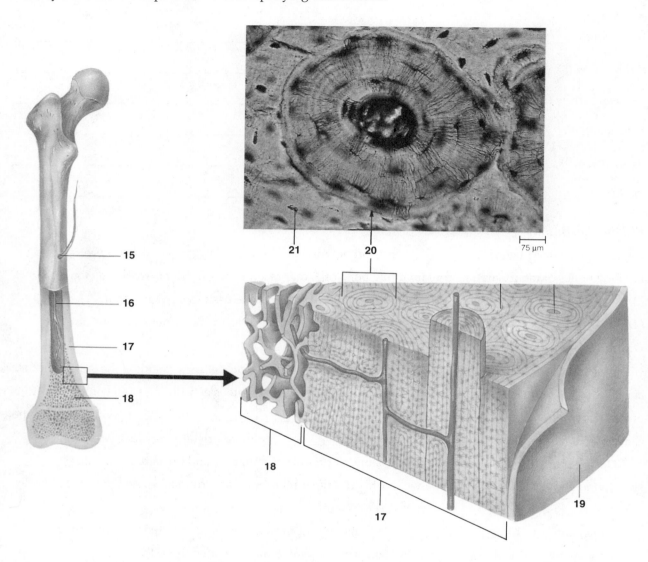

15. _____ _____

16. _____ _____

17. _____ _____

18. _____ _____

19. _____ _____ _____

20. _____ _____

21. _____

22. _____ _____
23. _____
24. _____
25. _____
26. _____
27. _____ _____
28. _____
29. _____
30. _____ _____
31. _____ _____

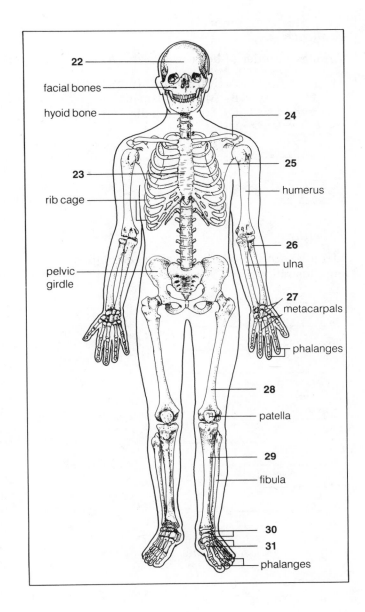

MUSCULAR SYSTEM (24-III, pp. 390–396)

Terms

skeletal muscle (391) _____

sarcomere (392) _____

myofibrils (392) _____

actin (392) _____

myosin (392) _____

Sequence

Arrange in order of decreasing size.

1. _____
2. _____
3. _____
4. _____
5. _____
6. _____

A. Muscle fiber (muscle cell)
B. Myosin filament
C. Muscle bundle
D. Muscle
E. Myofibril
F. Actin filament

Labeling

Identify each indicated part of the accompanying illustration.

7. _____
8. _____ _____
9. _____ _____
10. _____ _____
11. _____ _____
12. _____ _____
13. _____ _____
14. _____ _____
15. _____ _____
16. _____ _____

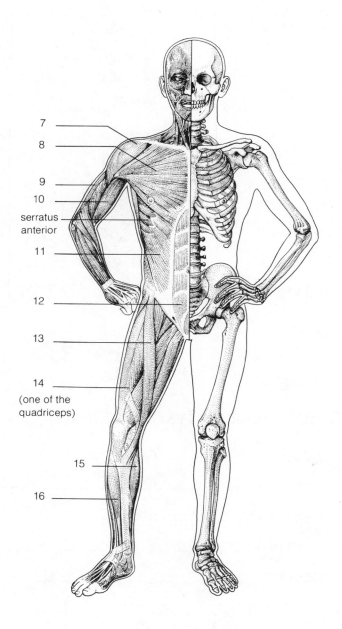

7
8
9
10
serratus
anterior
11
12
13
14
(one of the
quadriceps)
15
16

Label the numbered parts of the illustrations.

17._____ bundle

18._____ _____

19._____

20._____

21._____ _____

22._____ _____

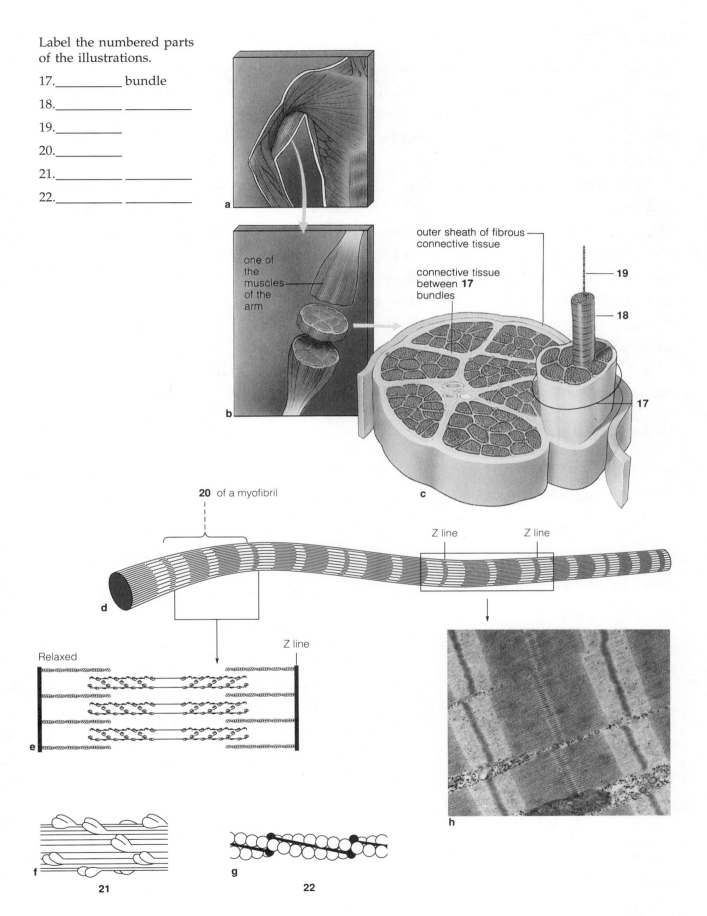

a

b

one of the muscles of the arm

outer sheath of fibrous connective tissue

connective tissue between **17** bundles

19

18

17

c

20 of a myofibril

Z line   Z line

d

h

Relaxed

Z line

e

f

21

g

22

Fill-in-the-Blanks

Together, the skeleton and its attached muscles are like a system of levers in which rigid rods,

(23)_____ , move about at fixed points, called (24)_____. Most attachments are close to joints, so a

muscle has to shorten only a small distance to produce a large movement of some body part. When the

(25)_____ contracts, the elbow joint bends (flexes). As it relaxes and as its partner, the (26)_____,

contracts, the forelimb extends and straightens.

There are three types of muscle tissue: (27)_____, which is striated, involuntary, and located in the

heart; (28)_____ , which is striated, largely voluntary, and generally attached to bones or cartilage; and

(29)_____ , which is involuntary, not striated, and mostly located in the wall of internal organs in ver-

tebrates. Of these three, (30)_____ muscle interacts with the skeleton to bring positional changes of

body parts and to move the animal through its environment.

The repetitive fundamental unit of muscle contraction in striated muscle cells is the (31)_____.

(32)_____ of an entire muscle is brought about by the combined decreases in length of the individ-

ual sarcomeres that make up the myofibrils of the muscle cells.

Each myofibril contains (33)_____ filaments, which have heads, and (34)_____ filaments,

which are thin and lack heads.

THE MUSCULAR SYSTEM (cont.) (pp. 393–396)

Terms

sliding-filament model (393) _____

cross-bridge formation (393) _____

tension (394) _____

motor neuron (394) _____

action potential (394) _____

sarcoplasmic reticulum (394) _____

muscle twitch (395) _____

tetanus (395) _____

muscle fatigue (395) _____

Fill-in-the-Blanks

Like a (35) _____ , a muscle cell is "excited" when a wave of electrical disturbance, a (an)
(36) _____ _____ , travels along its cell membrane. Neurons that send signals to muscles are
(37) _____ neurons. When excited by incoming signals from motor neurons, muscle cell membranes
cause (38) _____ ions to be released from the (39) _____ _____. These ions remove all obstacles that might interfere with myosin heads binding to the sites along (40) _____ filaments. According
to the (41) _____ - _____ model, (42) _____ filaments physically slide along actin filaments
and pull them toward the center of a (43) _____ during a contraction. The energy that drives the forming and breaking of the cross-bridges comes immediately from (44) _____ , which obtained its phosphate group from (45) _____ _____ , which is stored in muscle cells. When muscle cells
relax/rest, calcium ions are sent back to the sarcoplasmic reticulum by (46) _____ _____.

By controlling the (47)_____ _____ that reach the (48)_____ _____ in the first
place, the nervous system controls muscle contraction by controlling calcium ion levels in muscle tissue.

When supplies of available creatine phosphate run out, the muscle cells take (49)_____ from the
bloodstream and from the breakdown of glycogen stored in the muscle cells and send it through lactate
fermentation if the demand for muscle action is intense but brief. If the demand for muscle action is moderate, it can also be prolonged by muscle cells sending (50) _____ through the final stage of the aerobic
respiration pathway.

Labeling

Identify the numbered parts of the accompanying illustrations.

51. _____ _____

52. _____ _____

53. _____ _____

54. _____ _____

55. _____

56. _____ _____

57. _____ _____

58. _____ (_____)

59. _____ _____ _____

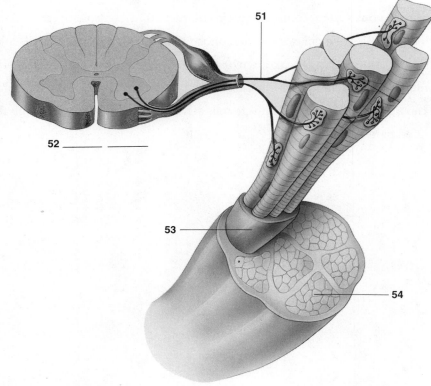

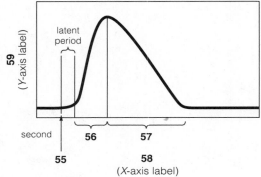

Complete the Table

For each item in the table below, state its specific role in muscle contraction.

aerobic respiration	60.
ATP	61.
calcium ions	62.
creatine phosphate	63.
glycogen	64.
lactate fermentation	65.
motor neuron	66.
myosin heads	67.
sarcomere	68.
sarcoplasmic reticulum	69.

Fill-in-the-Blanks

The larger the (70)_____ of a muscle, the greater its strength. A (71)_____ neuron sends signals to a group of muscle cells under its control. A (72)_____ _____ is a response in which a muscle contracts briefly when reacting to a single, brief stimulus and then relaxes. The strength of the muscular contraction depends on how far the (73)_____ response has proceeded by the time another signal arrives. (74)_____ is a large contraction in which a motor unit is being stimulated repeatedly and not allowed to relax. (75)_____ _____ are synthetic hormones that mimic the effects of testosterone in building greater (76)_____ mass in both men and women. It is illegal for competitive athletes to use them because of unfair advantage and because of the side effects. In men, (77)_____ , baldness, shrinking (78)_____ , and infertility are the first signs of damage. Aside from these physical side effects, some men experience uncontrollable (79)_____ , delusions, and wildly manic behavior.

Self-Quiz

For questions 1–7, choose from the following answers:

    a. bone
    b. cartilage
    c. epidermis
    d. dermis
    e. hypodermis

___ 1. Fat cells in adipose tissue are most likely to be located in this.

___ 2. Keratinized squamous cells are most likely to be located in this.

___ 3. Melanin in melanocytes is most likely to be here.

___ 4. Smooth muscles attached to hairs are probably here.

___ 5. This makes up the original "model" of the skeletal framework.

___ 6. Tips of sensory receptors are most likely here.

___ 7. This serves as a "bank" for withdrawing and depositing calcium and phosphate ions.

For questions 8–11, choose from the following answers:

    a. ligaments
    b. osteoblasts
    c. osteoclasts
    d. red marrow
    e. tendons

___ 8. _____ secrete bone-dissolving enzymes.

___ 9. Major site of blood cell formation.

___ 10. _____ remove Ca^{++} and $PO_4^{\equiv}$ ions from blood and build bone.

___ 11. _____ attach muscles to bone.

For questions 12–16, choose from the following answers:

    a. an action potential
    b. cross-bridge formation
    c. the sliding-filament model
    d. tension
    e. tetanus

___ 12. _____ is a mechanical force that causes muscle cells to shorten if it is not exceeded by opposing forces.

___ 13. A wave of electrical disturbance that moves along a neuron or muscle cell in response to a threshold stimulus

___ 14. _____ is a large contraction caused by repeated stimulation of motor units that are not allowed to relax.

___ 15. _____ is assisted by calcium ions and ATP.

___ 16. _____ explains how myosin filaments move to the centers of sarcomeres and back.

For questions 17–20, choose from the following answers:

    a. actin
    b. myofibril
    c. myosin
    d. sarcomere
    e. sarcoplasmic reticulum

___ 17. A(n) _____ contains many repetitive units of muscle contraction.

___ 18. The repetitive unit of muscle contraction.

___ 19. Thin filaments that depend upon calcium ions to clear their binding sites so that they can attach to parts of thick filaments.

___ 20. _____ stores calcium ions and releases them in response to an action potential.

Chapter Objectives/Review Questions

This section lists general and detailed chapter objectives that can be used as review questions. You can make maximum use of these items by writing answers on a separate sheet of paper. Fill in answers where blanks are provided. To check for accuracy, compare your answers with information given in the chapter or glossary.

Page   *Objectives/Questions*

(384)   1. Name the three functions of human skin.

(385)   2. Describe the two-layered structure of human skin and identify the items located in each layer.

(388)   3. Identify human bones by name and location.

(390)   4. Refer to Figure 24.9 of your main text and indicate (a) a muscle used in sit-ups, (b) another used in flexing the foot, and (c) another used in flexing the elbow joint.

(391)   5. Explain in detail the structure of muscles, from the molecular level to the organ systems level. Then explain how biochemical events occur in muscle contractions and how opposing muscle action refines movements.

(392)   6. Describe the fine structure of a muscle fiber; use terms such as *myofibril, sarcomere, motor unit, actin,* and *myosin.*

(393–394)   7. List, in sequence, the biochemical and fine structural events that occur during the contraction of a skeletal muscle fiber and explain how the fiber relaxes.

(395)   8. Distinguish twitch contractions from tetanic contractions.

Integrating and Applying Key Concepts

If humans had an exoskeleton rather than an endoskeleton, would they move differently from the way they do now? Name any advantages or disadvantages that having an exoskeleton instead of an endoskeleton would present in human locomotion.

Critical Thinking Exercise

A muscle can be clamped in an apparatus so that its sarcomeres are all limited to a certain minimum length. When the muscle is stimulated to contract, it generates tension, which can be measured, but it becomes no shorter than the apparatus allows. The graph below shows the typical behavior of a muscle set at various lengths. Which of the following statements is the best explanation of these data?

     a. There is an optimum length for muscle function.

     b. ATP is synthesized most rapidly at a certain muscle length.

     c. More calcium is released at a certain muscle length.

     d. More myosin heads can engage the actin filaments at a certain muscle length.

     e. The amount of force the arm can exert depends on the angle at which the elbow is bent.

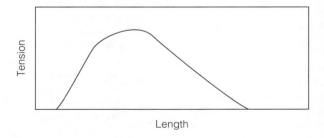

Note: Please see authors' note at end of Chapter 23 Critical Thinking Exercise.

Answers

Answers to Interactive Exercises

INTEGUMENTARY SYSTEM (24-I)
1. epidermis; 2. dermis; 3. hypodermis; 4. adipose;
5. hair; 6. sensory nerve ending; 7. sebaceous gland;
8. smooth muscle; 9. hair follicle; 10. sweat gland;
11. blood vessel; 12. epidermis; 13. dermis; 14. hypodermis; 15. a; 16. c; 17. d; 18. b; 19. c; 20. c; 21. e.

SKELETAL SYSTEMS/HUMAN SKELETAL SYSTEM (24-II)
1. muscles; 2. Haversian; 3. Red marrow; 4. calcium
(phosphate); 5. osteoblasts; 6. lack of exercise; 7. osteoporosis; 8. axial; 9. appendicular; 10. freely movable;
11. cartilage; 12. freely movable; 13. osteoarthritis;
14. rheumatoid arthritis; 15. nutrient canal; 16. yellow
marrow; 17. compact bone; 18. spongy bone; 19. connective tissue covering; 20. Haversian system; 21. osteocyte
(bone cell); 22. cranial bones; 23. sternum; 24. clavicle;
25. scapula; 26. radius; 27. carpal bones; 28. femur;
29. tibia; 30. tarsal bones; 31. metatarsal bones.

MUSCULAR SYSTEM (24-III)
1. D; 2. C; 3. A; 4. E; 5. B; 6. F; 7. deltoid; 8. pectoralis
major; 9. triceps brachii; 10. biceps brachii; 11. external
oblique; 12. rectus abdominis; 13. adductor longus; 14.
quadriceps femoris; 15. gastrocnemius; 16. tibalis anterior; 17. muscle; 18. muscle cell (muscle fiber); 19. myofibril; 20. sarcomere; 21. myosin filament; 22. actin
filament; 23. bones; 24. joints; 25. biceps brachii; 26. triceps brachii; 27. cardiac; 28. skeletal; 29. smooth;
30. skeletal; 31. sarcomere; 32. Contraction; 33. myosin;
34. actin; 35. neuron; 36. action potential; 37. motor;
38. calcium; 39. sarcoplasmic reticulum; 40. actin;
41. sliding-filament; 42. myosin; 43. sarcomere; 44. ATP;
45. creatine phosphate; 46. active transport; 47. action
potentials; 48. sarcoplasmic reticulum; 49. glucose;
50. glucose; 51. motor neuron; 52. spinal cord; 53. muscle
bundle; 54. muscle cell; 55. time that stimulus is applied;
56. contraction phase; 57. relaxation phase; 58. time (seconds); 59. strength of contraction; 60. Its oxygen-requiring reactions provide most of the ATP needed for muscle
contraction during prolonged, moderate exercise; 61. It
provides the energy to make myosin filaments slide
along actin filaments; 62. These are used to clear the actin
binding sites of any obstacles to cross-bridge formation
with myosin heads; 63. This supplies phosphate to ADP
→ ATP, which powers muscle contraction for a short
time because creatine phosphate supplies are limited;
64. Stored in muscles and in the liver, glucose is stored
by the animal body in this form of starch; 65. An anaerobic pathway in which glucose is broken down to lactate
with a small yield of ATP; this pathway operates during
intense exercise; 66. This supplies commands (signals) to
muscle cells to contract and relax; 67. Projections from
the thick filaments of myosin that bind to actin sites and
form temporary cross-bridges. Making and breaking
these cross-bridges cause myosin filaments to be pulled
to the center of a sarcomere; 68. The repetitive unit of
muscle contraction. Many sarcomeres constitute a
myofibril. Many myofibrils constitute a muscle cell;
69. This is the endoplasmic reticulum of a muscle cell. It
stores calcium ions and releases them in response to
incoming signals from motor neurons. It uses active
transport to bring the calcium ions back inside; 70. diameter; 71. motor; 72. muscle twitch; 73. twitch; 74. Tetanus;
75. Anabolic steroids; 76. muscle; 77. acne; 78. testes;
79. aggression.

Answers to Self-Quiz

1. e; 2. c; 3. c; 4. d; 5. b; 6. d; 7. a; 8. c; 9. d; 10. b; 11. e;
12. d; 13. a; 14. e; 15. b; 16. c; 17. b; 18. d; 19. a; 20. e.

Critical Thinking Analysis

a. This is a statement of the observations, a description
of the graph, not an explanation.

b. This statement assumes that the availability of ATP
limits the force that can be generated by a muscle.
This can be true when the contraction is prolonged,
but in a single twitch the number of ATP molecules
available is much greater than the number of
myosin heads that are activated. Furthermore, it is
very difficult to imagine a mechanism that would
couple the rate of ATP synthesis to the length of the
sarcomere.

c. Tension does seem to be proportional to calcium
concentration, but how changing cell length could
alter release of calcium from the sarcoplasmic reticulum is not at all clear. You might imagine membrane
permeability to calcium increasing as the sarcoplasmic reticulum is stretched, but this does not account
for the fact that calcium release is still dependent on
action potentials or for the drop in tension at high
sarcomere lengths.

d. The more individual forces applied to a rod, the
greater the total force. Thus, the more myosin heads
activated, the greater the tension generated. At very
short sarcomere lengths, the ends of the thin filaments extend over the central part of the thick filament where there are no myosin heads; hence, those
actin molecules do not function to generate more
tension. Furthermore, the ends of the actin filaments
begin to push against each other and on structures
in the center of the thick filaments, reducing the
total tension generated. Likewise, the ends of the
thick filaments can begin to push against the ends of
the sarcomere. At intermediate lengths, there is
maximum overlap between the thick and thin filaments, and the greatest number of myosin heads is

activated. As the length is increased, the overlap between the filaments decreases, and fewer active myosin heads generate less tension. At some point, the thick and thin filaments are pulled completely apart, and no tension can be generated.

e. This statement is a consequence of the observation, not an explanation. The angle of the elbow determines the length of the muscles that move it. If there is a length of maximum tension, there must be an angle of maximum tension.

Reference: Gordon, A. M., Huxley, A. F., and Julian, F. S. (1966) *J. Phys*. London 184:170-92.

25

DIGESTION AND HUMAN NUTRITION

Interactive Exercises

FUNCTIONS OF DIGESTIVE SYSTEMS (25-I, p. 399)

Terms

The page-referenced terms are important; they were in boldface type in the chapter. Refer to the instructions given in Chapter 1, p. 1 of this workbook.

nutrition (398)_____

digestive system (399) _____

Fill-in-the-Blanks

A digestive system is some form of body cavity or tube in which food is reduced first to (1)_____ and then to small (2)_____. Digested nutrients are then (3)_____ into the internal environment, and distributed to the body's cells. (4) _____ is a large concept that encompasses processes by which food is ingested, digested, absorbed and later converted to the body's own (5) _____, lipids, proteins, and nucleic acids.

    A(n) (6)_____ digestive system has only one opening, two-way traffic, and a highly branched gut cavity that serves both digestive and (7)_____ functions. A(n) (8)_____ digestive system has a tube or cavity with regional specializations and a(n) (9)_____ at each end.

HUMAN DIGESTIVE SYSTEM (25-II, pp. 400–405)

Terms

gut (400) _____

mouth (400) _____

tooth (400) _____

salivary glands (400) _____

pharynx (400) _____

esophagus (400) _____

stomach (402) _____

small intestine (403) _____

pancreas (403) _____

liver (403) _____

bile (403) _____

gallbladder (403) _____

emulsification (403) _____

villus, villi (404) _____

microvillus, microvilli (404) _____

micelle formation (404) _____

colon (405) _____

appendix (405) _____

bulk (405) _____

Complete the Table

1. Complete the following table by naming the organs described.

Organ	Main Functions
a.	Mechanically breaks down food, mixes it with saliva
b.	Moisten food; start polysaccharide breakdown; buffer acidic foods in mouth
c.	Stores, mixes, dissolves food; kills many microorganisms; starts protein breakdown; empties in a controlled way
d.	Digests and absorbs most nutrients
e.	Produces enzymes that break down all major food molecules; produces buffers against hydrochloric acid from stomach
f.	Secretes bile for fat emulsification; secretes bicarbonate, which buffers hydrochloric acid from stomach
g.	Stores, concentrates bile from liver
h.	Stores, concentrates undigested matter by absorbing water and salts
i.	Controls elimination of undigested and unabsorbed residues

Fill-in-the-Blanks

Saliva contains an enzyme, (2)_____ _____, that hydrolyzes starch. Contractions force the larynx against a cartilaginous flap called the (3)_____, which closes off the trachea. The (4)_____ is a muscular tube that propels food to the stomach. Any alternating progression of contracting and relaxing muscle movements along the length of a tube is known as a (5)_____ wave. The (6)_____'s most important function is to regulate the rate at which food reaches the intestine. Most digestion and absorption of nutrients occurs in the (7)_____ _____. (8)_____ is an enzyme that works in the stomach. (9)_____ is made by the liver, is stored in the gallbladder, and works in the (10)_____ _____. (11)_____ is an example of an enzyme that is made by the pancreas but works in the small intestine.

Labeling

Identify each numbered structure in the accompanying illustration.

12. _____ _____

13. _____ _____

14. _____

15. _____

16. _____

17. _____ _____

18. _____ _____

19. _____

20. _____

21. _____

22. _____

True/False

If the statement is true, write a T in the blank. If the statement is false, make it correct by changing the underlined word(s) and writing the correct word(s) in the answer blank.

_____ 23. Amylase digests starch, lipase digests lipids, and <u>proteases</u> break peptide bonds.

_____ 24. ATP is the end product of <u>digestion</u>.

_____ 25. The appendix has no known <u>digestive</u> functions.

_____ 26. Water and <u>sodium ions</u> are absorbed into the bloodstream from the lumen of the large intestine.

_____ 27. Fatty acids and monoglycerides recombine into fats inside epithelial cells lining the <u>colon</u>.

Fill-in-the-Blanks

(28)_____ involves the muscular movement of the gut wall, but (29)_____ is the release into the lumen of enzyme fluids and other substances required to carry out digestive functions.

HUMAN NUTRITIONAL REQUIREMENTS (25-III, pp. 406–411)
NUTRITION AND METABOLISM (25-IV, pp. 412–413)

Terms

essential fatty acid (407) _____

essential amino acid (407) _____

vitamin (408) _____

mineral (408) _____

kilocalorie (410) _____

obesity (410) _____

Complete the Table

1. Complete the following table by determining how many kilocalories the people described should take in daily, given the stated exercise level, in order to *maintain* their weight. Consult pages 410–411 of the text.

Height	Age	Sex	Level of Physical Activity	Present Weight (lbs.)	Number of Kilocalories
a. 5'6"	25	Female	Moderately active	138	
b. 5'10"	18	Male	Very active	145	
c. 5'8"	53	Female	Not very active	143	

Fill-in-the-Blanks

People who are (2)_____ percent heavier than "ideal" are considered to be obese. (3)_____ _____ are the body's main sources of energy; they should make up (4)_____ to _____ percent of the human daily caloric intake. (5)_____ and cholesterol are components of animal cell membranes. Fat deposits are used primarily as (6)_____ _____, but they also cushion many organs and provide insulation. Lipids should constitute less than (7)_____ percent of the human diet. One tablespoon a day of polyunsaturated oil supplies all (8)_____ _____ _____ that the body cannot synthesize. (9)_____ are digested to twenty common amino acids, of which eight are (10)_____, cannot be synthesized, and must be supplied by the diet. Animal proteins such as (11)_____ and (12)_____ contain high amounts of essential amino acids (that is, they are complete). (13)_____ are organic substances needed in small amounts in order to build enzymes or help them catalyze metabolic reactions. (14)_____ are inorganic substances needed for a variety of uses.

Amino acid conversions in the liver form (15)_____, which is potentially toxic to cells; the liver immediately converts this substance to (16)_____, a much less toxic waste product that is expelled by the urinary system from the body.

Short Answer

To answer the following questions, consult Table 25.1 and Figure 6.10.

17. What is the pool of amino acids used for in the human body? _____

_____

18. Which breakdown products result from carbohydrate and fat digestion (Table 25.1)?

_____

19. Monosaccharides, free fatty acids, and monoglycerides all have three uses; identify them.

_____

Review Problem

20. You are a 19-year-old male, very sedentary (TV, sleep, and computers), 6′1″ tall, medium frame, and you weigh 195 lbs.

   a. Use page 410, Fig. 25.11 to calculate the number of calories required to sustain your desired weight. Use the average of that range. Are you underweight, overweight, or just right? _____
   b. How many calories are you allowed to ingest every day? _____
   c. Use the new, improved food pyramid (p. 406) to construct a one-day diet that would eventually allow you to reach that weight if you ate a similar diet every day.

Complete the Table

Place your choices in the table below.

How many servings from each group below are you allowed to have daily?	What, specifically, could you choose to eat?
complex carbohydrates	d.
fruits	e.
vegetables	f.
dairy group	g.
assorted proteins	h.
The "sin" group at the top	i.

Self-Quiz

___1. The process that moves nutrients into the blood or lymph is _____.
   a. ingestion
   b. absorption
   c. assimilation
   d. digestion
   e. none of the above

___2. The enzymatic digestion of proteins begins in the _____.
   a. mouth
   b. stomach
   c. liver
   d. pancreas
   e. small intestine

___3. The enzymatic digestion of starches begins in the _____.
   a. mouth
   b. stomach
   c. liver
   d. pancreas
   e. small intestine

___4. The greatest amount of absorption of digested nutrients occurs in the
   _____.
   a. stomach
   b. pancreas
   c. liver
   d. colon
   e. duodenum

___5. Glucose moves through the membranes of the small intestine mainly by _____.
   a. peristalsis
   b. osmosis
   c. diffusion
   d. active transport
   e. bulk flow

___6. Which of the following is not found in bile?
   a. lecithin
   b. salts
   c. digestive enzymes

   d. cholesterol
   e. pigments

___7. The average American consumes approximately _____ pounds of sugar per year.
   a. 25
   b. 50
   c. 75
   d. 100
   e. 125

___8. Of the following, _____ has (have) the highest net protein utilization.
   a. milk
   b. eggs
   c. fish
   d. meat
   e. bread

___9. Obesity is defined as being _____ percent above the ideal weight, which has been defined by insurance companies.
   a. 5
   b. 10
   c. 15
   d. 20
   e. 25

___10. A deficiency of vitamin _____ causes rickets in children and osteomalacia in adults.
   a. A
   b. B
   c. C
   d. D
   e. E

___11. The element needed by humans for blood clotting, nerve impulse transmission, and bone and tooth formation is _____.
   a. magnesium
   b. iron
   c. calcium
   d. iodine
   e. zinc

Chapter Objectives/Review Questions

This section lists general and detailed chapter objectives that can be used as review questions. You can make maximum use of these items by writing answers on a separate sheet of paper. To check for accuracy, compare your answers with information given in the chapter or glossary.

Page	Objectives/Questions
(399)	1. Distinguish between incomplete and complete digestive systems and tell which is characterized by (a) specialized regions and (b) two-way traffic.
(400)	2. Define and distinguish among motility, secretion, digestion, and absorption.
(400–405)	3. List all parts (in order) of the human digestive system through which food actually passes. Then list the auxiliary organs that contribute one or more substances to the digestive process.
(400–405)	4. Explain how, during digestion, food is mechanically broken down. Then explain how it is chemically broken down.
(403–404)	5. Describe how the digestion and absorption of fats differ from the digestion and absorption of carbohydrates and proteins.
(404–405)	6. List the items that leave the digestive system and enter the circulatory system during the process of absorption.
(403)	7. Tell which foods undergo digestion in each of the following parts of the human digestive system and state what the food is broken into: oral cavity, stomach, small intestine, large intestine.
(403)	8. List the enzyme(s) that act in (a) the oral cavity, (b) the stomach, and (c) the small intestine. Then tell where each enzyme was originally made.
(404–405)	9. Describe the cross-sectional structure of the small intestine and explain how its structure is related to its function.
(405)	10. State which processes occur in the colon (large intestine).
(406–409)	11. Compare the contributions of carbohydrates, proteins, and fats to human nutrition with the contributions of vitamins and minerals.
(406)	12. Reproduce from memory the Food Pyramid Diagram as revised in 1992. Identify each of the six components, list the numerical range of servings permitted from each group and also list some of the choices available.
(406–410)	13. Summarize the 1979 U.S. Surgeon General's report that presented ideas for promoting health by eating properly.
(406–410)	14. Summarize the daily nutritional requirements of a 25-year-old man who works at a desk job and exercises very little. State what he needs in energy, carbohydrates, proteins, and lipids and name at least six vitamins and six minerals that he needs to include in his diet every day.
(406)	15. Construct an ideal diet for yourself for one 24-hour period. Calculate the number of calories necessary to maintain your weight (see p. 410) and then use the Food Pyramid (p. 406) to choose exactly what to eat and how much.
(407–409)	16. Distinguish vitamins from minerals, and state what is meant by net protein utilization.
(409)	17. Name four minerals that are important in human nutrition and state the specific role of each.
(413)	18. List four functions of the liver.
(412–413)	19. Explain how the human body manages to meet the energy and nutritional needs of the various body parts even though the person may be feasting sometimes and fasting at other times.

Integrating and Applying Key Concepts

Suppose you could not eat solid food for two weeks and you had only water to drink. List in correct sequential order the measures your body would take to try to preserve your life. Mention the command signals that are given as one after another critical point is reached, and tell which parts of the body are the first and the last to make up for the deficit.

Critical Thinking Exercise

Tadpoles, which are herbivorous, have a much longer intestine than adult frogs, which are carnivorous. Which of the following predictions would be most likely from this observation?

a. Tadpoles require less energy than adult frogs.
b. Cellulose-digesting enzymes work more slowly than proteases.
c. The intestinal epithelium is more permeable in tadpoles than in adult frogs.
d. Tadpoles are smaller than frogs.
e. The caloric content of cellulose is greater than that of protein.

Answers

Answers to Interactive Exercises

FUNCTIONS OF DIGESTIVE SYSTEMS (25-I)
1. particles; 2. molecules; 3. absorbed; 4. Nutrition; 5. carbohydrates; 6. incomplete; 7. circulatory; 8. complete; 9. opening.

HUMAN DIGESTIVE SYSTEM (25-II)
1. a. Mouth; b. Salivary glands; c. Stomach; d. Small intestine; e. Pancreas; f. Liver; g. Gallbladder; h. Large intestine; i. Rectum; 2. salivary amylase; 3. epiglottis; 4. esophagus; 5. peristaltic; 6. stomach; 7. small intestine; 8. Pepsin; 9. Bile; 10. small intestine; 11. Amylase (Lipase, Trypsin, or Chymotrypsin); 12. salivary glands; 13. oral cavity; 14. liver; 15. stomach; 16. gallbladder; 17. small intestine; 18. large intestine; 19. anus; 20. pancreas; 21. esophagus; 22. pharynx; 23. T; 24. F, respiration; 25. T; 26. T; 27. F, small intestine; 28. Motility; 29. secretion.

HUMAN NUTRITIONAL REQUIREMENTS (25-III) NUTRITION AND METABOLISM (25–IV)
1. a. 2,070; b. 2,900; c. 1,230; 2. 25; 3. Complex carbohydrates; 4. 50 to 60; 5. Phospholipids; 6. energy reserves; 7. 30; 8. essential fatty acids; 9. Proteins; 10. essential; 11. milk (eggs); 12. eggs (milk); 13. Vitamins; 14. Minerals; 15. ammonia; 16. urea; 17. constructing hormones, nucleotides, proteins, and enzymes; 18. monosaccharides, free fatty acids, and glycerol; 19. The three uses are (a) to construct components of cells and storage forms (such as glycogen) and specialized derivatives such as steroids and acetylcholine; (b) to convert to amino acids as needed; and (c) to serve as a source of energy; 20. a. Consult Figure 25.11 to discover from the men's chart (6'1", medium frame) the 160–174 ideal weight range. Subtracting these values from 195 tells us this person is overweight by 21–35 pounds. b. Multiply 160 and 174 times 10 (see p. 410) to obtain 1600–1740 as the correct range of kilocalories. The above diet is a diet that *maintains* weight in the correct size range. The excess 21–35 lbs should be lost gradually by adopting an everyday exercise program that over many months will gradually eliminate the excess kilocalories that are stored mostly in the form of fat. c. The smallest range of serving sizes shown in Fig. 25.9 will help keep the total caloric intake to about 1,600 kcal:

- 6 servings - bread, cereal, rice, pasta
- 2 servings - fruits
- 3 servings - vegetables
- 2 servings - milk, yogurt, or cheese
- 2 servings - legumes, nuts, poultry, fish, or meats
- Scarcely any added fats and simple sugars

d.–i. Choose from Fig. 25.9.

Answers to Self-Quiz

1. b; 2. b; 3. a; 4. e; 5. d; 6. c; 7. e; 8. b; 9. e; 10. d; 11. c.

Critical Thinking Analysis

a. If tadpoles require less energy, then, assuming that the diets have about the same energy content per gram, the tadpole would have to digest less food. This would lead to the prediction that the tadpole gut would be shorter, not longer.

b. This would mean that vegetable food would have to be exposed to the appropriate enzymes for a longer time than the same amount of meat. Assuming that food moves through the gut at the same speed in both cases, the herbivore (the tadpole) would be expected to have a longer gut than the carnivore.

c. If the epithelium is more permeable, the necessary amount of digested food could be absorbed more quickly and thus would require a shorter gut. In addition, most of the monomers are absorbed by carrier-mediated processes, not by simple diffusion, so permeability is not a factor.

d. Having a longer gut would lead to the prediction of a larger body. However, there is no reason to assume a constant proportion of body size devoted to any single organ. A larger gut could be accommodated even in a smaller body.

e. If cellulose in plant cell walls had more calories per gram than meat, an herbivore would require fewer total grams of food and would be expected to have a smaller gut.

26

CIRCULATION

Interactive Exercises

CIRCULATORY SYSTEMS: AN OVERVIEW (26-I, pp. 416–417)

Terms

The page-referenced terms are important; they were in boldface type in the chapter. Refer to the instructions given in Chapter 1, p. 1 of this workbook.

circulatory system (416) _____

_____

interstitial fluid (416) _____

blood (416) _____

heart (416) _____

capillary beds (417) _____

lymphatic system (417) _____

_____

Fill-in-the-Blanks

Cells survive by taking in from their surroundings what they need, (1) _____, and giving back to their surroundings materials that they don't need, (2) _____. In most animals, substances move rapidly to and from living cells by way of a (3) _____ circulatory system. (4) _____, a fluid connective tissue within the (5) _____ and blood vessels, is the transport medium.

Most of the cells of animals are bathed in a(n) (6) _____ _____; blood is constantly delivering nutrients and removing wastes from that fluid. The (7) _____ generates the pressure that keeps blood flowing. Blood flows (8) [choose one] ( ) rapidly, ( ) slowly through large diameter vessels to and from the heart, but where the exchange of nutrients and wastes occurs, in the (9) _____ beds, the blood is divided up into vast numbers of smaller-diameter vessels with tremendous surface area that enables the exchange to occur by diffusion. An elaborate network of drainage vessels attracts excess interstitial fluid and reclaimable (10) _____ and returns them to the circulatory system. This network is part of the (11) _____ system, which also helps clean the blood of disease agents.

Nutrients are absorbed into the blood from the (12) _____ and (13) _____ systems. Carbon dioxide is given to the (14) _____ system for elimination, and excess water, solutes, and wastes are eliminated by the (15) _____ system.

Labeling

Label the numbered parts in the illustrations below.

16. _____

17. _____ _____

18. _____

Describe the kind of circulatory system in:

19. Creature A. _____

20. Creature B. _____

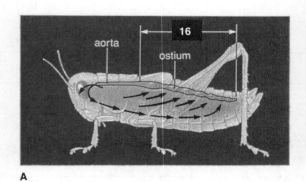

A

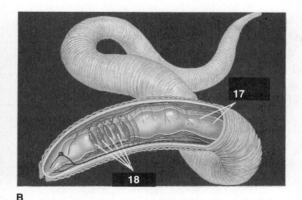

B

CHARACTERISTICS OF BLOOD (26-II, pp. 418–420)

Terms

plasma (418) _____

red blood cells (418) _____

white blood cells (418) _____

platelets (418) _____

stem cells (419) _____

cell count (419) _____

Complete the Table

1. Complete the following table, which describes the components of blood.

Components	Relative Amounts	Functions
Plasma Portion (50%–60% of total volume):		
Water	91%–92% of plasma volume	Solvent
a. (albumin, globulins, fibrinogen, etc.)	7%–8%	Defense, clotting, lipid transport, roles in extracellular fluid volume, etc.
Ions, sugars, lipids, amino acids, hormones, vitamins, dissolved gases	1%–2%	Roles in extracellular fluid volume, pH, etc.
Cellular Portion (40%–50% of total volume):		
b.	4,800,000–5,400,000 per microliter	O_2, CO_2 transport
White blood cells:		
c.	3,000–6,750	Phagocytosis
d.	1,000–2,700	Immunity
Monocytes (macrophages)	150–720	Phagocytosis
Eosinophils	100–360	Roles in inflammatory response, immunity
Basophils	25–90	Roles in inflammatory response, anticlotting
e.	250,000–300,000	Roles in clotting

Fill-in-the-Blanks

Blood is a highly specialized fluid (2)_____ tissue that helps stabilize internal (3)_____ and equalize internal temperature throughout an animal's body. Organisms with (4)_____ circulatory systems generally also have a supplementary (5)_____ _____ _____ that recovers and purifies interstitial fluid and returns it to the major blood vessels. Oxygen binds with the (6)_____ atom in a hemoglobin molecule. The red blood (7)_____ _____ in males is about 5.4 million cells per microliter of blood; in females, it is 4.8 million per microliter. The plasma portion constitutes approximately (8) _____ to _____ percent of the total blood volume. Erythrocytes are produced in the (9)_____ _____ _____. (10)_____ _____ are immature cells not yet fully differentiated. (11)_____ and monocytes are highly mobile and phagocytic; they chemically detect, ingest, and destroy bacteria, foreign matter, and dead cells. (12)_____ (thrombocytes) are cell fragments that aid in forming blood clots.

In humans, mature red blood cells lack their (13)_____, but they contain enough proteins to sustain them for about (14)_____ months. Platelets also have no (15)_____, but they last a maximum of (16)_____ days in the human bloodstream.

Label-Match

Identify the numbered cell types in the illustration below. Complete the exercise by matching and entering the letter of the appropriate function in the parentheses following the given cell types. A letter may be used more than once. Cell types 17 and 23 have no matching letters.

17. _____

18. _____ _____ ( )

19. _____ ( )

20. _____ ( )

21. _____ ( )

22. _____ ( )

23. _____

24. _____ ( )

A. Phagocytosis
B. Plays a role in the inflammatory response
C. Plays a role in clotting
D. Immunity
E. O_2, CO_2 transport

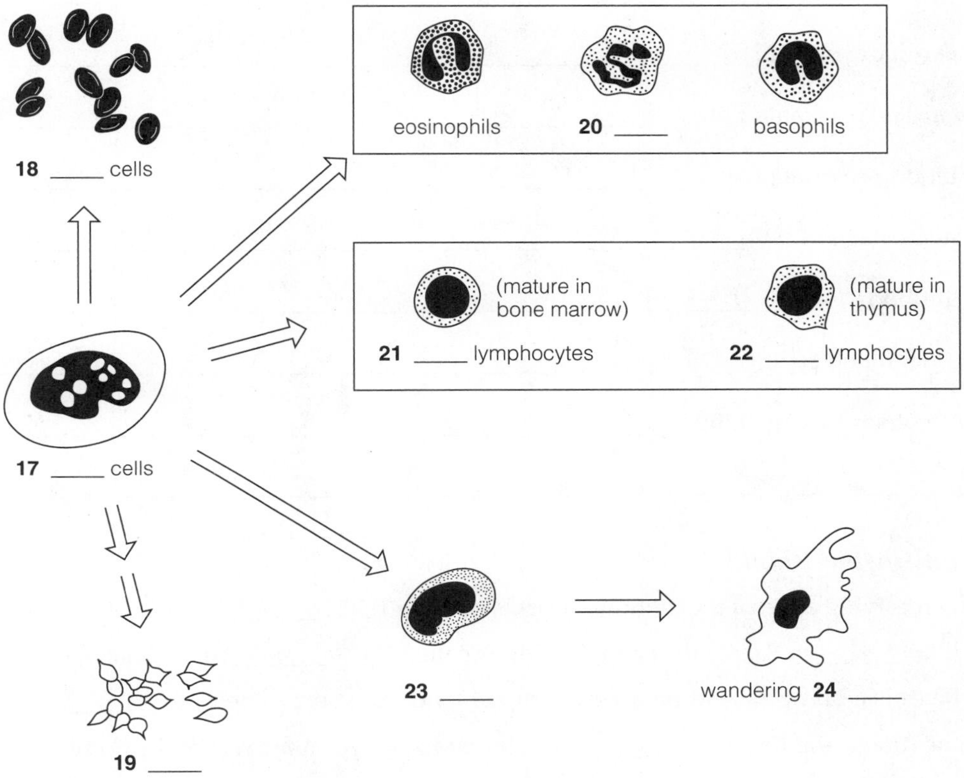

18 ____ cells

eosinophils   20 ____   basophils

(mature in bone marrow)   (mature in thymus)

21 ____ lymphocytes   22 ____ lymphocytes

17 ____ cells

19 ____

23 ____

wandering 24 ____

HUMAN CARDIOVASCULAR SYSTEM (26-III, pp. 421–429)

Terms

artery (421) _____

arteriole (421) _____

capillary (421) _____

venule (421) _____

vein (421) _____

pulmonary circuit (421) _____

_____

systemic circuit (421) _____

_____

atrium (422) _____

ventricle (422) _____

aorta (422) _____

cardiac cycle (422) _____

cardiac pacemaker (423) _____

blood pressure (424) _____

hemostasis (428) _____

antibody (428) _____

ABO blood typing (428) _____

Rh blood typing (429) _____

_____

Fill-in-the-Blanks

A receiving zone of a vertebrate heart is called a(n) (1)_____; a departure zone is called a(n)
(2)_____. Each contraction period is called (3)_____; each relaxation period is (4)_____. The
heart is a pumping station for two major blood transport routes: the (5)_____ circulation to and from
the lungs, and the (6)_____ circulation to and from the rest of the body.

In the pulmonary circuit, the heart pumps (7)_____-poor blood to the lungs; then the
(8)_____-enriched blood flows back to the (9)_____. The (10)_____ _____ is the cardiac
pacemaker. During a cardiac cycle, contraction of the (11)_____ is the driving force for blood circulation; (12)_____ contraction helps fill the ventricles.

A(n) (13)_____ carries blood away from the heart. A(n) (14)_____ is a blood vessel with such
a small diameter that red blood cells must flow through it single-file; its wall consists of no more than a
single layer of (15)_____ cells resting on a basement membrane. In each (16)_____ _____,
small molecules move between the bloodstream and the (17)_____ fluid. (18)_____ are in the
walls of veins and prevent backflow. Both (19)_____ and (20)_____ serve as temporary reservoirs
for blood volume.

(21)_____ are pressure reservoirs that keep blood flowing smoothly away from the heart while the (22)_____ are relaxing. (23)_____ are control points where adjustments can be made in the volume of blood flow to be delivered to different capillary beds. They offer great resistance to flow, so there is a major drop in (24)_____ in these tubes.

One cause of (25)_____ is the rupture of one or more blood vessels in the brain. A (26)_____ _____ blocks a coronary artery. (27)_____ is a term for a formation that can include cholesterol, calcium salts, and fibrous tissue. It is not healthful to have a high concentration of (28)_____-density lipoproteins in the bloodstream. A clot that stays in place is a (29) _____, but a clot that travels in the bloodstream is an embolus. Bleeding is stopped by several mechanisms that are referred to as (30)_____; the mechanisms include blood vessel spasm, (31)_____ _____ _____, and blood (32)_____. Once the platelets reach a damaged vessel, through chemical recognition they adhere to exposed (33)_____ fibers in damaged vessel walls. Reactions cause rod-shaped proteins to assemble into long (34)_____ fibers. These trap blood cells and components of plasma. Under normal conditions, a clot eventually forms at the damaged site.

If you are blood type (35)_____, you have no antibodies against A or B markers in your plasma. If you are type (36) _____, you have antibodies against A and B markers in your plasma. People who are (37) [choose one] RH$^+$/Rh$^-$ (38) [choose one] women/men have to be careful so that they don't develop erythroblastosis fetalis.

True/False

If the statement is true, write a T in the blank. If the statement is false, make it correct by changing the underlined word(s) and writing the correct word(s) in the answer blank.

_____ 39. The pulse <u>rate</u> is the difference between the systolic and the diastolic pressure readings.

_____ 40. Because the total volume of blood remains constant in the human body, blood pressure must <u>also remain constant</u> throughout the circuit.

Labeling

Identify each indicated part of the accompanying illustrations.

41. _____

42. _____ _____ _____

43. _____ _____

44. _____

45. _____ _____ _____

46. _____ _____

47. _____ _____ _____

48. _____ _____ _____

49. _____

50. _____

51. _____

52. _____

53. _____ _____ , _____ _____

54. _____

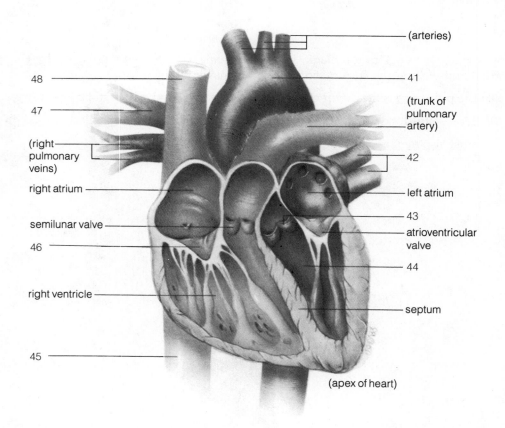

(arteries)

48

41

47

(trunk of
pulmonary
artery)

(right
pulmonary
veins)

42

right atrium

left atrium

semilunar valve

43

46

atrioventricular
valve

44

right ventricle

septum

45

(apex of heart)

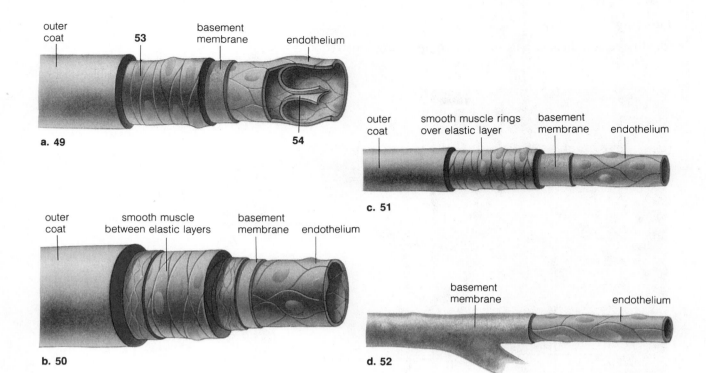

a. 49 — outer coat, **53**, basement membrane, endothelium, **54**

b. 50 — outer coat, smooth muscle between elastic layers, basement membrane, endothelium

c. 51 — outer coat, smooth muscle rings over elastic layer, basement membrane, endothelium

d. 52 — basement membrane, endothelium

LYMPHATIC SYSTEM (26-IV, pp. 430–431)

Terms

lymph (430) _____

lymph capillary (430) _____

lymph vessel (430) _____

lymph node (431) _____

spleen (431) _____

thymus (431) _____

Labeling

Identify each indicated part of the accompanying illustration.

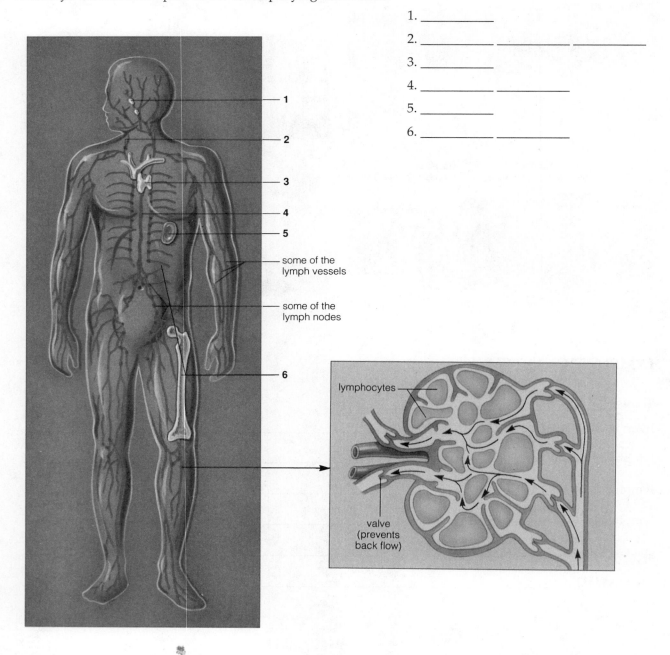

1. _____

2. _____ _____ _____

3. _____

4. _____ _____

5. _____

6. _____ _____

- 1
- 2
- 3
- 4
- 5

some of the lymph vessels

some of the lymph nodes

- 6

lymphocytes

valve (prevents back flow)

Fill-in-the-Blanks

(7)_____ vessels reclaim fluid lost from the bloodstream, purify the blood of microorganisms, and transport (8)_____ from the (9)_____ _____ to the bloodstream.

Self-Quiz

___1. Most of the oxygen in human blood is transported by _____.
   a. plasma
   b. serum
   c. platelets
   d. hemoglobin
   e. leukocytes

___2. Of all the different kinds of white blood cells, two classes of _____ are the ones that respond to *specific* invaders and confer *immunity* to a variety of disorders.
   a. basophils
   b. eosinophils
   c. monocytes
   d. neutrophils
   e. lymphocytes

___3. Open circulatory systems generally lack _____.
   a. a heart
   b. arterioles
   c. capillaries
   d. veins
   e. arteries

___4. Red blood cells originate in the _____.
   a. liver
   b. spleen
   c. yellow bone marrow
   d. thymus gland
   e. red bone marrow

___5. Hemoglobin contains _____.
   a. copper
   b. magnesium
   c. sodium
   d. calcium
   e. iron

___6. The pacemaker of the human heart is the _____.
   a. sinoatrial node
   b. semilunar valve

c. inferior vena cava
d. superior vena cava
e. atrioventricular node

___7. During systole, _____.
   a. oxygen-rich blood is pumped to the lungs
   b. the heart muscle tissues contract
   c. the atrioventricular valves suddenly open
   d. oxygen-poor blood from all parts of the human body, except the lungs, flows toward the right atrium
   e. none of the above

___8. _____ are reservoirs of blood pressure in which resistance to flow is low.
   a. Arteries
   b. Arterioles
   c. Capillaries
   d. Venules
   e. Veins

___9. Begin with a red blood cell located in the superior vena cava and travel with it in proper sequence as it goes through the following structures. Which will be *last* in sequence?
   a. aorta
   b. left atrium
   c. pulmonary artery
   d. right atrium
   e. right ventricle

___10. The lymphatic system is the principal avenue in the human body for transporting _____.
   a. fats
   b. wastes
   c. carbon dioxide
   d. amino acids
   e. interstitial fluids

Chapter Objectives/Review Questions

This section lists general and detailed chapter objectives that can be used as review questions. You can make maximum use of these items by writing answers on a separate sheet of paper. Fill in answers where blanks are provided. To check for accuracy, compare your answers with information given in the chapter or glossary.

Page   *Objectives/Questions*

(418–420)   1. Describe the composition of human blood, using percentages of volume.
(418–419)   2. State where erythrocytes, leukocytes, and platelets are produced.
(422–423)   3. Explain what causes a heart to beat. Then describe how the rate of heartbeat can be slowed down or speeded up.
(422–424)   4. List the factors that cause blood to leave the heart and the factors that cooperate to return blood to the heart.
(424–425)   5. Explain what causes high pressure and low pressure in the human circulatory system. Then show where major drops in blood pressure occur in humans.
(424–425)   6. Describe how the structures of arteries, capillaries, and veins differ.
(425)   7. Explain how veins and venules can act as reservoirs of blood volume.
(426–427)   8. Distinguish a stroke from a coronary artery blockage.
(426)   9. Describe how hypertension develops, how it is detected, and whether it can be corrected.
(427)   10. State the significance of high- and low-density lipoproteins to cardiovascular disorders.
(428–429) 11. Describe how blood is typed for the ABO blood group and for the Rh factor.
(430–431) 12. Describe the composition and function of the lymphatic system.

Integrating and Applying Key Concepts

You observe that some people appear as though fluid had accumulated in their lower legs and feet. Their lower extremities resemble those of elephants. You inquire about what is wrong and are told that the condition is caused by the bite of a mosquito that is active at night. Construct a testable hypothesis that would explain (1) why the fluid was not being returned to the torso, as normal, and (2) what the mosquito did to its victims.

Critical Thinking Exercise

1. In the mammalian circulatory system, blood passes through four cardiac chambers in a series. Which of the following would be the most likely measurements of oxygen concentration in each of the four chambers?

	Right Atrium	Left Atrium	Right Ventricle	Left Ventricle
a.	40	40	40	40
b.	100	100	40	40
c.	100	40	100	40
d.	40	100	40	100
e.	40	40	100	100

Answers

Answers to Interactive Exercises

CIRCULATORY SYSTEMS: AN OVERVIEW (26-I)
1. nutrients (food); 2. wastes; 3. closed; 4. Blood; 5. heart; 6. interstitial fluid; 7. heart; 8. rapidly; 9. capillary; 10. solutes; 11. lymphatic; 12. digestive (respiratory); 13. respiratory (digestive); 14. respiratory; 15. urinary; 16. heart(s); 17. blood vessels; 18. hearts; 19. open circulatory system; blood is pumped into short tubes that open into spaces in the body's tissues, mingles with tissue fluids, then is reclaimed by open-ended tubes that lead back to the heart; 20. closed circulatory system; blood flow is confined within blood vessels that have continuously connected walls and is pumped by 5 pairs of "hearts."

CHARACTERISTICS OF BLOOD (26-II)
1. a. Plasma proteins; b. Red blood cells; c. Neutrophils; d. Lymphocytes; e. Platelets; 2. connective; 3. pH; 4. closed; 5. lymph vascular system; 6. iron; 7. cell count; 8. 50 to 60; 9. red bone marrow; 10. Stem cells; 11. Neutrophils; 12. Platelets; 13. nucleus; 14. four; 15. nucleus; 16. nine; 17. stem; 18. red blood (E); 19. platelets (C); 20. neutrophils (A); 21. B (D); 22. T (D); 23. monocytes; 24. macrophages (A).

HUMAN CARDIOVASCULAR SYSTEM (26-III)
1. atrium; 2. ventricle; 3. systole; 4. diastole; 5. pulmonary; 6. systemic; 7. oxygen; 8. oxygen; 9. heart; 10. SA node; 11. ventricles; 12. atrial; 13. artery; 14. capillary; 15. endothelial; 16. capillary bed (diffusion zone); 17. interstitial; 18. Valves; 19. veins (venules); 20. venules (veins); 21. Arteries; 22. ventricles; 23. Arterioles; 24. pressure; 25. stroke; 26. coronary occlusion; 27. Plaque; 28. low; 29. thrombus; 30. hemostasis;

31. platelet plug formation; 32. coagulation; 33. collagen; 34. insoluble; 35. AB; 36. O; 37. -; 38. women; 39. F, pressure; 40. F, blood pressure cannot remain constant because it passes through various kinds of vessels that have varied structures; 41. aorta; 42. left pulmonary veins; 43. semilunar valve; 44. left ventricle; 45. inferior vena cava; 46. atrioventricular valve; 47. right pulmonary artery; 48. superior vena cava; 49. vein; 50. artery; 51. arteriole; 52. capillary; 53. smooth muscle, elastic fibers; 54. valve.

LYMPHATIC SYSTEM (26-IV)
1. tonsils; 2. right lymphatic duct; 3. thymus; 4. thoracic duct; 5. spleen; 6. bone marrow; 7. Lymph; 8. fats; 9. small intestine.

Answers to Self-Quiz
1. d; 2. e; 3. c; 4. e; 5. e; 6. a; 7. b; 8. a; 9. a; 10. a.

Critical Thinking Analysis

Blood flows through the right side of the heart and then to the lungs, where oxygen concentration rises. Then blood returns to the left side of the heart and flows through it to the systemic organs, which consume oxygen, lowering the oxygen concentration in the blood. Thus, blood flowing through the two chambers of one side of the heart has the same oxygen concentration in both chambers. Furthermore, the blood in the left side of the heart has high oxygen concentration, and the blood in the right side has low oxygen concentration. The only choice that meets these criteria is (d).

27

IMMUNITY

FIRST LINE OF DEFENSE—SURFACE BARRIERS TO INVASION

SECOND LINE OF DEFENSE—NONSPECIFIC RESPONSES
    First on the Front Lines
    Inflammation

THIRD LINE OF DEFENSE—THE IMMUNE SYSTEM
    Overview of Immune Responses
    Immune Functions of Macrophages
    Focus on Health: Cancer and Immunotherapy
    Helper T Cells and Cytotoxic T Cells

B Cells and Antibodies
Control of Immune Responses
Battlegrounds for the Immune System
The Basis of Specificity and Memory

IMMUNIZATION

ABNORMAL OR DEFICIENT IMMUNE RESPONSES
    Allergies
    Autoimmune Disorders
    Deficient Immune Responses
    Focus on Health: AIDS —The Immune System Compromised

Interactive Exercises

FIRST LINE OF DEFENSE—SURFACE BARRIERS TO INVASION/SECOND LINE OF DEFENSE—NONSPECIFIC RESPONSES (27-I, pp. 434–437)

Terms

The page-referenced terms are the important terms in the chapter. Refer to the instructions given in Chapter 1, p. 1 of this workbook.

pathogen (434) _____

lysozyme (434) _____

neutrophil (435) _____

eosinophil (435) _____

basophil (435) _____

macrophages (435) _____

complement system (436) _____

lysis (436) _____

acute inflammation (437) _____

interleukins (437) _____

fever (437) _____

Fill-in-the-Blanks

Several barriers prevent pathogens from crossing the boundaries of your body. Intact skin and (1)_____ membranes are effective barriers. (2)_____ is an enzyme that destroys the cell wall of many bacteria. (3)_____ fluid destroys many food-borne pathogens in the gut. Normal (4)_____ residents of the skin, gut, and vagina outcompete pathogens for resources and help keep their numbers under control.

When a sharp object cuts through the skin and foreign microbes enter, some plasma proteins come to the rescue and seal the wound with a (5) _____ mechanism. (6) _____ white blood cells engulf bacteria soon thereafter. Also, there are about twenty plasma proteins (collectively referred to as the (7)_____ _____) that are activated one after another in a "cascade" of reactions to help you destroy invading microorganisms. (8)_____ are Y-shaped proteins that lock onto specific foreign targets and thereby tag them for destruction by phagocytes or by activating the complement system.

The inflammatory response engages in battle both specific and nonspecific invaders. When the complement system is activated, circulating basophils and mast cells in tissues release (9)_____, which dilates (10)_____ and makes them "leaky," so fluid seeps out and causes the inflamed area to become swollen and warm.

Complete the Table

Complete the table by providing the specific functions carried out by each of the four different kinds of white blood cells listed below.

Basophils	11.
Eosinophils	12.
Neutrophils	13.
Macrophages	14.

THIRD LINE OF DEFENSE—THE IMMUNE SYSTEM (27-II, pp. 438–445)

Terms

B lymphocytes (B cells) (438) _____

_____

T lymphocytes (T cells) (438) _____

_____

antigens (438) _____

antibodies (438) _____

helper T cells (438) _____

_____

cytotoxic T cells (438) _____

MHC markers (438) _____

antigen-presenting cells (438) _____

_____

perforins (440) _____

natural killer cells, NK cells (441) _____

_____

plasma cells (442) _____

immunoglobulins (442) _____

_____

lymphoid organs (444) _____

memory cells (445) _____

Fill-in-the-Blanks

If the (1) _____ defenses (fast-acting white blood cells such as neutrophils, eosinophils, and basophils; slower-acting macrophages; and the plasma proteins involved in clotting and complement) fail to repel the microbial invaders, then the body calls on its (2) _____ _____, which identifies *specific* targets to kill and *remembers* the identities of its targets. Your own unique (3) _____ _____ patterns identify your cells as "self" cells. Any other surface pattern is, by definition, (4) _____, and doesn't belong in your body.

The principal actors of the immune system are (5) _____ descended from stem cells (consult Fig. 27.6) in the bone marrow, which have two different strategies of action to deal with their different kinds of enemies. (6) _____ _____ clones secrete antibodies and act principally against the extracellular (ones that stay outside the body cells) enemies that are pathogens in blood or on the cell surfaces of body tissues. (7) _____ _____ clones descend from lymphocytes that matured in the (8) _____ where they acquired specific markers on their cell surfaces; they defend principally against intracellular pathogens such as (9) _____, and against any cells ((10) _____ cells and grafts of foreign tissue) that are perceived as abnormal or foreign.

Labeling

Identify each numbered part in the accompanying illustration.

11._____-_____ _____   15._____ _____ _____

12._____ _____ _____ _____   16._____ _____ _____

13._____ _____ _____   17._____

14._____   18._____

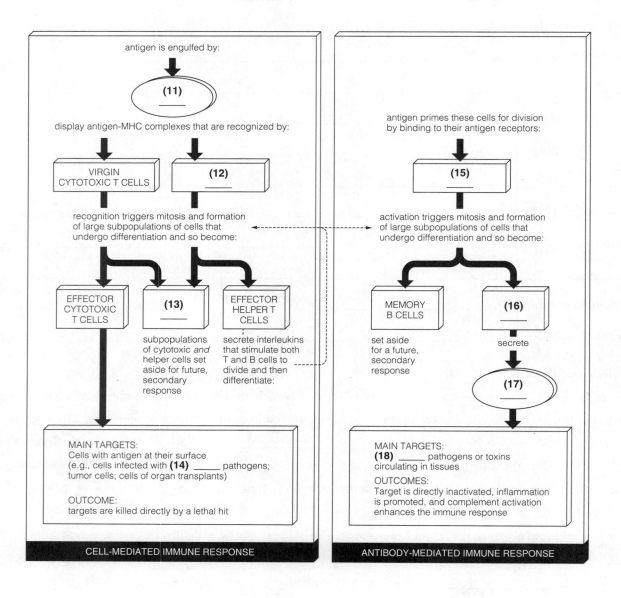

antigen is engulfed by:

(11)
____

display antigen-MHC complexes that are recognized by:

| VIRGIN CYTOTOXIC T CELLS | **(12)** ____ |

recognition triggers mitosis and formation of large subpopulations of cells that undergo differentiation and so become:

| EFFECTOR CYTOTOXIC T CELLS | **(13)** ____ | EFFECTOR HELPER T CELLS |

subpopulations of cytotoxic *and* helper cells set aside for future, secondary response

secrete interleukins that stimulate both T and B cells to divide and then differentiate:

MAIN TARGETS:
Cells with antigen at their surface (e.g., cells infected with **(14)** ____ pathogens; tumor cells; cells of organ transplants)

OUTCOME:
targets are killed directly by a lethal hit

CELL-MEDIATED IMMUNE RESPONSE

antigen primes these cells for division by binding to their antigen receptors:

(15)
____

activation triggers mitosis and formation of large subpopulations of cells that undergo differentiation and so become:

| MEMORY B CELLS | **(16)** ____ |

set aside for a future, secondary response

secrete

(17)
____

MAIN TARGETS:
(18) ____ pathogens or toxins circulating in tissues

OUTCOMES:
Target is directly inactivated, inflammation is promoted, and complement activation enhances the immune response

ANTIBODY-MEDIATED IMMUNE RESPONSE

Choice

Match each of the white blood cell types (a–e) with the appropriate function (19–23).

(a) effector cytotoxic T cells   (b) effector helper T cells
(c) macrophages   (d) memory cells   (e) effector B cells

___ 19. Lymphocytes that directly destroy body cells already infected by certain viruses or by parasitic fungi

___ 20. A portion of B and T cell populations that were set aside as a result of a first encounter, now circulate freely and respond rapidly to any later attacks by the same type of invader

___ 21. Lymphocytes and their progeny that produce antibodies

___ 22. Lymphocytes that serve as master switches of the immune system; stimulate the rapid division of B cells and cytotoxic T cells

___ 23. Nonlymphocytic white blood cells that develop from monocytes, engulf anything perceived as foreign, and alert helper T cells to the presence of specific foreign agents

Fill-in-the-Blanks

In addition to effector B cells and cytotoxic T cells (executioner lymphocytes that mature in the thymus), (24) _____ _____ cells mature in other lymphoid tissues and search out any cell that is either coated with complement proteins or antibodies, *or* bears any foreign molecular pattern. When cytotoxic T cells find foreign cells, they secrete (25) _____ and other toxic substances to poison and lyse the offenders.

Antibodies are plasma proteins that are part of the (26) _____ group of proteins. Some of these circulate in blood; others are present in other body fluids or bound to B cells.

The (27)_____ _____ _____ is the route taken during a first-time contact with an antigen; the secondary immune response is more rapid because patrolling battalions of (28) _____ _____ are in the bloodstream on the lookout for enemies they have conquered before. When these meet up with recognizable (29) _____, they divide at once, producing large clones of B or T cells within 2–3 days.

The (30)_____ _____ _____ explains how an individual has immunological memory, which is the basis of a secondary immune response; the theory also explains in part how self cells are distinguished from (31)_____ cells in the vertebrate immune response.

(32)_____ refers to cells that have lost control over cell division. Milstein and Kohler developed a means of producing large amounts of (33)_____ _____.

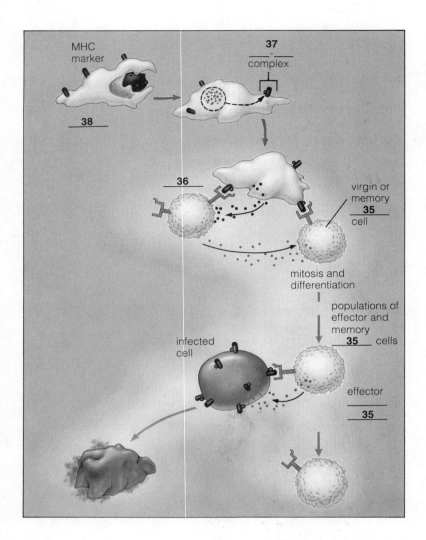

Example of an (34) _____-mediated immune response as carried out by activated (35) _____ _____ cells.

Labeling

Identify each numbered part in the accompanying illustrations.

34. _____

35. _____ _____

36. _____ _____

37. _____ - _____

38. _____

Note: If the same number is used more than once, that is because the label is the same in all such situations. Indeed, you may find numbered structures that are on this page on one or more of the pages that follow. If you have labeled it correctly on this page, then its identification should be correct on subsequent pages also.

Labeling

Identify each numbered part in the accompanying illustration.

36. _____ _____

37. _____ - _____

38. _____

39. _____

40. _____ _____

41. _____ _____

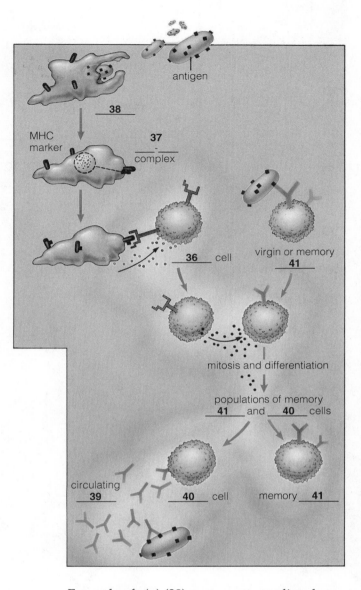

antigen

38

MHC marker

37 - complex

36 cell

virgin or memory 41

mitosis and differentiation

populations of memory 41 and 40 cells

circulating 39

40 cell

memory 41

Example of a(n) (39) _____-mediated immune response to a bacterial invasion.

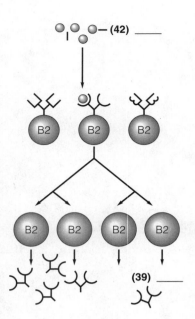

Labeling

Identify each numbered part in the accompanying illustration and its legend.

39. _____

42. _____

43. _____

44. _____   _____

45. _____   _____

Clonal selection of a (43) _____ cell, the descendants of which produced the specific (39) _____ that can combine with a specific antigen.

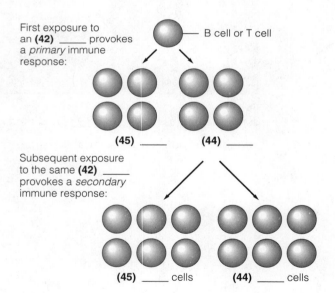

First exposure to an **(42)** _____ provokes a *primary* immune response:

B cell or T cell

(45) _____   **(44)** _____

Subsequent exposure to the same **(42)** _____ provokes a *secondary* immune response:

(45) _____ cells   **(44)** _____ cells

IMMUNIZATION/ABNORMAL OR DEFICIENT
IMMUNE RESPONSES (27-III, pp. 446–450)

Terms

immunization (446) _____

vaccine (446) _____

primary immune response (446) _____

allergy (447) _____

autoimmune response (447) _____

AIDS (448) _____

Fill-in-the-Blanks

Deliberately provoking the production of memory lymphocytes is known as (1)_____. In a(n)
(2) _____ immunization, a vaccine containing antigens is injected into the body or taken orally. The
first one elicits a (3) _____ _____ _____, and the second one (a booster shot) elicits a secondary immune response, which causes the body to produce more antibodies and (4) _____
_____ to provide long-lasting protection.

  (5)_____ is an altered secondary response to a normally harmless substance that may actually
cause injury to tissues. (6)_____ _____ is a disorder in which the body mobilizes its forces against
certain of its own tissues. (7) _____ _____ is an example of this kind of disorder in which antibodies tag acetylcholine receptors on skeletal muscle cells and cause progressive weakness. (8) _____
_____ is a similar kind of disorder in which skeletal joints are chronically inflamed. AIDS is a constel-
lation of disorders that follow infection by the (9)_____ _____ _____. In the United States,
transmission has occurred most often among intravenous drug abusers who share needles and among
(10)_____ _____. HIV is a (11)_____; its genetic material is RNA rather than DNA, and it
has several copies of an enzyme (12)_____ _____, which uses the viral RNA as a template for
making DNA, which is then inserted into a host chromosome. By 1992, an estimated (13) _____ million
people in the United States were HIV-I carriers and more than (14) _____ had full-blown AIDS.

Matching

Choose the most appropriate description for each term.

15. ___ allergy

16. ___ antibody

17. ___ antigen

18. ___ macrophage

19. ___ clone

20. ___ complement

21. ___ histamine

22. ___ MHC marker

23. ___ effector B cell

24. ___ T cell

A. Begins its development in bone marrow, but matures in the thymus gland
B. Cell that has directly or indirectly descended from the same parent cell
C. A potent chemical that causes blood vessels to dilate and let protein pass through the vessel walls
D. Y-shaped immunoglobulin
E. A nonself marker
F. The progeny of turned-on B cells
G. A group of about fifteen proteins that participate in the inflammatory response
H. An altered secondary immune response to a substance that is normally harmless to other people
I. The basis for self-recognition at the cell surface
J. Principal perpetrator of phagocytosis

Complete the Table

Indicate with a check (√) the type of vaccine used (oral vs. combined injected vaccine) and the age(s) of vaccination.

			Age Vaccination is Administered		
	(a)	(b)	(c)	(d)	(e)
Disease	CIV	O	0–8 months after birth	12–20 months after birth	11–12 years after birth
25. Diphtheria					
26. Hemophilus influenzae					
27. Hepatitis B					
28. Measles					
29. Mumps					
30. Polio					
31. Rubella					
32. Tetanus					
33. Whooping cough					

Self-Quiz

___ 1. All the body's phagocytes are derived from stem cells in the _____.
   a. spleen
   b. liver
   c. thymus
   d. bone marrow
   e. thyroid

___ 2. The plasma proteins that are activated when they contact a bacterial cell are collectively known as the _____ system.
   a. shield
   b. complement
   c. Ig G
   d. MHC
   e. HIV

___ 3. _____ are divided into two groups: T cells and B cells.
   a. Macrophages
   b. Lymphocytes
   c. Platelets
   d. Complement cells
   e. Cancer cells

___ 4. _____ produce and secrete antibodies that set up bacterial invaders for subsequent destruction by macrophages.
   a. B cells
   b. Phagocytes
   c. T cells
   d. Bacteriophages
   e. Thymus cells

___ 5. Antibodies are shaped like the letter _____.
   a. Y
   b. W
   c. Z
   d. H
   e. E

___ 6. The markers for every cell in the human body are referred to by the letters _____.
   a. HIV
   b. MBC
   c. RNA
   d. DNA
   e. MHC

___ 7. Plasma cells _____.
   a. fight against extracellular bacteria, viruses, some fungal parasites, and some protozoans
   b. develop from B cells
   c. manufacture and secrete antibodies
   d. do not divide and form clones
   e. all of the above

___ 8. Clones of B or T cells are _____.
   a. being produced continually
   b. sometimes known as memory cells if they keep circulating in the bloodstream
   c. only produced when their surface proteins recognize other specific proteins previously encountered
   d. produced and mature in the bone marrow
   e. both (b) and (c)

___ 9. Whenever the body is reexposed to a specific sensitizing agent, IgE antibodies cause _____.
   a. prostaglandins and histamine to be produced
   b. clonal cells to be produced
   c. histamine to be released
   d. the immune response to be suppressed
   e. none of the above

___ 10. The leading cause of death among transplant patients is _____.
   a. failure of the MHC in plasma cells to do its work
   b. pneumocystis infections
   c. loss of a vital organ when the transplant fails
   d. an excessive number of antigens being released into the bloodstream
   e. a transplant reaction similar to that which causes blood transfusion deaths

Chapter Objectives/Review Questions

This section lists general and detailed chapter objectives that can be used as review questions. You can make maximum use of these items by writing answers on a separate sheet of paper. Fill in answers where blanks are provided. To check for accuracy, compare your answers with information given in the chapter or glossary.

Page   *Objectives/Questions*

(434–435)   1. Describe typical external barriers that organisms present to invading organisms.
(435–437)   2. List and discuss four nonspecific defense responses that serve to exclude microbes from the body.
(436–437)   3. Explain how the complement system is related to an inflammatory response.
(438–442)   4. Distinguish between the antibody-mediated response pattern and the cell-mediated response pattern.
(439, 445)   5. Explain what is meant by primary immune pathway as contrasted with secondary immune pathway.
(440, 446)   6. Explain what monoclonal antibodies are and tell how they are currently being used in passive immunization and cancer treatment.
(444–445)   7. Describe the clonal selection theory and tell what it helps to explain.
(446)   8. Describe two ways that people can be immunized against specific diseases.
(447)   9. Distinguish allergy from autoimmune disease.
(448–449) 10. Describe how AIDS specifically interferes with the human immune system.

Integrating and Applying Key Concepts

Suppose you wanted to get rid of forty-seven warts that you have on your hands by treating them with monoclonal antibodies. Outline the steps you would have to take.

Critical Thinking Exercise

Among adult humans, 1 to 5 percent show an allergic response to penicillin. The response is caused by the presence of a specific protein antibody in their blood. One hypothesis is that penicillin causes a mutation in each individual and the mutant allele codes for the antibody. Which of the following is the best alternative explanation?

    a. The allele for the antibody is dominant.
    b. Penicillin treatment exerts a strong selective pressure against the mutant allele.
    c. Allergic responses lead to penicillin resistance.
    d. The allele for the antibody is already present in some individuals before penicillin exposure.
    e. Penicillin exposure destroys the allele for the antibody.

Note: Please see authors' note on page 287 of this workbook.

Answers

Answers to Interactive Exercises

FIRST LINE OF DEFENSE—SURFACE BARRIERS TO INVASION/SECOND LINE OF DEFENSE—NONSPECIFIC RESPONSES (27-I)

1. mucous; 2. Lysozyme; 3. Gastric; 4. bacterial; 5. clotting; 6. Phagocytic; 7. complement system; 8. Antibodies; 9. histamine; 10. capillaries;

11. *Basophils* secrete histamine and prostaglandins that change permeability of blood vessels in damaged or irritated tissues.

12. *Eosinophils* attack parasitic worms by secreting corrosive enzymes.

13. *Neutrophils* are the most abundant white blood cells; they quickly phagocytize bacteria and reduce them to molecules that can be used for other purposes.

14. Monocytes mature into *macrophages*, which slowly go about engulfing foreign agents and cleaning out dead and damaged cells.

THIRD LINE OF DEFENSE—THE IMMUNE SYSTEM (27-II)

1. nonspecific; 2. immune system; 3. MHC marker; 4. nonself; 5. lymphocytes; 6. B cell; 7. T cell; 8. thymus; 9. viruses; 10. cancer; 11. antigen-presenting cells; 12. virgin helper T cells; 13. memory T cells; 14. intracellular; 15. virgin B cells; 16. effector B cells; 17. antibodies; 18. extracellular; 19. a; 20. d; 21. e; 22. b; 23. c; 24. natural killer; 25. perforins; 26. immunoglobulin; 27. primary immune response; 28. memory cells; 29. antigens; 30. clonal selection theory; 31. nonself; 32. Cancer; 33. monoclonal (pure) antibodies; 34. cell-; 35. cytotoxic T; 36. helper T; 37. antigen-MHC; 38. macrophage; 39. antibody; 40. effector B; 41. B cell; 42. antigen; 43. B; 44. memory cells; 45. effector cells.

IMMUNIZATION/ABNORMAL OR DEFICIENT IMMUNE RESPONSES (27-III)

1. immunization; 2. active; 3. primary immune response; 4. memory cells; 5. Allergy; 6. Autoimmune disease; 7. Myasthenia gravis; 8. Rheumatoid arthritis; 9. human immunodeficiency virus; 10. male homosexuals; 11. retrovirus; 12. reverse transcriptase; 13. 1; 14. 200,000; 15. H; 16. D; 17. E; 18. J; 19. B; 20. G; 21. C; 22. I; 23. F; 24. A; 25. a, c, d; 26. a, c, d; 27. a, c, d; 28. a, d, e; 29. a, d, e; 30. b, c, d; 31. a, d, e; 32. a, c, d; 33. a, c, d.

Answers to Self-Quiz

1. d; 2. b; 3. b; 4. a; 5. a; 6. e; 7. e; 8. e; 9. a; 10. b.

Critical Thinking Analysis

a. The allele would be dominant, because even in an individual heterozygous for an inactive protein the active antibody would be present. However, this is not an alternative explanation for the source of the allele.

b. Because allergic reactions to penicillin can be lethal, this is a valid statement, but again it is not an alternative explanation.

c. This statement is meaningless, because the allergic response is a phenomenon in humans and penicillin resistance is a property of some bacteria.

d. This would explain why some individuals are allergic and others are not. The ultimate source of the allele would have to have been mutation some time in the past, not caused by penicillin exposure.

e. If penicillin destroyed the allele for the antibody, penicillin exposure would prevent, not trigger, the allergic response.

28

RESPIRATION

Interactive Exercises

THE NATURE OF RESPIRATION (28-I, pp. 453–455)

Terms

The page-referenced terms are the important terms in the chapter. Refer to the instructions given in Chapter 1, p. 1 of this workbook.

respiratory system (453) _____

respiratory surface (453) _____

_____

hemoglobin (454) _____

integumentary exchange (454) _____

tracheal respiration (454) _____

_____

gills (454) _____

lungs (455) _____

Fill-in-the-Blanks

A (1)_____ is an outfolded, thin, moist membrane endowed with blood vessels. Gas transfer is enhanced by (2)_____ _____, in which water flows past the bloodstream in the opposite direction. Insects have (3)_____ (chitin-lined air tubes leading from the body surface to the interior). At sea level, atmospheric pressure is approximately 760 mm Hg, and oxygen represents about (4)_____ percent of the total volume.

The energy to drive animal activities comes mainly from (5)_____ _____, which uses (6)_____ and produces (7)_____ _____ wastes. In a process called (8)_____, animals move (6) into their internal environment and give up (7) to the external environment.

All respiratory systems make use of the tendency of any gas to diffuse down its (9)_____ _____. Such a (9) exists between (6) in the atmosphere [(10)_____ pressure] and the metabolically active cells in body tissues [where (6) is used rapidly; pressure is (11) [choose one] ( ) highest, ( ) lowest here]. Another (9) exists between (7) in body tissues (12) [choose one] ( ) high, ( ) low pressure and the atmosphere, with its (13) [choose one] ( ) higher, ( ) lower amount of (7).

The more extensive the (14) _____ _____ of a respiratory surface membrane and the larger the differences in (15) _____ _____ across it, the faster a gas diffuses across the membrane. (16)_____ is an important transport pigment, each molecule of which can bind loosely with as many as four O_2 molecules in the lungs.

A(n) (17)_____ is an internal respiratory surface in the shape of a cavity or sac. In all lungs, (18)_____ carry gas to and from one side of the respiratory surface, and (19)_____ in blood vessels carries gas to and from the other side.

Labeling

Identify the numbered parts of the accompanying illustration, which shows the respiratory system found in many fishes.

20. _____

21. _____

    _____

22. _____-

    _____

    _____

23. _____-

    _____

    _____

24. _____

25. _____

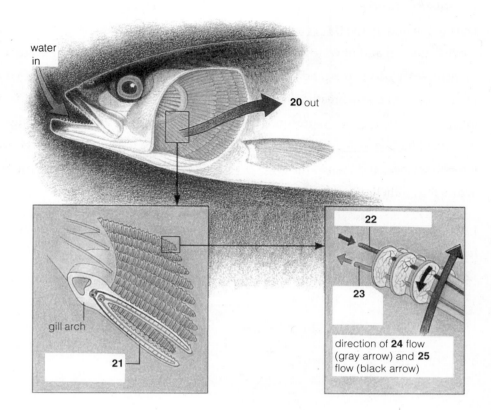

water in

20 out

gill arch

21

22

23

direction of **24** flow (gray arrow) and **25** flow (black arrow)

HUMAN RESPIRATORY SYSTEM (28-II, pp. 456–458)

Terms

nasal cavities (456) _____

diaphragm (457) _____

bronchioles (457) _____

alveolus, -oli (457) _____

pharynx (457) _____

larynx (457) _____

vocal cords (457) _____

epiglottis (457) _____

trachea (457) _____

bronchus (457) _____

Fill-in-the-Blanks

During inhalation, the (1)_____ moves downward and flattens, and the (2)_____ _____ moves outward and upward; when these things happen, the chest cavity volume (3) [choose one] ( ) increases, ( ) decreases, and the internal pressure (4) [choose one] ( ) rises, ( ) drops, ( ) stays the same. Every time you take a breath, you are (5)_____ the respiratory surfaces of your lungs. The (6)_____ _____ surrounds each lung. In succession, air passes through the nasal cavities, pharynx, and (7)_____, past the epiglottis into the (8)_____ (the space between the true vocal cords), into the trachea, and then to the (9)_____, (10)_____, and alveolar ducts. Exchange of gases occurs across the epithelium of the (11)_____.

Labeling

Identify each indicated part of the accompanying illustration.

12. _____ _____   16. _____ _____   20. _____ _____   24. _____

13. _____   17. _____   21. _____ _____

14. _____   18. _____   22. _____

15. _____   19. _____   23. _____ (sing.), _____ (plural)

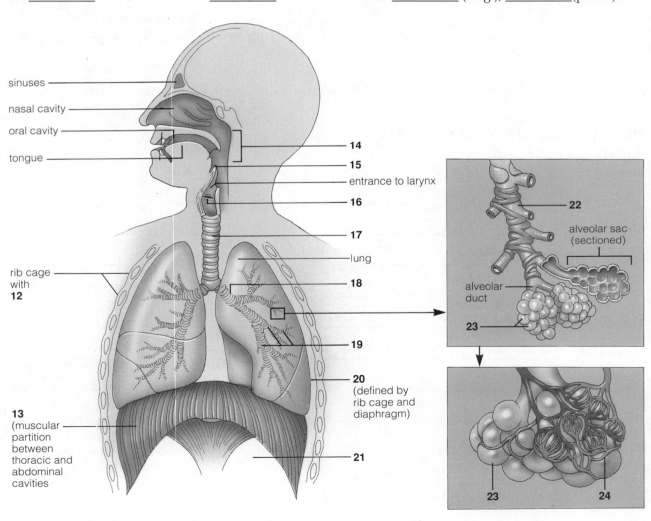

sinuses

nasal cavity

oral cavity

tongue

rib cage with **12**

13 (muscular partition between thoracic and abdominal cavities)

14

15

entrance to larynx

16

17

lung

18

19

20 (defined by rib cage and diaphragm)

21

22

alveolar sac (sectioned)

alveolar duct

23

23   **24**

GAS EXCHANGE AND TRANSPORT/RESPIRATION IN UNUSUAL ENVIRONMENTS (28-III, pp. 458–462)

Fill-in-the-Blanks

Oxygen is said to exert a (1)_____ _____ 160 mm Hg. (2)_____ alone moves oxygen from the alveoli into the bloodstream, and it is enough to move (3)_____ _____ in the reverse direction. (4)_____ is the medical name for oxygen deficiency; it is characterized by faster breathing, faster heart rate, and anxiety at altitudes of 8,000 feet above sea level. About 70 percent of the carbon dioxide in the blood is transported as (5)_____. Without (6)_____, the plasma would be able to carry only about 2 percent of the oxygen that whole blood carries. When oxygen-rich blood reaches a (7)_____ tissue capillary bed, oxygen diffuses outward, and carbon dioxide moves from tissues into the capillaries. With the assistance of (8)_____ _____, carbonic acid dissociates to form water and carbon dioxide.

(9)_____ results from persistent bronchitis episodes that damage alveoli and replace them with scar tissue. The loss of gas exchange efficiency is such that running, walking, and even exhaling are painful experiences. At least 90 percent of all (10)_____ _____ deaths are the result of cigarette smoking; only about 10 percent of afflicted individuals will survive.

When a diver ascends, (11)_____ tends to move out of the tissues and into the bloodstream. If the ascent is too rapid, many bubbles of nitrogen gas collect at the (12)_____, hence the common name, "the bends," for what is otherwise known as (13)_____ sickness.

The rate of breathing is governed by a respiratory center in the (14)_____, which monitors signals coming in from arterial walls, from blood vessels, and from other brain regions.

Self-Quiz

___1. Most forms of life depend on _____ to obtain oxygen and eliminate carbon dioxide.
a. active transport
b. bulk flow
c. diffusion
d. osmosis
e. muscular contractions

___2. _____ is the most abundant gas in Earth's atmosphere.
a. Water vapor
b. Oxygen
c. Carbon dioxide
d. Hydrogen
e. Nitrogen

___3. With respect to respiratory systems, countercurrent flow is a mechanism that explains how _____.
a. oxygen uptake by blood capillaries in the lamellae of fish gills occurs
b. ventilation occurs
c. intrapleural pressure is established
d. sounds originating in the vocal cords of the larynx are formed
e. all of the above

___4. _____ have the most efficient respiratory system.
a. Amphibians
b. Reptiles
c. Birds
d. Mammals
e. Humans

___5. Immediately before reaching the alveoli, air passes through the _____.
   a. bronchioles
   b. glottis
   c. larynx
   d. pharynx
   e. trachea

___6. During inhalation, _____.
   a. the pressure in the thoracic cavity is less than the pressure within the lungs
   b. the pressure in the chest cavity is greater than the pressure within the lungs
   c. the diaphragm moves upward and becomes more curved
   d. the thoracic cavity volume decreases
   e. all of the above

___7. Hemoglobin _____.
   a. releases oxygen more readily in tissues with high rates of cellular respiration
   b. tends to release oxygen in places where the temperature is lower
   c. tends to hold on to oxygen when the pH of the blood drops
   d. tends to give up oxygen in regions where partial pressure of oxygen exceeds that in the lungs
   e. all of the above

___8. Oxygen moves from alveoli to the bloodstream _____.
   a. whenever the concentration of oxygen is greater in alveoli than in the blood
   b. by means of active transport
   c. by using the assistance of carbamino-hemoglobin
   d. principally due to the activity of carbonic anhydrase in the red blood cells
   e. by all of the above

___9. Oxyhemoglobin releases O_2 when _____.
   a. carbon dioxide concentrations are high
   b. body temperature is lowered
   c. pH values are high
   d. CO_2 concentrations are low
   e. all of the above occur

___10. Nonsmokers live an average of _____ longer than people in their mid-twenties who smoke two packs of cigarettes each day.
   a. 6 months
   b. 1–2 years
   c. 3–5 years
   d. 7–9 years
   e. over 12 years

Chapter Objectives/Review Questions

This section lists general and detailed chapter objectives that can be used as review questions. You can make maximum use of these items by writing answers on a separate sheet of paper. Fill in answers where blanks are provided. To check for accuracy, compare your answers with information given in the chapter or glossary.

Page   *Objectives/Questions*

(435–459, 461)   1. Understand how the human respiratory system is related to the circulatory system, to cellular respiration, and to the nervous system. As the rate of cellular respiration increases, what must happen to the human respiratory and circulatory systems?

(454–455)   2. Describe how incoming oxygen is distributed to the tissues of insects and contrast this process with the process that occurs in mammals.

(456–457)   3. List all the principal parts of the human respiratory system and explain how each structure contributes to transporting oxygen from the external world to the bloodstream.

(458–459)   4. Explain why oxygen diffuses from alveolar air spaces, through interstitial fluid, and across capillary epithelium. Then explain why carbon dioxide diffuses in the reverse direction.

(458–459)   5. Explain why oxygen diffuses from the bloodstream into the tissues far from the lungs. Then explain why carbon dioxide diffuses into the bloodstream from the same tissues.

(460)   6. Distinguish bronchitis from emphysema. Then explain how lung cancer differs from emphysema.

(462)   7. List some of the things that go awry with the respiratory system when it attempts to operate in divers diving in very deep water or in people living at high altitudes, and describe the characteristics of the breakdown.

Integrating and Applying Key Concepts

Consider the amphibians—animals that generally have aquatic larval forms (tadpoles) and terrestrial adults. Outline the respiratory changes that you think might occur as an aquatic tadpole metamorphoses into a land-going juvenile.

Critical Thinking Exercise

Consider the following measurements of carbon dioxide concentrations:

Inhaled air	35
Exhaled air	28
Pulmonary arterial blood	47

Why would you most likely suspect these data to be in error?

a. The concentration of carbon dioxide in blood should be higher than in exhaled air.
b. The concentration of carbon dioxide in exhaled air should be higher than in the inhaled air.
c. The concentration of carbon dioxide in blood should be lower than in inhaled air.
d. The concentration of carbon dioxide in inhaled air should be higher than in exhaled air.
e. The concentration of carbon dioxide should be equal in all three compartments.

Answers

Answers to Interactive Exercises

THE NATURE OF RESPIRATION (28-I)
1. gill; 2. countercurrent flow; 3. tracheas; 4. 21; 5. aerobic metabolism; 6. oxygen (O_2) ; 7. carbon dioxide (CO_2); 8. respiration; 9. pressure gradient; 10. high; 11. lowest; 12. high; 13. lower; 14. surface area; 15. partial pressure; 16. Hemoglobin; 17. lung; 18. airways; 19. blood; 20. water; 21. blood vessel in gill filament; 22. oxygen-poor blood; 23. oxygen-rich blood; 24. water; 25. blood.

HUMAN RESPIRATORY SYSTEM (28–II)
1. diaphragm; 2. rib cage; 3. increases; 4. drops; 5. ventilating; 6. pleural sac; 7. larynx; 8. glottis; 9. bronchi; 10. bronchioles; 11. alveoli; 12. intercostal muscles; 13. diaphragm; 14. pharynx; 15. epiglottis; 16. vocal cords; 17. trachea; 18. bronchus; 19. bronchioles; 20. thoracic cavity; 21. abdominal cavity; 22. bronchiole; 23. alveolus, alveoli; 24. capillary.

GAS EXCHANGE AND TRANSPORT/ RESPIRATION IN UNUSUAL ENVIRONMENTS (28-III)
1. partial pressure; 2. Diffusion; 3. carbon dioxide; 4. Hypoxia; 5. bicarbonate; 6. oxyhemoglobin (hemoglobin); 7. systemic (low-pressure); 8. carbonic anhydrase; 9. Emphysema; 10. lung cancer; 11. N_2 (nitrogen gas); 12. joints; 13. decompression; 14. brain.

Answers to Self-Quiz

1. c; 2. e; 3. a; 4. c; 5. a; 6. a; 7. a; 8. a; 9. a; 10. d.

Critical Thinking Analysis

There is net movement of carbon dioxide out of the blood through the lungs. This means that the blood entering the lungs should contain the highest concentration of carbon dioxide. At most, the concentration of carbon dioxide in exhaled air could be equal to that in blood, if the two compartments are in contact long enough to reach equilibrium. If diffusion does not reach equilibrium, the blood concentration will still be larger than that in the exhaled air. Inhaled air should have the lowest concentration. This will allow net diffusion of carbon dioxide from blood into the alveolar lumen and raise the carbon dioxide concentration there. Choice (b) is the only one that fits this set of predictions.

29

WATER-SOLUTE BALANCE

MAINTAINING EXTRACELLULAR FLUID
    Water Gains and Losses
    Solute Gains and Losses
URINARY SYSTEM OF MAMMALS
    Kidney Structure
    Nephron Structure
URINE FORMATION
    Urine-Forming Processes

Factors Influencing Filtration
Reabsorption of Water and Sodium
Focus on Health: Kidney Failure and Dialysis
Kidney Malfunctions
ACID-BASE BALANCE
ON FISH, FROGS, AND KANGAROO RATS

Interactive Exercises

MAINTAINING EXTRACELLULAR FLUID/
URINARY SYSTEM OF MAMMALS (29-I, pp. 465–467)

Terms

The page-referenced terms are important; they were in boldface type in the chapter. Refer to the instructions given in Chapter 1, p. 1 of this workbook.

extracellular fluid (465) _____

urinary excretion (465) _____

_____

kidneys (466) _____

urine (466) _____

ureter (466) _____

urinary bladder (467) _____

urethra (467) _____

urinary system (467) _____

nephrons (467) _____

glomerulus (467) _____

Bowman's capsule (467) _____

_____

proximal tubule (467) _____

loop of Henle (467) _____

distal tubule (467) _____

Fill-in-the-Blanks

The body gains water by absorbing water from the slurry in the lumen of the small intestine and from
(1)_____ during condensation reactions. The mammalian body loses water mostly by excretion of
(2)_____, evaporation through the skin and (3)_____ , elimination of feces from the gut, and
(4)_____ as the body is cooled. (5)_____ behavior, in which the brain compels the individual to
seek liquids, influences the gain of water.

The body gains solutes by absorption of substances from the gut, by the secretion of hormones and
other substances, and by (6)_____, which produces CO_2 and other waste products of degradative reactions. Besides CO_2, there are several major metabolic wastes that must be eliminated: (7)_____, formed
when amino groups are detached from amino acids; (8)_____, which is produced in the liver during
reactions that link two ammonia molecules to CO_2 and release a molecule of water; and (9)_____
_____, which is formed in reactions that break down nucleic acids.

Labeling

Identify each indicated part of the accompanying illustrations.

10. _____   13. _____

11. _____   14. _____

12. _____ _____   15. _____

16. _____

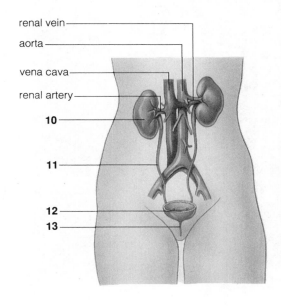

renal vein

aorta

vena cava

renal artery

10

11

12
13

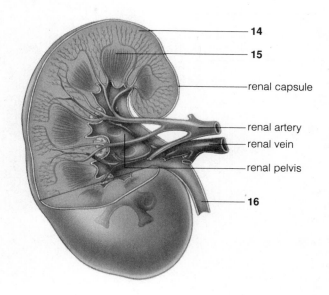

14

15

renal capsule

renal artery

renal vein

renal pelvis

16

Labeling

Identify each indicated part of the accompanying illustration.

17. _____ _____

18. _____ _____

19. _____ _____

20. _____ _____

21. _____ _____

22. _____ _____

_____

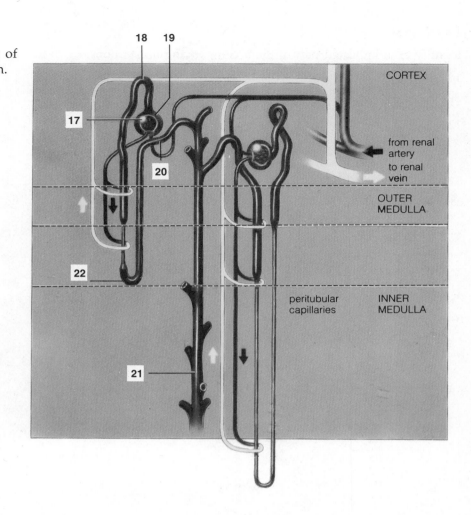

URINE FORMATION (29-II, pp. 468–470)

Terms

filtration (468) _____

_____

reabsorption (468) _____

_____

secretion (468) _____

ADH (469) _____

aldosterone (469) _____

adrenal cortex (469) _____

True/False

If the statement is true, write a T in the blank. If the statement is false, make it correct by changing the underlined word(s) and writing the correct word(s) in the answer blank.

_____ 1. When the body rids itself of excess water, urine becomes more <u>dilute.</u>

_____ 2. Water reabsorption into capillaries is achieved by <u>diffusion and active transport.</u>

Fill-in-the-Blanks

In mammals, urine formation occurs in a pair of (3)_____. Each contains about a million tubelike blood-filtering units called (4)_____. The function of (3) depends on intimate links between the (4) and the (5)_____. In every (4) blood flows from a(n) (6)_____ into a set of capillaries inside the (7)_____ _____, then into a second set of capillaries that thread around the tubular parts of the nephron, then back to the bloodstream, leaving the kidney. Urine composition and volume depend on three processes: filtration of blood at the (8)_____ of a nephron, with (9)_____ _____ providing the force for filtration; *reabsorption*, in which water and (10)_____ move out of tubular parts of the nephron and back into adjacent (11)_____ _____; and (12)_____, in which excess ions and a few foreign substances move out of those capillaries and back into the nephron so that they are disposed of in the urine. (13)_____ carry urine away from the kidney to the (14)_____ _____, where it is stored until it is released via a tube called the (15)_____, which carries urine to the outside.

Two hormones, ADH and (16)_____, adjust the reabsorption of water and (17)_____ along the distal tubules and collecting ducts. An increase in the secretion of aldosterone causes (18) [choose one] ( ) more, ( ) less sodium to be excreted in the urine. When the body cannot rid itself of excess sodium, it inevitably retains excess water, and this leads to a rise in (19)_____ _____. Abnormally high blood pressure is called (20)_____; it can damage the kidneys, the vascular system, and the brain. One way to control hypertension is to restrict the intake of (21)_____ _____. Increased secretion of (22)_____ enhances water reabsorption at distal tubules and collecting ducts when the body must conserve water. When excess water must be excreted, ADH secretion is (23) [choose one] ( ) stimulated, ( ) inhibited.

ACID-BASE BALANCE/ON FISH, FROGS, AND KANGAROO RATS
(29-III, pp. 470–471)

Fill-in-the-Blanks

The (1) _____ control the acid-base balance of body fluids by controlling the levels of dissolved ions, especially (2) _____ ions. The extracellular pH of humans must be maintained between 7.37 and (3) _____. (4) [choose one] ( ) Acids, ( ) bases lower the pH and (5) [choose one] ( ) acids, ( ) bases raise it. If you were to drink a gallon of orange juice, the pH would be (6) [choose one] ( ) raised, ( ) lowered, but the effect is minimized when excess (7) _____ ions are neutralized by (8) _____ ions in the bicarbonate–carbon dioxide buffer system. Only the (9) _____ system eliminates excess H^+ and restores buffers. Desert-dwelling kangaroo rats have very long (10) _____ _____ _____ so that nearly all (11) _____ that reaches their very long collecting ducts is reabsorbed. In freshwater, bony fishes and amphibians tend to gain (12) _____ and lose (13) _____; they produce (14) [choose one] ( ) very dilute, ( ) very concentrated urine.

Self-Quiz

___1. The most toxic waste product of metabolism is _____.
a. water
b. uric acid
c. urea
d. ammonia
e. carbon dioxide

___2. An entire subunit of a kidney that purifies blood and restores solute and water balance is called a _____.
a. glomerulus
b. loop of Henle
c. nephron
d. ureter
e. none of the above

___3. In humans, the thirst center is located in the _____.
a. adrenal cortex
b. thymus
c. heart
d. adrenal medulla
e. hypothalamus

___4. The longer the _____, the greater an animal's capacity to conserve water and to concentrate solutes to be excreted in the urine.
a. loop of Henle
b. proximal tubule
c. ureter
d. Bowman's capsule
e. collecting tubule

___5. During reabsorption, sodium ions cross the proximal tubule walls into the interstitial fluid principally by means of _____.
a. phagocytosis
b. countercurrent multiplication
c. bulk flow
d. active transport
e. all of the above

___6. Filtration of the blood in the kidney takes place in the _____.
a. loop of Henle
b. proximal tubule
c. distal tubule
d. Bowman's capsule
e. all of the above

___7. _____ primarily controls the concentration of solutes in urine.
a. Insulin
b. Glucagon
c. Antidiuretic hormone
d. Aldosterone
e. Epinephrine

___8. Hormonal control over excretion primarily affects _____.
a. Bowman's capsules
b. distal tubules
c. proximal tubules
d. the urinary bladder
e. loops of Henle

___9. The last portion of the excretory system passed by urine before it is eliminated from the body is the _____.
a. renal pelvis
b. bladder
c. ureter
d. collecting ducts
e. urethra

___10. Desert animals excrete _____ as their principal nitrogenous waste in a highly concentrated urine.
a. urea
b. uric acid
c. ammonia
d. amino acids
e. ADH

Chapter Objectives/Review Questions

This section lists general and detailed chapter objectives that can be used as review questions. You can make maximum use of these items by writing answers on a separate sheet of paper. Fill in answers where blanks are provided. To check for accuracy, compare your answers with information given in the chapter or glossary.

Page   *Objectives/Questions*

(465)   1. List some of the factors that can change the composition and volume of body fluids.
(465)   2. List three soluble by-products of animal metabolism that are potentially toxic.
(466–467) 3. List successively the parts of the human urinary system that constitute the path of urine formation and excretion.
(468)   4. Locate the processes of filtration, reabsorption, and tubular secretion along a nephron and tell what makes each process happen.
(469)   5. State explicitly how the hypothalamus, posterior pituitary, and distal tubules of the nephrons are interrelated in regulating water and solute levels in body fluids.
(470)   6. Describe the role of the kidney in maintaining the pH of the extracellular fluids between 7.35 and 7.45.
(470)   7. List two kidney disorders and explain what can be done if kidneys become too diseased to work properly.

Integrating and Applying Key Concepts

The hemodialysis machine used in hospitals is expensive and time-consuming. So far, artificial kidneys capable of allowing people who have nonfunctional kidneys to purify their blood by themselves, without having to go to a hospital or clinic, have not been developed. Which aspects of the hemodialysis procedure do you think have presented the most problems in development of a method of home self-care? If you had an unlimited budget and were appointed head of a team to develop such a procedure and its instrumentation, what strategy would you pursue?

Critical Thinking Exercise

One type of evidence for the existence of renal secretion comes from experiments with dye molecules. Slices of kidney tissue are incubated in a medium that contains the dye. After incubation, the concentration of dye in the tissue slice is higher than in the medium. In order to conclude that this accumulation of dye is evidence for renal secretion, which of the following assumptions must be made?

a. The dye is chemically changed by the cells in the slice.
b. The dye is accumulated in the lumen of the renal tubule.
c. The dye is present in the filtrate formed during the incubation.
d. The dye is not filtered during the incubation.
e. Cell respiration does not occur during the incubation.

Crossword Puzzle

ACROSS

1. Connects urinary bladder to exterior
4. _____ enhances sodium reabsorption, and is produced by the adrenal cortex.
6. Extracellular _____ bathes the body's cells.
7. _____ acid is the least toxic nitrogenous waste and is constructed from nucleic acid breakdown.
8. Bowman's _____ encloses the glomerulus.
10. _____ is an enzyme secreted by kidney cells that detaches part of a protein circulating in the blood so that the new protein can be made into a hormone that acts on the adrenal cortex.
11. A cluster of capillaries enclosed by a Bowman's capsule that filters blood
13. Connects a kidney to the urinary bladder
14. An organ that adjusts the volume and composition of blood and helps maintain the composition of the extracellular fluid
16. A small tube
17. Relating to kidney function
19. _____ of Henle
20. Most toxic nitrogenous waste
21. Nitrogenous waste that is formed in the liver and is relatively harmless
22. Water and solutes move out of the nephron tubule then into adjacent capillaries.
24. The urinary _____ includes three organs and three larger waste-containing tubes.
25. More than a million of these units are packed inside each fist-sized kidney.
26. The elimination of fluid wastes
27. The _____ tubule is the part of the nephron that is farthest from the Bowman's capsule.

DOWN

1. Product of kidneys
2. _____ fluid bathes the body's cells.
3. Loop of _____
4. _____ glands are perched on top of the kidneys.
5. _____ moves excess H^+ and a few other substances by active transport from the capillaries into the cells of the nephron wall and then into the urine.
6. Water and small-molecule solutes are forced from the blood into the Bowman's capsule.
9. Urinary _____ stores urine.
12. The middle region of the kidney
13. _____ excretion is a process that dumps excess mineral ions and metabolic wastes.
15. The _____ tubule is nearest to its Bowman's capsule.
18. The outer region of kidney, adrenal gland, or brain
23. The "water conservation" hormone produced by the posterior pituitary

Answers

Answers to Interactive Exercises

MAINTAINING EXTRACELLULAR FLUID/ URINARY SYSTEM OF MAMMALS (29-I)

1. metabolism; 2. urine; 3. lungs; 4. sweating; 5. Thirst; 6. metabolism; 7. ammonia; 8. urea; 9. uric acid; 10. kidney; 11. ureter; 12. urinary bladder; 13. urethra; 14. cortex; 15. medulla; 16. ureter; 17. glomerular capillaries; 18. proximal tubule; 19. Bowman's capsule; 20. distal tubule; 21. collecting duct; 22. loop of Henle.

URINE FORMATION (29-II)

1. T; 2. T; 3. kidneys; 4. nephrons; 5. bloodstream; 6. arteriole; 7. Bowman's capsule; 8. glomerulus; 9. blood pressure; 10. solutes; 11. blood capillaries (peritubular capillaries); 12. secretion; 13. Ureters; 14. urinary bladder; 15. urethra; 16. aldosterone; 17. sodium; 18. less; 19. blood pressure; 20. hypertension; 21. table salt (sodium chloride); 22. ADH; 23. inhibited.

ACID-BASE BALANCE/ON FISH, FROGS, AND KANGAROO RATS (29-III)

1. kidneys; 2. H^+; 3. 7.45; 4. Acids; 5. bases; 6. lowered; 7. H^+; 8. bicarbonate (HCO_3^-); 9. urinary; 10. loops of Henle; 11. water; 12. water; 13. solutes; 14. very dilute.

Answers to Self-Quiz

1. d; 2. c; 3. e; 4. a; 5. d; 6. d; 7. d; 8. b; 9. e; 10. b.

Critical Thinking Analysis

a. Because the observations and conclusions concerned the location of the unchanged dye itself, this assumption would be irrelevant to the conclusion.

b. Secretion is a process of active transport of material into the tubular lumen. The observation does not show where the dye is localized within the tissue. If the dye was accumulated inside the epithelial cells or some other compartment, it would not be evidence of secretion. This assumption is necessary for the conclusion.

c. This assumption would not be made, because the tissue is a slice removed from the animal; hence, it has no blood pressure or flow and no filtration.

d. This statement is a valid conclusion, not an assumption, because without blood circulation there can be no filtration. It is also a necessary condition for the interpretation, because if there was filtration the accumulation of dye could result from filtration of dye and water and reabsorption of water, leaving a concentrated dye solution.

e. The opposite assumption must be made. If respiration stops, ATP becomes depleted, and secretion by active transport cannot continue.

Answers to Crossword Puzzle

Answers appear on next page.

Crossword puzzle (completed):

Across:
1. URETHRA
4. ALDOSTERONE
6. FLUID
7. URIC
8. CAPSULE
10. RENIN
11. GLOMERULUS
13. URETER
14. KIDNEY
16. TUBULE
17. RENAL
19. LOOP
20. AMMONIA
21. UREA
22. REABSORPTION
24. SYSTEM
25. NEPHRON
26. EXCRETION
27. DISTAL

Down:
1. URINARY
2. EXCRETA
3. HENLE
5. SECRETION
9. BLADDER
12. MEDULLA
15. PROXIMAL
18. CORTEX
23. BEAD

30

NEURAL CONTROL AND THE SENSES

Interactive Exercises

CELLS OF THE NERVOUS SYSTEM (30-I, pp. 474–480)

Terms

The page-referenced terms are important; they were in boldface type in the chapter. Refer to the instructions given in Chapter 1, p. 1 of this workbook.

nervous system (474) _____

_____

neuron (474) _____

dendrite (474) _____

axon (474) _____

resting membrane potential (475) _____

_____

action potential (475) _____

_____

sodium-potassium pump (476) _____

_____

neurotransmitters (478) _____

chemical synapses (478) _____

synaptic integration (479) _____

nerve (480) _____

Fill-in-the-Blanks

Nerve cells that conduct messages are called (1)_____. (2)_____ cells, which support and nurture the activities of neurons, make up less than half the volume of the nervous system. (3)_____ neurons respond to specific kinds of environmental stimuli, (4)_____ connect different neurons in the spinal cord and brain, and (5)_____ neurons are linked with muscles or glands. All neurons have a (6)_____ _____ that contains the nucleus and the metabolic means to carry out protein synthesis. (7)_____ are short, slender extensions of (6), and together these two neuronal parts are the neurons' "input zone" for receiving (8)_____. The (9)_____ is a single long cylindrical extension away from the (10)_____ _____; in motor neurons, the (11)_____ has finely branched (12)_____ that terminate on muscle or gland cells and are "output zones," where messages are sent on to other cells.

A neuron at rest establishes unequal electric charges across its plasma membrane, and a (13)_____ _____ is maintained. Another name for (13) is the (14)_____ _____ _____; it represents an ability for the membrane to be disturbed. Weak disturbances of the neuronal membrane might set off only slight changes across a small patch, but strong disturbances can cause a(n) (15)_____ _____, which is an abrupt, short-lived reversal in the polarity of charge across the plasma membrane of the neuron. For a fraction of a second, the cytoplasmic side of a bit of membrane becomes positive with respect to the outside. The (16)_____ that travels along the neural membrane is nothing more than short-lived changes in the membrane potential. When action potentials reach the end of a motor neuron, they cause (17)_____ to be released that serve as chemical signals to adjacent muscle cells. Muscles (18)_____ in response to the signals.

How is the resting membrane potential established, and what restores it between action potentials? The concentrations of (19)_____ ions (K^+), sodium ions (20) (___$^+$), and other charged substances are not the same on the inside and outside of the neuronal membrane. (21)_____ proteins that span the membrane affect the diffusion of specific types of ions across it. (22)_____ proteins that span the membrane pump sodium and potassium ions against their concentration gradients across it by using energy stored in ATP. A neuronal membrane has many more positively charged (23)_____ ions inside than out and many more positively charged (24)_____ ions outside than inside. There are about (25)_____ times more potassium ions on the cytoplasmic side as outside, and there are about (26)_____ times more sodium ions outside as inside. Some channel proteins leak ions through them all the time; others have (27)_____ that open only when stimulated. Transport proteins called (28)_____-_____ _____ counter the leakage of ions across the neuronal membrane and maintain the resting membrane potential. In all neurons, stimulation at an input zone produces (29)_____ signals that do not spread very far (half a millimeter or less). (30)_____ means that signals can vary in magnitude—small or large—depending on the intensity and (31)_____ of the stimulus. When stimulation is intense or prolonged, graded signals can spread into an adjacent (32)_____ _____ of the membrane—the site where action potentials can be initiated.

A(n) (33)_____ _____ is an all-or-nothing, brief reversal in membrane potential; it is also known as a(n) (34)_____ _____. Once an action potential has been achieved, it is an (35)_____-_____-_____ event; its amplitude will not change even if the strength of the stimulus changes. The minimum change in membrane potential needed to achieve an action potential is the (36)_____ value. Some narrow-diameter neurons are wrapped in lipid-rich (37)_____ produced by specialized neuroglial cells called Schwann cells; each of these is separated from the next by a (38)_____ _____ _____—a small gap where the axon is exposed to extracellular fluid.

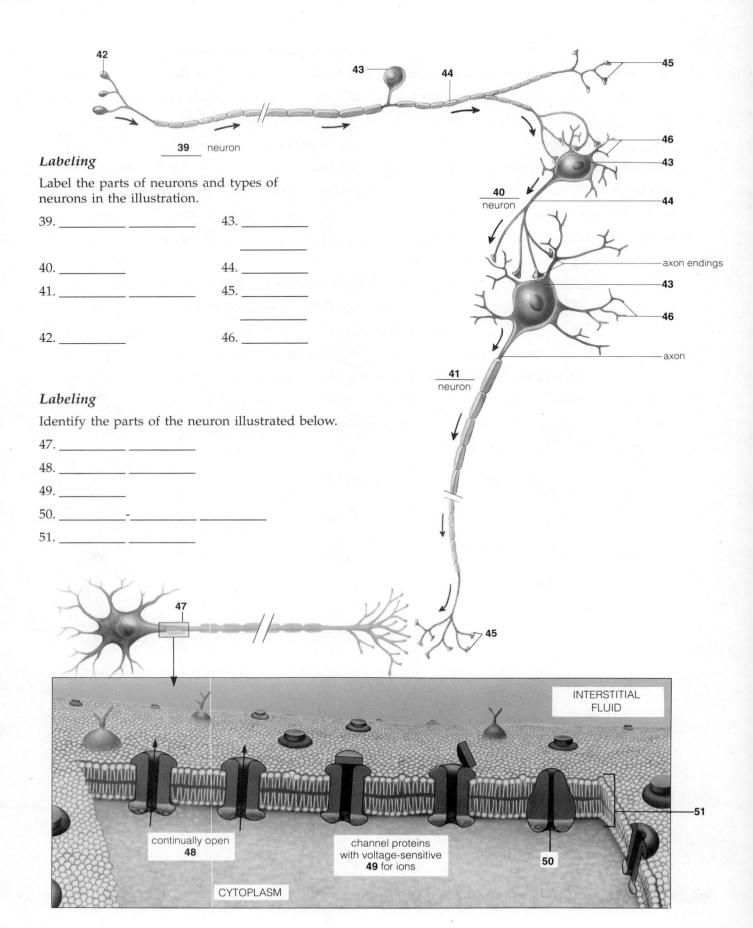

39 _____ _____ neuron

40 _____ neuron

41 _____ neuron

43

44

45

46

axon endings

43

46

axon

Labeling

Label the parts of neurons and types of neurons in the illustration.

39. _____ _____   43. _____

    _____

40. _____   44. _____

41. _____ _____   45. _____

    _____

42. _____   46. _____

Labeling

Identify the parts of the neuron illustrated below.

47. _____ _____

48. _____ _____

49. _____

50. _____ - _____ _____

51. _____ _____

continually open
48

channel proteins with voltage-sensitive **49** for ions

50

51

INTERSTITIAL FLUID

CYTOPLASM

Labeling

Identify the numbered parts of the accompanying illustrations.

52. _____ _____

53. _____

54. _____ _____ _____

55. _____

56. _____

57. _____ _____ _____

58. _____ _____

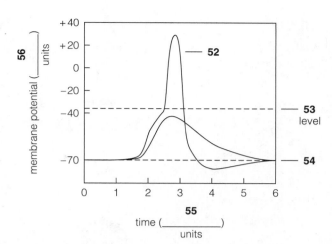

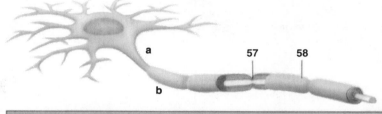

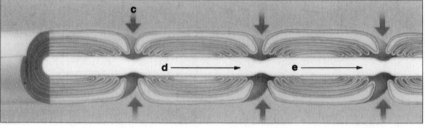

Labeling

Label the parts of the nerve illustrated at the right.

59. _____

60. _____ _____

61. _____ _____

62. _____

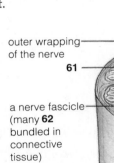

outer wrapping of the nerve

61

a nerve fascicle (many **62** bundled in connective tissue)

Matching

Match the choices below with the correct number in the diagram. (Two of the numbers match with two lettered choices.)

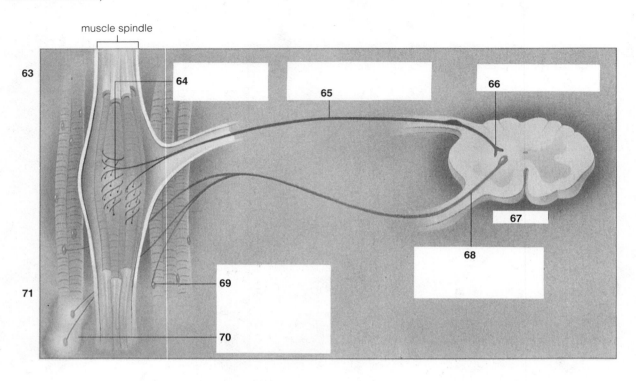

63. _____   A. Response
64. _____   B. Action potentials generated in motor neuron and propagated along its axon toward
                      muscle
65. _____   C. Motor neuron synapses with muscle cells
66. _____   D. Muscle cells contract
                  E. Local signals in receptor endings of sensory neuron
67. _____   F. Muscle spindle stretches
68. _____   G. Action potentials generated in all muscle cells innervated by motor neuron
                  H. Stimulus
69. _____   I. Axon endings synapse with motor neuron
70. _____   J. Spinal cord
                  K. Action potential propagated along sensory neuron toward spinal cord
71. _____

Fill-in-the-Blanks

The junction specialized for transmission between a neuron and another cell is called a (72)_____

_____. Usually, the signal being sent to the receiving cell is carried by chemical messengers called

(73)_____. (74)_____ is an example of this type of chemical messenger that diffuses across the

synaptic cleft, combines with protein receptor molecules on the muscle cell membrane, and soon thereafter

is rapidly broken down by enzymes. At an (75)_____ synapse, the membrane potential is driven

toward the threshold value and increases the likelihood that an action potential will occur. At an

(76)_____ synapse, the membrane potential is driven away from the threshold value, and the receiving neuron is less likely to achieve an action potential. A specific transmitter substance can have either excitatory or inhibitory effects depending on which type of protein channel it opens up in the (77)_____ membrane.

(78)_____ are neuromodulators that inhibit perceptions of pain and may have roles in memory and learning, emotional states, temperature regulation, and sexual behavior. (79)_____ _____ at the cellular level is the moment-by-moment tallying of all excitatory and inhibitory signals acting on a neuron. Incoming information is (80)_____ by cell bodies, and the charge differences across the membranes are either enhanced or inhibited. Cordlike communication lines within the brain and spinal cord that connect neurons from one region to neurons in different regions are called (81) _____

_____.

A (82)_____ is an involuntary sequence of events elicited by a stimulus. During a (83)_____ _____, a muscle contracts involuntarily whenever conditions cause a stretch in length; many of these help you maintain an upright posture despite small shifts in balance.

Imbalances can occur at chemical synapses; a neurotoxin produced by *Clostridium tetani* interferes with the effect of (84)_____ on motor neurons, which may cause tetanus—a prolonged, spastic paralysis that can lead to death.

VERTEBRATE NERVOUS SYSTEMS (30-II, pp. 481–486)

Terms

reflex pathway (481) _____

_____

central nervous system (481) _____

peripheral nervous system (481) _____

_____

parasympathetic nerve (482) _____

_____

sympathetic nerves (482) _____

_____

spinal cord (483) _____

brain (483) _____

cerebral cortex (484) _____

_____

memory (484) _____

_____

limbic system (485) _____

_____

drug addiction (486) _____

_____

Fill-in-the-Blanks

All motor-nerves-to-skeletal-muscle pathways and all sensory pathways make up the (1)_____ nervous system. The remaining nerve tissue, which generally is not under conscious control, is collectively known as the (2)_____ nervous system; it is subdivided into two parts: (3)_____ nerves, which respond to emergency situations, and (4)_____ nerves, which oversee the restoration of normal body functioning. The (5)_____ _____ _____ consists of the brain and spinal cord. A skull encloses the brain, and the (6)_____ _____ encloses and protects the spinal cord in vertebrates. In humans, thirty-one pairs of spinal nerves connect with the spinal cord and are grouped by anatomical region; twelve pairs of (7)_____ nerves connect parts of the head and neck with brain centers.

Matching

Match the most appropriate letter with each numbered item.

8. ___ cervical   A. Chest
                         B. Neck
9. ___ coccygeal   C. Pelvic
10. ___ lumbar   D. Tail
                         E. Waist
11. ___ sacral

12. ___ thoracic

Fill-in-the-Blanks

The storage of individual bits of information somewhere in the brain is called (13)_____. Experiments suggest that at least two stages are involved in its formation. One is a (14)_____-_____ _____ period, lasting at most only a few hours; information is limited to a few bits that temporarily activate neural pathways. The other is a (15)_____-_____ _____; information then is put in a different neural representation and is permanently filed in the brain.

(16)_____ are analgesics produced by the brain that inhibit regions concerned with our emotions and perception of (17)_____. Imbalances in (18)_____ can produce emotional disturbances.

Labeling

Identify the divisions of the nervous system in the posterior view below.

19. _____ _____

20. _____ _____

21. _____ _____

22. _____ _____

23. _____ _____

24. _____ _____

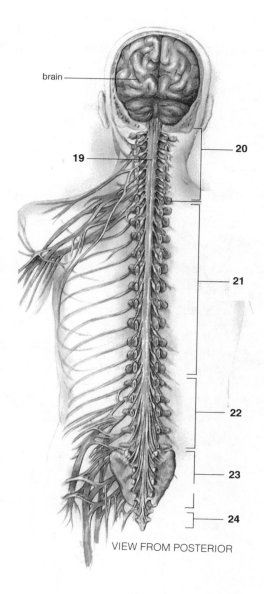

brain

19

20

21

22

23

24

VIEW FROM POSTERIOR

Labeling

Label each numbered part of the accompanying illustration.

25. _____

26. _____

27. _____

28. _____

29. _____

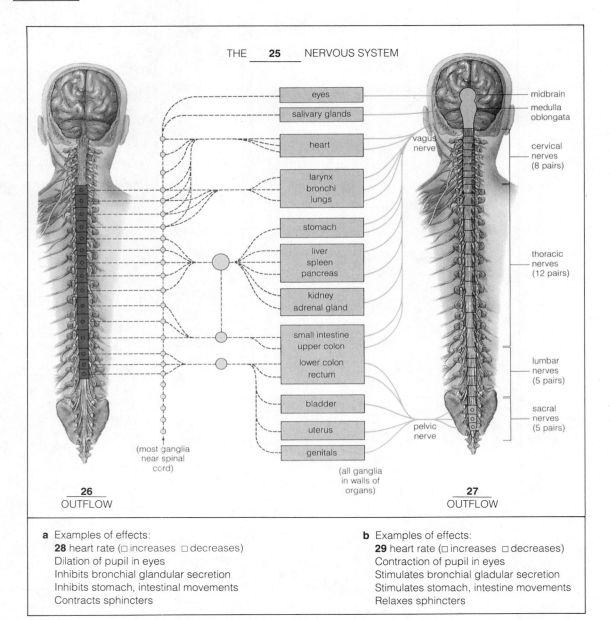

THE ___25___ NERVOUS SYSTEM

eyes

salivary glands

heart — vagus nerve

larynx
bronchi
lungs

stomach

liver
spleen
pancreas

kidney
adrenal gland

small intestine
upper colon

lower colon
rectum

bladder

uterus — pelvic nerve

genitals

(most ganglia near spinal cord)

(all ganglia in walls of organs)

midbrain
medulla oblongata
cervical nerves (8 pairs)
thoracic nerves (12 pairs)
lumbar nerves (5 pairs)
sacral nerves (5 pairs)

___26___
OUTFLOW

___27___
OUTFLOW

a Examples of effects:
28 heart rate (□ increases □ decreases)
Dilation of pupil in eyes
Inhibits bronchial glandular secretion
Inhibits stomach, intestinal movements
Contracts sphincters

b Examples of effects:
29 heart rate (□ increases □ decreases)
Contraction of pupil in eyes
Stimulates bronchial gladular secretion
Stimulates stomach, intestine movements
Relaxes sphincters

Fill-in-the-Blanks

The (30)_____ _____ is a region of local integration and reflex connections with nerve pathways leading to and from the brain; its (31)_____ _____, which contains myelinated sensory and motor axons, is the expressway that connects the brain with the peripheral nervous system. The (32)_____ _____ includes neuronal cell bodies, dendrites, nonmyelinated axon terminals, and neuroglial cells; this is the integrative zone that deals mainly with (33) _____ for limb movements (such as walking) and organ activity (such as bladder emptying).

The hindbrain is an extension and enlargement of the upper spinal cord; it consists of the (34)_____ _____, which contains the control centers for the heartbeat rate, blood pressure, and breathing reflexes. The hindbrain also includes the (35)_____, which helps coordinate motor responses associated with refined limb movements, maintenance of posture, and spatial orientation. The (36)_____ is a major routing station for nerve tracts passing between the cerebellum and cerebral cortex.

The (37)_____ evolved as a center that coordinates reflex responses to images and sounds. The midbrain, pons, and medulla oblongata make up the brain stem; within its core, the (38)_____ _____ is a major network of interneurons that extends its entire length and helps govern an organism's level of nervous system functioning. The (39)_____ contains two cerebral hemispheres, the surfaces of which are composed of gray matter. Underlying the cerebral hemispheres is the (40)_____, a region that monitors internal organs and influences hunger, thirst, and (41)_____ behaviors, and the (42)_____, the coordinating center where some motor pathways converge and relay signals to and from the cerebrum. Both the spinal cord and the brain are bathed in (43)_____ _____.

Labeling

Identify the numbered parts of the illustration below.

44. _____ _____

45. _____

46. _____

47. _____ _____

48. _____

49. _____ _____

50. _____ _____

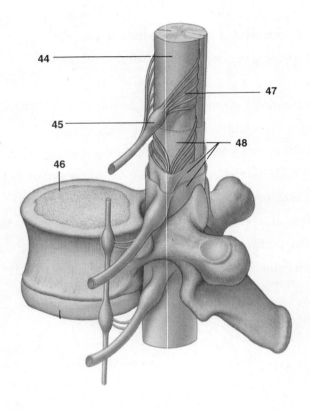

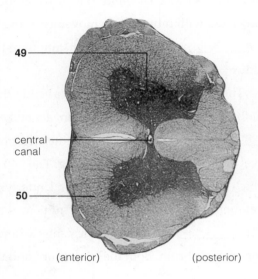

central canal

(anterior)   (posterior)

Matching

Match the named part with the letter that describes its function.

51. _____ Broca's area

52. _____ cerebellum

53. _____ corpus callosum

54. _____ hypothalamus

55. _____ limbic system

56. _____ medulla oblongata

57. _____ motor cortex

58. _____ olfactory lobes

59. _____ primary auditory cortex

60. _____ primary visual cortex

61. _____ primary somatic sensory cortex

62. _____ thalamus

A. Monitors visceral activities; influences behaviors related to thirst, hunger, reproductive cycles, and temperature control
B. Coordinates muscles required for speech
C. Receives inputs from cochleas of inner ears
D. Issues commands to muscles
E. Relays and coordinates sensory signals to the cerebrum
F. Receives inputs from receptors in nasal epithelium
G. Receives and processes input from body-feeling areas
H. Broad channel of white matter that keeps the two cerebral hemispheres communicating with each other
I. Coordinates nerve signals for maintaining balance, posture, and refined limb movements
J. Receives inputs from retinas of eyeballs
K. Connects pons and spinal cord; contains reflex centers involved in respiration, stomach secretion, and cardiovascular function
L. Contains brain centers that coordinate activities underlying emotional expression

Labeling

Identify each numbered part of the accompanying illustration.

63. _____

64. _____ _____

65. _____

66. _____

67. _____

68. _____ _____

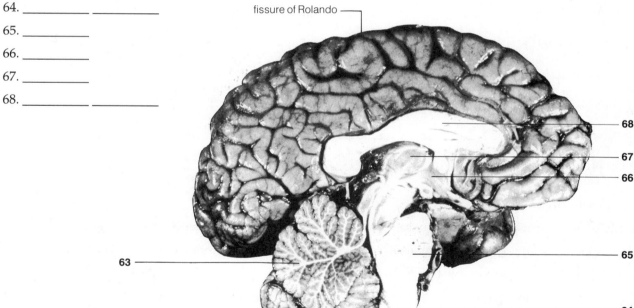

fissure of Rolando

68

67

66

65

63

64

(to spinal cord)

Matching

Match the most appropriate letter with each numbered item.

69. _____ amphetamines   A. Depressant or hypnotic drug

70. _____ caffeine

     B. Narcotic analgesic drug

     C. Psychedelic or hallucinogenic drug

71. _____ cocaine   D. Stimulant

72. _____ ethyl alcohol

73. _____ heroin

74. _____ LSD

75. _____ marijuana

76. _____ nicotine

77. _____ Quaalude

78. _____ Valium

SENSORY SYSTEMS (30-III, pp. 487–492)

Terms

sensory system (487) _____

_____

sensation (487) _____

chemoreceptor (487) _____

mechanoreceptor (487) _____

_____

photoreceptor (487) _____

thermoreceptor (487) _____

nociceptor (487) _____

Fill-in-the-Blanks

Finely branched peripheral endings of sensory neurons that detect specific kinds of stimuli are
(1)_____. A (2)_____ is any form of energy change in the environment that the body actually
detects. (3)_____ detect substances dissolved in water or air; (4)_____ detect pressure, stretching,
and vibrational changes; (5)_____ detect the energy of visible and ultraviolet light; (6)_____
detect radiant energy associated with temperature changes. A (7)_____ is conscious awareness of
change in internal or external conditions; this is not to be confused with (8)_____, which is an understanding of what sensation means. A sensory system consists of sensory receptors for specific stimuli,

(9)_____ _____ that conduct information from those receptors to the brain, and (10)_____ _____ where information is evaluated.

Matching

Select the best match for each item below.

11. ___ Vision is associated with _____ .

12. ___ Pain is associated with _____ .

13. ___ Odors are detected by _____ .

14. ___ Hearing is detected by _____ .

15. ___ CO_2 concentration in the blood is detected by _____ .

16. ___ Environmental temperature is detected by _____ .

17. ___ Internal body temperature is detected by _____ .

18. ___ Touch is detected by _____ .

19. ___ Rods and cones

20. ___ Hair cells in the ear's organ of Corti

21. ___ Pacinian corpuscles in the skin

22. ___ Olfactory receptors

23. ___ Any stimulus that causes tissue damage

24. ___ The movement of fluid in the inner ear is associated with _____ .

A. chemoreceptors
B. mechanoreceptors
C. nociceptors
D. photoreceptors
E. thermoreceptors

Fill-in-the-Blanks

The somatic sensations (awareness of (25)_____ , pressure, heat, (26)_____ , and pain) start with receptor endings that are embedded in (27)_____ and other tissues at the body's surfaces, in (28)_____ muscles, and in the walls of internal organs. All skin (29)_____ are easily deformed by pressure on the skin's surface; these make you aware of touch, vibrations, and pressure. (30)_____ nerve endings serve as "heat" receptors, and their firing of action potentials increases with increases in temperature. (31)_____ is the perception of injury to some body region. The brain sometimes gets confused and may associate perceived pain with a tissue some distance from the damaged area; this phenomenon is called (32)_____ _____ . Usually the nerve pathways to both the injured and the mistaken area pass through the same segment of spinal cord. (33)_____ in skeletal muscle, joints, tendons, ligaments, and (34)_____ are responsible for awareness of the body's position in space and of limb movements.

Label-Match

Identify each indicated part of the illustrations to the right. Complete the exercise by entering the appropriate letter in the parentheses that follow the labels.

35. _____ _____ _____ ( )

36. _____ _____ ( )

37. _____ _____ ( )

38. _____

39. _____

40. _____ _____ ( )

A. React continually to ongoing stimuli
B. Contribute to sensations of vibrations
C. Involved in sensing heat, light, pressure, and pain
D. Stimulated at the beginning and end of sustained pressure

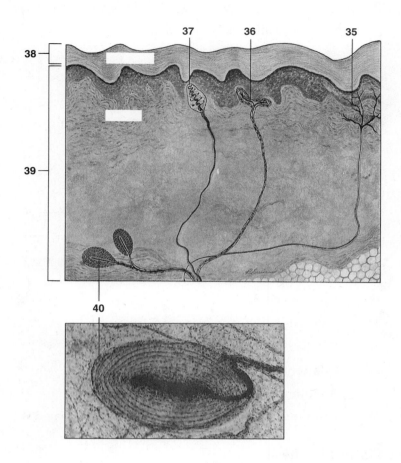

Fill-in-the-Blanks

(41)_____ receptors detect molecules that become dissolved in fluid next to some body surface. Receptors are the modified dendrites of (42)_____ neurons. In the case of taste, these receptors, when located on animal tongues, are often part of sensory organs, (43)_____ _____, which are enclosed by circular papillae. Sensory nerve pathways lead from the nasal cavity to the region of the brain where odors are identified and associated with their sources—the (44)_____ bulb and nerve tract. The (45)_____ (perceived loudness) of sound depends on the height of the sound wave. The (46)_____ (perceived pitch) of sound depends on how fast the wave changes occur. The faster the vibrations, the (47) [choose one] ( ) higher, ( ) lower the sound. Hair cells are (48) [choose one] ( ) nociceptors, ( ) mechanoreceptors, ( ) thermoreceptors that detect vibrations. The hammer, anvil, and stirrup are located in the (49) [choose one] ( ) inner, ( ) middle ear. The (50)_____ is a coiled tube that resembles a snail shell and contains the (51)_____ _____ _____—the organ that changes vibrations into electrochemical impulses. Structures that detect rotational acceleration in humans are (52)_____

_____.

Light is a stream of (53)_____—discrete energy packets. (54)_____ is a process in which photons are absorbed by pigment molecules and photon energy is transformed into the electrochemical energy of a nerve signal. (55)_____ requires precise light focusing onto a layer of photoreceptive cells that are dense enough to sample details of the light stimulus, followed by image formation in the brain. (56)_____ are simple clusters of photosensitive cells, usually arranged in a cuplike depression in the epidermis. (57)_____ are well-developed photoreceptor organs that allow at least some degree of image formation. The (58)_____ is a transparent cover of the lens area, and the (59)_____ consists of tissue containing densely packed photoreceptors. In the vertebrate eye, lens adjustments assure that the (60)_____ _____ for a specific group of light rays lands on the retina. (61)_____ refers to the lens adjustments that bring about precise focusing onto the retina. (62)_____ people focus light from nearby objects posterior to the retina. (63)_____ cells are concerned with daytime vision and, usually, color perception. A (64)_____ is a funnel-shaped pit on the retina that provides the greatest visual acuity.

Labeling

Identify each indicated part of the accompanying illustrations.

65. _____ _____

66. _____

67. _____ _____

68. _____ _____

69. _____ _____

70. _____ _____

71. _____ _____

72. _____ _____

73. _____

74. _____

75. _____

76. _____ _____

77. _____ _____

78. _____

79. _____

80. _____ _____

81. _____ _____

82. _____

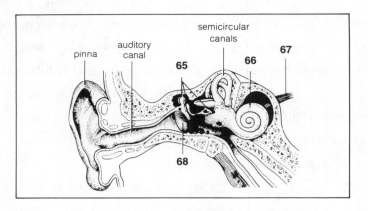

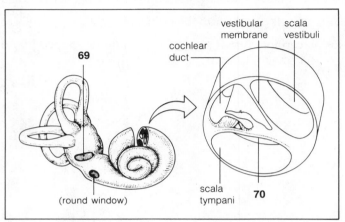

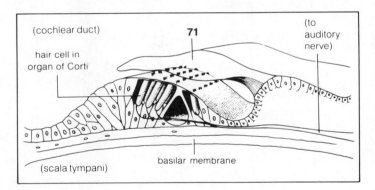

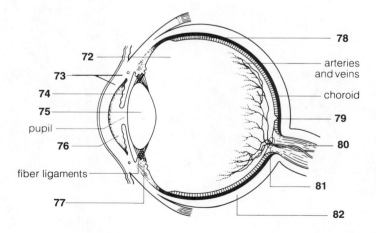

Self-Quiz

___1. Which of the following is *not* true of an action potential?
a. It is a short-range message that can vary in size.
b. It is an all-or-none brief reversal in membrane potential.
c. It doesn't decay with distance.
d. It is self-propagating.

___2. The principal place in the human ear where sound waves are amplified is _____.
a. the pinna
b. the ear canal
c. the middle ear
d. the organ of Corti
e. none of the above

___3. The place where vibrations are translated into patterns of nerve impulses is _____.
a. the pinna
b. the ear canal
c. the middle ear
d. the organ of Corti
e. none of the above

___4. The resting membrane potential _____.
a. exists as long as a charge difference sufficient to do work exists across a membrane
b. occurs because there are more potassium ions outside the neuronal membrane than there are inside
c. occurs because of the unique distribution of receptor proteins located on the dendrite exterior
d. is brought about by a local change in membrane permeability caused by a greater-than-threshold stimulus

___5. An action potential is brought about by _____.
a. a sudden membrane impermeability
b. the movement of negatively charged proteins through the neuronal membrane

c. the movement of lipoproteins to the outer membrane
d. a local change in membrane permeability caused by a greater-than-threshold stimulus

___6. The _____ are the protective coverings of the brain.
a. ventricles
b. meninges
c. tectums
d. olfactory bulbs
e. pineal glands

___7. The left hemisphere of the brain is responsible for _____.
a. music
b. mathematics
c. language skills
d. abstract abilities
e. artistic ability and spatial relationships

___8. To produce a split-brain individual, an operation would need to sever the _____.
a. pons
b. fissure of Rolando
c. hypothalamus
d. reticular formation
e. corpus callosum

___9. Accommodation involves the ability to _____.
a. change the sensitivity of the rods and cones by means of transmitters
b. change the width of the lens by relaxing or contracting certain muscles
c. change the curvature of the cornea
d. adapt to large changes in light intensity
e. all of the above

___10. Nearsightedness is caused by _____.
a. eye structure that focuses an image in front of the retina
b. uneven curvature of the lens
c. eye structure that focuses an image posterior to the retina
d. uneven curvature of the cornea
e. none of the above

Matching

11. __ axon

12. __ Broca's area

13. __ cell body

14. __ cerebellum

15. __ cerebrum

16. __ cornea

17. __ fovea

18. __ hypothalamus

19. __ interneuron

20. __ iris

21. __ medulla oblongata

22. __ parasympathetic

23. __ retina

24. __ sclera

25. __ sympathetic

A. The part of the brain that controls the basic responses necessary to maintain life processes (breathing, heartbeat) is _____.

B. The adjustable ring of contractile and connective tissues that controls the amount of light entering the eye is the _____.

C. Rods and cones are located in the _____.

D. The integrative zone of a neuron is the _____.

E. The white protective fibrous tissue of the eye is the _____.

F. The center for balance and coordination in the human brain is the _____.

G. The conducting zone of a neuron is the _____.

H. The outer transparent protective covering of part of the eyeball is the _____.

I. The highest concentration of cones is in the _____.

J. _____ are responsible for integration in the nervous system.

K. The center of consciousness and intelligence is the _____.

L. _____ nerves generally dominate internal events when environmental conditions permit normal body functioning.

M. _____ is concerned with speech.

N. Monitors internal organs; acts as gatekeeper to the limbic system; helps the reasoning centers of the brain to dampen rage and hatred; and governs hunger, thirst, and sex drives.

O. _____ nerves generally dominate internal events when "fight or flight" situations occur.

Chapter Objectives/Review Questions

This section lists general and detailed chapter objectives that can be used as review questions. You can make maximum use of these items by writing answers on a separate sheet of paper. Fill in answers where blanks are provided. To check for accuracy, compare your answers with information given in the chapter or glossary.

Page	Objectives/Questions
(474)	1. Outline some of the ways by which information flow is regulated and integrated in the human body.
(474)	2. Draw a neuron and label it according to its three general zones, its specific structures, and the specific function(s) of each structure.
(475)	3. Define resting membrane potential; explain what establishes it and how it is used by the cell neuron.
(475–477)	4. Describe the distribution of the invisible array of large proteins, ions, and other molecules in a neuron, both at rest and as a neuron experiences a change in potential.
(476)	5. Define sodium-potassium pump and state how it helps maintain the resting membrane potential.
(477–478)	6. Define action potential by stating its three main characteristics.
(477–478)	7. Explain the chemical basis of the action potential. Look at Figure 30.7 in your text and determine which part of the curve represents the following:

a. the point at which the stimulus was applied;

b. the events prior to achievement of the threshold value;

c. the opening of the ion gates and the diffusing of the ions;

d. the change from net negative charge inside the neuron to net positive charge and back again to net negative charge; and

e. the active transport of sodium ions out of and potassium ions into the neuron.

(476–478)   8. Explain how graded signals differ from action potentials.
(478–479)   9. Understand how a nerve impulse is received by a neuron, conducted along a neuron, and transmitted across a synapse to a neighboring neuron, muscle, or gland.
(480–481) 10. Explain what a reflex is by drawing and labeling a diagram and telling how it functions.
(486)   11. List the major classes of psychoactive drugs and provide an example of each class.
(481–483) 12. Describe the basic structural and functional organization of the spinal cord. In your answer, distinguish spinal cord from vertebral column.
(481–483) 13. Contrast the central and peripheral nervous systems.
(482)   14. Explain how parasympathetic nerve activity balances sympathetic nerve activity.
(484–485) 15. For each part of the brain state how the behavior of a normal person would change if he or she suffered a stroke in that part of the brain.
(484)   16. Describe how the cerebral hemispheres are related to the other parts of the forebrain.
(488, 490) 17. Distinguish the types of stimuli detected by tactile and stretch receptors from those detected by hearing and equilibrium receptors.
(490)   18. Follow a sound wave from pinna to organ of Corti; mention the name of each structure it passes and state where the sound wave is amplified and where the pattern of pressure waves is translated into electrochemical impulses.
(491–492) 19. Explain what a visual system is and list four of the five aspects of a visual stimulus that are detected by different components of a visual system.
(491–492) 20. Contrast the structure of compound eyes with the structures of invertebrate eyespots and of the human eye.
(492)   21. Define nearsightedness and farsightedness and relate each to eyeball structure.
(492)   22. Describe how the human eye perceives color and black-and-white.
(492)   23. Explain the general principles that affect how light is detected by photoreceptors and changed into electrochemical messages.

Integrating and Applying Key Concepts

Suppose that anger is eventually determined to be caused by excessive amounts of specific transmitter substances in the brains of angry people. Also suppose that an inexpensive antidote to anger that neutralizes these anger-producing transmitter substances is readily available. Can violent murderers now argue that they have been wrongfully punished because they were victimized by their brain's transmitter substances and could not have acted in any other way? Suppose an antidote is prescribed to curb violent tempers in an easily angered person. Suppose also that the person forgets to take the pill and subsequently murders a family member. Can the murderer still claim to be victimized by transmitter substances?

Critical Thinking Exercise

A fine electrode was inserted into a single cell in the olfactory epithelium of an anaesthetized frog. Action potentials were recorded as vapors of different chemicals were blown over the epithelium. Recordings from two different cells with the compounds menthol and menthone are shown below.

For each of the two cells, indicate whether it can do the following:

a. detect the presence of menthol
b. detect the presence of menthone
c. detect the difference between menthol and menthone

Answers

Answers to Interactive Exercises

CELLS OF THE NERVOUS SYSTEM (30-I)
1. neurons; 2. Neuroglial; 3. Sensory; 4. interneurons; 5. motor; 6. cell body; 7. Dendrites; 8. signals (stimuli); 9. axon; 10. cell body; 11. axon; 12. endings; 13. voltage differential; 14. resting membrane potential; 15. action potential (nerve impulse); 16. disturbance; 17. neurotransmitters; 18. contract; 19. potassium; 20. Na; 21. Channel; 22. Transport; 23. potassium; 24. sodium; 25. 30; 26. 10; 27. gates; 28. sodium-potassium pumps; 29. localized; 30. Graded; 31. duration; 32. trigger zone; 33. action potential; 34. nerve impulse; 35. all-or-nothing; 36. threshold; 37. myelin; 38. node of Ranvier; 39. sensory neuron; 40. inter; 41. motor neuron; 42. receptor; 43. cell body; 44. axon; 45. axon endings; 46. dendrites; 47. axonal membrane; 48. channel proteins; 49. gates; 50. sodium-potassium pump; 51. lipid bilayer; 52. action potential; 53. threshold; 54. resting membrane potential; 55. milliseconds; 56. millivolts; 57. node of Ranvier; 58. Schwann cell (myelin sheath); 59. axon; 60. myelin sheath; 61. blood vessels; 62. axons; 63. H, F; 64. E; 65. K; 66. I; 67. J; 68. B; 69. C; 70. G; 71. A, D; 72. chemical synapse; 73. neurotransmitters; 74. Acetylcholine (ACh); 75. excitatory; 76. inhibitory; 77. postsynaptic; 78. Endorphins; 79. Synaptic integration; 80. summed; 81. nerve tracts; 82. reflex; 83. stretch reflex; 84. acetylcholine.

VERTEBRATE NERVOUS SYSTEMS (30-II)
1. somatic; 2. autonomic; 3. sympathetic; 4. parasympathetic; 5. central nervous system; 6. vertebral column; 7. cranial; 8. B; 9. D; 10. E; 11. C; 12. A; 13. memory; 14. short-term memory; 15. long-term memory; 16. Endorphins (Enkephalins); 17. pain; 18. neurotransmitters; 19. spinal cord; 20. cervical nerves; 21. thoracic nerves; 22. lumbar nerves; 23. sacral nerves; 24. coccygeal nerves; 25. autonomic; 26. sympathetic; 27. parasympathetic; 28. increases; 29. decreases; 30. spinal cord; 31. white matter; 32. gray matter; 33. reflexes; 34. medulla oblongata; 35. cerebellum; 36. pons; 37. midbrain; 38. reticular formation; 39. forebrain; 40. hypothalamus; 41. sexual; 42. thalamus; 43. cerebrospinal fluid; 44. spinal cord; 45. ganglion; 46. vertebra; 47. spinal nerve; 48. meninges; 49. gray matter; 50. white matter; 51. B; 52. I; 53. H; 54. A; 55. L; 56. K; 57. D; 58. F; 59. C; 60. J; 61. G; 62. E; 63. cerebellum; 64. medulla oblongata; 65. pons; 66. hypothalamus; 67. thalamus; 68. corpus callosum; 69. D; 70. D; 71. D; 72. A; 73. B; 74. C; 75. C; 76. D; 77. A; 78. A.

SENSORY SYSTEMS (30-III)
1. receptors; 2. stimulus; 3. Chemoreceptors; 4. mechanoreceptors; 5. photoreceptors; 6. thermoreceptors; 7. sensation; 8. perception; 9. nerve pathways; 10. brain regions; 11. D; 12. C; 13. A; 14. B; 15. A; 16. E; 17. E; 18. B; 19. D; 20. B; 21. B; 22. A; 23. C; 24. B; 25. touch; 26. cold; 27. skin; 28. skeletal; 29. mechanoreceptors; 30. Free; 31. Pain; 32. referred pain; 33. Mechanoreceptors; 34. skin; 35. free nerve endings (C); 36. Ruffini endings (A); 37. Meissner corpuscle (D); 38. epidermis; 39. dermis; 40. Pacinian corpuscle (B); 41. Taste; 42. sensory; 43. taste buds; 44. olfactory; 45. amplitude; 46. frequency; 47. higher; 48. mechanoreceptors; 49. middle; 50. cochlea; 51. organ of Corti; 52. semicircular canals; 53. photons; 54. Photoreception; 55. Vision; 56. Eyespots; 57. Eyes; 58. cornea; 59. retina; 60. focal point; 61.Accommodation; 62. Farsighted; 63. Cone; 64. fovea; 65. middle earbones (malleus, incus, stapes); 66. cochlea; 67. auditory nerve; 68. tympanic membrane/eardrum; 69. oval window; 70. basilar membrane; 71. tectorial membrane; 72. vitreous body; 73. cornea; 74. iris; 75. lens; 76. aqueous body; 77. ciliary muscle; 78. retina; 79. fovea; 80. optic nerve; 81. blind spot/optic disk; 82. sclera.

Answers to Self-Quiz

1. a; 2. c; 3. d; 4. a; 5. d; 6. b; 7. c; 8. e; 9. b; 10. a; 11. G; 12. M; 13. D; 14. F; 15. K; 16. H; 17. I; 18. N; 19. J; 20. B; 21. A; 22. L; 23. C; 24. E; 25. O.

Critical Thinking Analysis

Cell 1 generates action potentials at the same frequency whether either substance is present or not. It can detect neither substance. The firing frequency of Cell 2 is depressed when menthol is present and increased when menthone is present. This cell can detect both substances and the difference between them. These data illustrate two forms of specificity of olfactory reception: (1) cells that are responsive to a substance versus cells that are not responsive to it and (2) cells that respond differently to two different substances. The data also show that information can be encoded in either an increase or a decrease in action potential frequency.

Reference: Gesteland, R. C. (1966) *Discover* 27(2).

31

ENDOCRINE CONTROL

Interactive Exercises

"THE ENDOCRINE SYSTEM" (31-I, p. 496)

Term

The page-referenced terms are important; they were in boldface type in the chapter. Refer to the instructions given in Chapter 1, p. 1 of this workbook.

endocrine system (496):_____

Complete the Table

1. Complete the table below by identifying the numbered components of the endocrine system shown in the illustration on the facing page as well as the hormones produced by each gland.

Gland Name	Number	Hormones Produced
a. Hypothalamus		
b. Pituitary, anterior lobe		
c. Pituitary, posterior lobe		
d. Adrenal glands (cortex)		
e. Adrenal glands (medulla)		
f. Ovaries (two)		
g. Testes (two)		
h. Pineal		
i. Thyroid		
j. Parathyroids (four)		
k. Thymus		
l. Pancreatic islets		

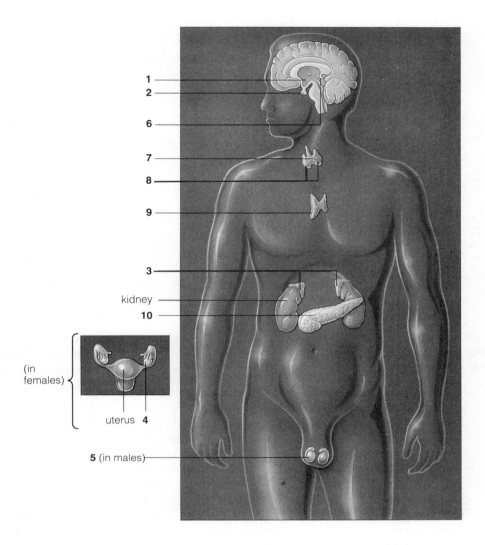

1
2
6
7
8
9
3
kidney
10
(in females) {
uterus   4
5 (in males)

HORMONES AND OTHER SIGNALING MOLECULES (31-II, pp. 496–497)

Terms

target cells (496):_____

hormones (496): _____

Fill-in-the-Blanks

Chemical substances are taken up and released by cells in response to (1) _____ in their surroundings.
In vertebrates, the responses of countless numbers of cells must be (2) _____ in ways that benefit the
entire body. This integration is carried out by way of (3) _____ molecules, which include hormones,
local signaling molecules, transmitter substances, and pheromones. All of these act on (4) _____ cells,

which are any cells that possess receptors for a specific signaling molecule and that may alter their behavior as a response to it. (5) _____ are chemical products of endocrine glands, endocrine cells, and some neurons; they are picked up by the bloodstream, which distributes them to nonadjacent target cells.

THE HYPOTHALAMUS AND PITUITARY GLAND (31-III, pp. 498–500)

Terms

hypothalamus (498):_____

pituitary gland (498): _____

hypothalamic triggers (499):_____

anterior pituitary hormones (498):_____

Choice-Match

Label each hormone given below with an "A" if it is secreted by the anterior lobe of the pituitary, a "P" if it is released from the posterior pituitary, and an "I" if it is secreted by the intermediate lobe. Complete the exercise by entering the letter of the corresponding action in the parentheses following each label.

1. ___ ( ) ACTH
2. ___ ( ) ADH
3. ___ ( ) FSH
4. ___ ( ) GH (STH)
5. ___ ( ) LH
6. ___ ( ) MSH
7. ___ ( ) oxytocin
8. ___ ( ) PRL
9. ___ ( ) TSH

A. Acts on ovaries and testes to produce gametes
B. In many cases, it governs reversible changes in skin or fur color
C. Acts on mammary glands to develop milk supplies
D. Acts on ovaries, testes to release gametes and to stimulate the production of testosterone in males and formation of corpus luteum in females
E. Induces uterine contractions and milk movement into secretory ducts
F. Acts on the thyroid gland
G. Acts on the kidneys to conserve water
H. Acts on the adrenal cortex to increase adrenal steroid hormone production
I. Acts on most cells to induce protein synthesis and cell division; also plays a role in the metabolism of glucose and protein in adults

Dichotomous Choice

Circle one of two possible answers given between parentheses in each statement.

10. The (hypothalamus/pituitary gland) region of the brain monitors internal organs and activities related to their functioning, such as eating and sexual behavior; it also secretes some hormones.
11. The (posterior/anterior) lobe of the pituitary stores and secretes two of the hypothalamic hormones.
12. The (posterior/anterior) lobe of the pituitary produces and secretes its own hormones, which govern the release of hormones from other endocrine glands.
13. Most hypothalamic hormones acting in the posterior pituitary lobe are (releasers/inhibitors) and cause target cells there to secrete hormones of their own.
14. Some hypothalamic hormones slow down secretion from their targets; these are classed as (releasers/inhibitors).
15. Pituitary (dwarfism/gigantism) results when not enough somatotropin was produced during childhood.

16. Production of excessive amounts of somatotropin during childhood results in pituitary (dwarfism/gigantism).
17. Excess somatotropin production during adulthood results in thicker bone, cartilage, and connective tissues of hands, feet, jaws, and epithelia; this condition is known as (gigantism/acromegaly).
18. (ACTH/TSH) is an anterior pituitary hormone acting on adrenal glands.
19. In addition to LH the anterior pituitary hormone having a role in reproduction is (FSH/TSH).

SELECTED EXAMPLES OF HORMONAL CONTROL (31-IV, pp. 500–505)

Terms

homeostatic feedback loops (500):_____

adrenal cortex (502):_____

adrenal medulla (502):_____

thyroid gland (502): _____

parathyroid glands (503): _____

gonads (503):_____

pancreatic islet (504): _____

thymus gland (503):_____

pineal gland (505): _____

Complete the Table

1. Complete the table below by matching the gland/organ and the hormone(s) it produces it to the descriptions of hormone action.

Gland/Organ

A. adrenal cortex
B. adrenal medulla
C. thyroid
D. parathyroids
E. testes
F. ovaries
G. pancreas (alpha cells)
H. pancreas (beta cells)
I. pancreas (delta cells)
J. thymus
K. pineal

Hormones

(a) thyroxine and triiodothyronine
(b) glucagon
(c) PTH
(d) androgens (includes testosterone)
(e) somatostatin
(f) thymosins
(g) glucocorticoids
(h) estrogens (includes progesterone)
(i) epinephrine and norepinephrine
(j) melatonin
(k) insulin

Gland/Organ	Hormones	Hormone Action
a.		stimulates bone cells to release calcium and phosphate, and the kidneys to conserve it; also helps activate vitamin D
b.		influences carbohydrate metabolism by control of food digestion; can block secretion of insulin and glucagon
c.		required in egg maturation and release; prepares and maintains the uterine lining for pregnancy; influences growth and development
d.		helps maintain the blood level of glucose, e.g., cortisol, and also suppresses inflammatory responses
e.		lowers blood sugar level by stimulating glucose uptake by liver, muscle, and adipose cells; promotes protein and fat synthesis; inhibits protein conversion to glucose
f.		required in sperm formation, genital development, and maintenance of sexual traits; influences growth and development
g.		regulates metabolism, roles in growth and development
h.		influences daily biorhythms and influences gonad development and reproductive cycles
i.		helps adjust blood circulation and carbohydrate metabolism when the body is excited or stressed, the "fight-flight" response
j.		roles in immunity
k.		raises blood sugar level by causing glycogen and amino acid conversion to glucose in the liver

Dichotomous Choice

Circle one of two possible answers given between parentheses in each statement.

2. In hypoglycemia, blood levels of glucose fall below a set point and a (positive/negative) feedback mechanism operates.
3. In hypoglycemia, the hypothalamus detects the glucose decrease and initiates a stress response by secreting (CRH/ACTH).
4. The releasing hormone operating in hypoglycemia makes the anterior pituitary secrete (CRH/ACTH), which makes the adrenal cortex secrete cortisol—which inhibits the uptake of more blood glucose by muscle cells.
5. (Hyperthyroidism/Hypothyroidism) results from insufficient concentration of thyroid hormones; adults are overweight, sluggish, dry-skinned, and intolerant of cold.
6. In vitamin D deficiency, not enough calcium and phosphorus are absorbed and bones do not develop properly; this ailment is called (goiter/rickets).
7. After a meal, blood glucose rises; pancreatic beta cells secrete (glucagon/insulin). Targets use glucose or store it as glycogen.
8. Blood glucose levels decrease between meals. (Glucagon/Insulin) is secreted by stimulated pancreas alpha cells. Targets convert glycogen back to glucose, which then enters the blood.
9. (Insulin/Glucagon) deficiency can lead to diabetes mellitus.
10. In ("type 1 diabetes"/"type 2 diabetes") the body mounts an immune response against its own insulin-secreting beta cells and destroys them.
11. In ("type 1 diabetes,"/"type 2 diabetes") insulin levels are close to or above normal, but target cells fail to respond to insulin.
12. In the (thymus/pineal) gland, white blood cells multiply, differentiate, and mature.
13. The (thymus/pineal) gland is a photosensitive organ in the brain that secretes melatonin, a hormone that influences gonad development and reproductive cycles.

SIGNALING MECHANISMS (31-V, pp. 506–507)

Terms

steroid hormones (506):_____

nonsteroid hormones (506): _____

second messengers (507): _____

Fill-in-the-Blanks

Hormones and other signaling molecules induce (1) _____ cells to take up substances. Hormones

cause cells to alter rates of (2) _____ synthesis, modify existing proteins, change cell shape, and modify

internal structures. Different hormones activate (3) _____ cellular mechanisms; not all cells can

respond to a given signal. Many cells have receptors for cortisol, so this hormone has (4) _____ effects.

Two different types of hormones have effects on cells; they are categorized as steroid and (5) _____

hormones.

Choice

For questions 6–12, choose from the following.

a. steroid hormones   b. nonsteroid hormones

___ 6. Hormones synthesized from cholesterol, are lipid-soluble, and diffuse directly across the lipid bilayer of a target cell's plasma membrane

___ 7. Include amines, peptides, proteins, and glycoproteins

___ 8. An example is testosterone, defective receptors, and a condition called testicular feminization

___ 9. Hormones that activate second messengers

___ 10. Hormones first bind to receptors at the plasma membrane; the complex then moves into the cell by endocytosis to cause further action

___ 11. Lipid-soluble hormones that move through the target cell's plasma membrane to the nucleus where the hormone-receptor complex interacts with DNA to stimulate or inhibit transcription of mRNA

___ 12. A hormone-receptor complex that activates transport proteins or triggers the opening of channel proteins across the membrane

---

Self-Quiz

___1. The _____ governs the release of hormones from other endocrine glands, it is controlled by the _____.
a. pituitary, hypothalamus
b. pancreas, hypothalamus
c. thyroid, parathyroid glands
d. hypothalamus, pituitary
e. pituitary, thalamus

___2. Neurons of the _____ produce ADH and oxytocin that are stored within axon endings of the _____.
a. anterior pituitary, posterior pituitary
b. adrenal cortex, adrenal medulla
c. posterior pituitary, hypothalamus
d. posterior pituitary, thyroid
e. hypothalamus, posterior pituitary

___3. If you were lost in the desert and had no fresh water to drink, the level of _____ in your blood would increase as a means to conserve water.
a. insulin
b. corticotropin
c. oxytocin
d. antidiuretic hormone
e. salt

For questions 4–6, choose from the following answers:
a. estrogen
b. PTH
c. FSH
d. somatotropin
e. prolactin

___4. _____ stimulates bone cells to release calcium and phosphate and the kidneys to conserve it.

___5. _____ stimulates and sustains milk production in mammary glands.

___6. Protein synthesis and cell division are activities stimulated by _____.

For questions 7–9, choose from the following answers:
a. adrenal medulla
b. adrenal cortex
c. thyroid
d. anterior pituitary
e. posterior pituitary

___7. The _____ produces glucocorticoids that help maintain the blood level of glucose and suppress inflammatory responses.

___8. The gland that is most closely associated with emergency situations is the _____.

___9. The _____ gland regulates the basic metabolic rate.

___10. If all sources of calcium were eliminated from your diet, your body would secrete more _____ in an effort to release calcium stored in your body and send it to the tissues that require it.
a. parathyroid hormone
b. aldosterone
c. calcitonin
d. mineralocorticoids
e. none of the above

Matching

11. ___ ACTH

12. ___ ADH

13. ___ thymosins

14. ___ oxytocin

15. ___ cortisol

16. ___ epinephrine and norepinephrine

17. ___ estrogen

18. ___ glucagon

19. ___ insulin

20. ___ melatonin

21. ___ parathyroid hormone

22. ___ STH (GH)

23. ___ calcitonin

24. ___ testosterone

25. ___ thyroxine

26. ___ progesterone

27. ___ TSH

A. Raises the glucose level in the blood
B. Influences daily biorhythms, gonad development, and reproductive cycles
C. Affects development of male sexual traits
D. Increases heart rate and controls blood volume; the "emergency hormones"
E. Produced by gonad; essential for egg maturation and maintenance of secondary sex characteristics in the female
F. The water conservation hormone; released from posterior pituitary
G. Lowers blood sugar by encouraging cells to take in glucose; responsible for synthesis of proteins and fats
H. Stimulates adrenal cortex to secrete cortisol
I. Elevates calcium levels in blood by stimulating calcium reabsorption from bone and kidneys and calcium absorption from gut
J. Influences overall metabolic rate, growth, and development
K. Roles in immunity
L. Triggers uterine contractions during labor and causes milk release during nursing
M. Prepares, maintains uterine lining for pregnancy; stimulates breast development
N. Inhibits uptake of more blood glucose by muscle cells
O. Lowers calcium levels in blood; bone is the target
P. Secreted by anterior pituitary; stimulates release of thyroid hormones
Q. Secreted by anterior pituitary; enhances growth in young animals, especially of cartilage and bone

Chapter Objectives/Review Questions

This section lists general and detailed chapter objectives that can be used as review questions. You can make maximum use of these items by writing answers on a separate sheet of paper. Fill in answers where blanks are provided. To check for accuracy, compare your answers with information given in the chapter or glossary.

Page	Objectives/Questions
(496)	1. Hormones, transmitter substances, local signaling molecules, and pheromones are all known as _____ molecules that carry out integration.
(496)	2. _____ cells are any cells that have receptors for a specific signaling molecule and that may alter their behavior in response to it.
(498)	3. State how, even though the anterior and posterior lobes of the pituitary are compounded as one gland, the tissues of each part differ in character.
(498)	4. The _____ and the pituitary gland interact as a major neural-endocrine control center.
(498)	5. Identify the hormones produced by the anterior lobe of the pituitary and tell which target tissues or organs each acts on.
(498)	6. Identify the hormones released from the posterior lobe of the pituitary and state their target tissues.
(499)	7. Most hypothalamic hormones acting in the anterior pituitary lobe are _____; they cause target cells to secrete hormones of their own. Some are _____; they slow down secretion from their targets.
(500)	8. Pituitary dwarfism, gigantism, and acromegaly are all associated with abnormal secretion of _____ by the pituitary gland.
(501)	9. The adrenal _____ secretes glucocorticoids.
(501)	10. The adrenal _____ contains neurons that secrete epinephrine and norepinephrine.
(501)	11. List all the hormones (and their secretory tissues) involved in regulating carbohydrate metabolism.
(503)	12. Describe the characteristics of hypothyroidism and hyperthyroidism.
(503)	13. Name the glands that secrete PTH and give the function of this hormone.
(503)	14. A woman reports that she has absolutely no interest in sex. Before you recommend that she consult a psychologist, state the items for which you (as a practicing endocrinologist) would test first.
(504)	15. Be able to name the hormones secreted by alpha, beta, and delta pancreatic cells; list the effect of each.
(505)	16. Describe the symptoms of diabetes mellitus and distinguish between type 1 and type 2.
(505)	17. Discuss the connection between melatonin and puberty.
(506–507)	18. Contrast the proposed mechanisms of hormonal action on target cell activities by (a) steroid hormones and (b) hormones that are proteins or are derived from proteins.

Integrating and Applying Key Concepts

Suppose you suddenly quadruple your already high daily consumption of calcium. State which body organs would be affected and tell how they would be affected. Name two hormones whose levels would most probably be affected and tell whether your body's production of them would increase or decrease. Suppose you continue this high rate of calcium consumption for ten years. Can you predict the organs that would be subject to the most stress as a result?

Critical Thinking Exercise

1. The concentration of iodine inside the cells of the thyroid gland is much higher than in any other tissue or in the extracellular fluid. Which of the following mechanisms would best explain the accumulation?

    a. Iodine is synthesized within the cells of the thyroid.

    b. Iodine is bound to TSH and carried on it into the cells.

    c. Iodine is moved into the cells by active transport.

    d. Iodine moves into the cells by diffusion, but the membrane is permeable to iodine only in the inward direction.

    e. Iodine moves into the cells by diffusion and is attached to impermeable molecules in the cells.

Answers

Answers to Interactive Exercises

"THE ENDOCRINE SYSTEM" (31-I)

1. a. 1, six releasing and inhibiting hormones, synthesizes ADH, oxytocin; b. 2, ACTH, TSH, FSH, LH, GSH; c. 2, stores and secretes two hypothalamic hormones, ADH and oxytocin; d. 3, sex hormones of opposite sex, cortisol, aldosterone; e. 3, epinephrine, norepinephrine; f. 4, estrogens, progesterone; g. 5, testosterone; h. 6, melatonin; i. 7, thyroxine and triiodothyronine; j. 8, parathyroid hormone (PTH); k. 9, thymosins; l. 10, insulin, glucagon, somatostatin.

HORMONES AND OTHER SIGNALING MOLECULES (31-II)

1. changes; 2. integrated; 3. signaling; 4. target; 5. Hormones.

THE HYPOTHALAMUS AND PITUITARY GLAND (31-III)

1. A(H); 2. P(G); 3. A(A); 4. A(I); 5. A(D); 6. I(B); 7. P(E); 8. A(C); 9. A(F); 10. hypothalamus; 11. posterior; 12. anterior; 13. releasers; 14. inhibitors; 15. dwarfism; 16. gigantism; 17. acromegaly; 18. ACTH; 19. FSH.

SELECTED EXAMPLES OF HORMONAL CONTROL (31-IV)

1. a. D(c); b. I(e); c. F(h); d. A(g); e. H(k); f. E(d); g. C(a); h. K(j); i. B(i); j. J(f); k. G(b); 2. negative; 3. CRH ; 4. ACTH; 5. Hypothyroidism; 6. rickets; 7. insulin; 8. Glucagon; 9. Insulin; 10. "type 1 diabetes"; 11. "type 2 diabetes"; 12. thymus; 13. pineal.

SIGNALING MECHANISMS (31-V)

1. target; 2. protein; 3. different; 4. widespread; 5. nonsteroid; 6. a; 7. b; 8. a; 9. b; 10. b; 11. a; 12. b.

Answers to Self-Quiz

1. a; 2. e; 3. d; 4. b; 5. e; 6. d; 7. b; 8. a; 9. c; 10. a; 11. H; 12. F; 13. K; 14. L; 15. N; 16. D; 17. E; 18. A; 19. G; 20. B; 21. I; 22. Q; 23. O; 24. C; 25. J; 26. M; 27. P.

Critical Thinking Analysis

a. Although compounds do tend to have higher concentrations where they are synthesized than elsewhere, iodine is an element, not a compound, and thus is not synthesized.

b. This mechanism is akin to an active transport mechanism and would be a way to accumulate iodine if TSH were present in extracellular fluid at a higher concentration than iodine. TSH is a protein and belongs to the class of impermeable hormones.

c. This is the obvious mechanism to maintain a substance at a higher concentration on one side of a membrane than on the other side.

d. This is not an acceptable mechanism. The membrane is essentially a lipid bilayer, the same on both sides. A substance is equally permeable or impermeable in both directions.

e. This is a common mechanism and operates in the thyroid. Thyroid hormone is a modified amino acid with iodine covalently bonded to it. The hormone is stored by being incorporated into a protein with normal peptide bonds. The stimulus to secrete causes hydrolysis (digestion) of the protein and release of the iodinated amino acid hormone.

32

REPRODUCTION AND DEVELOPMENT

Interactive Exercises

THE BEGINNING: REPRODUCTIVE MODES (32-I, p. 510)

Fill-in-the-Blanks

New sponges budding from parent sponges and a flatworm dividing into two flatworms represent examples of (1) _____ reproduction. This type of reproduction is useful when gene-encoded traits are strongly adapted to a limited set of (2) _____ conditions. Separation into male and female sexes comes at high cost; this cost is offset by a selective advantage: (3) _____ in traits among the offspring.

BASIC PATTERNS OF DEVELOPMENT (32-II, pp. 510–513)

Terms

The page-referenced terms are important; they were in boldface type in the chapter. Refer to the instructions given in Chapter 1, p. 1 of this workbook.

embryos (510):_____

gamete formation (510):_____

fertilization (510): _____

cleavage (510): _____

gastrulation (511): _____

endoderm (511): _____

mesoderm (511):_____

ectoderm (511):_____

organ formation (511):_____

growth, tissue specialization (511): _____

cell differentiation (513): _____

morphogenesis (513): _____

Sequence

Arrange the following events in correct chronological sequence. Write the letter of the first step next to 1, the letter of the second step next to 2, and so on.

1. _____   A. Gastrulation
2. _____   B. Fertilization
3. _____   C. Cleavage
4. _____   D. Growth, tissue specialization
5. _____   E. Organ formation
6. _____   F. Gamete formation

Complete the Table

7. Complete the table below by entering the correct germ layer (ectoderm, mesoderm, or endoderm) that forms the tissues and organs listed.

Tissues/Organs	Germ Layer
a. Muscle, circulatory organs	
b. Nervous tissues	
c. Inner lining of the gut	
d. Circulatory organs (blood vessels, heart)	
e. Outer layer of the integument	
f. Reproductive and excretory organs	
g. Organs derived from the gut	
h. Most of the skeleton	
i. Connective tissues of the gut and integument	

HUMAN REPRODUCTIVE SYSTEM (32-III, pp. 514–519)

Terms

ovaries (514):_____

testes (514): _____

secondary sexual traits (514):_____

sperm (514):_____

testosterone (515): _____

LH (515): _____

FSH (515): _____

oocyte (516): _____

menstrual cycle (516): _____

estrogens (516): _____

progesterone (516): _____

follicle (517): _____

polar body (517):_____

ovulation (517):_____

corpus luteum (518): _____

Fill-in-the-Blanks

The numbered items on the illustration below represent missing information; complete the numbered blanks in the narrative below to supply the missing information on the illustration. Some illustrated structures are numbered more than once to aid identification.

Within each testis and following repeated (1) _____ divisions of undifferentiated diploid cells just inside the (2) _____ tubule walls, (3) _____ occurs to form haploid, mature (4) _____. Males produce sperm continuously from puberty onward. Sperm leaving a testis enter a long coiled duct, the (5) _____; the sperm are stored in the last portion of this organ. When a male is sexually aroused, muscle contractions quickly propel the sperm through a thick-walled tube, the (6) _____ _____ , then to ejaculatory ducts and finally the (7) _____, which opens at the tip of the penis. During the trip to the urethra, glandular secretions become mixed with the sperm to form semen. (8) _____ _____ secrete fructose to nourish the sperm and prostaglandins to induce contractions in the female reproductive tract. (9) _____ _____ secretions help neutralize vaginal acids. (10) _____ glands secrete mucus to lubricate the penis, aid vaginal penetration, and improve sperm motility.

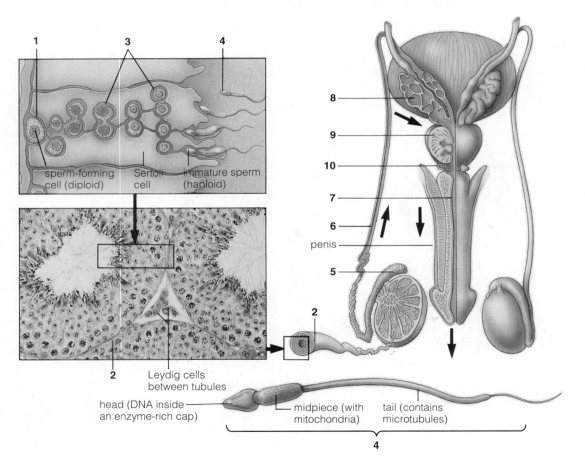

Dichotomous Choice

Circle one of two possible answers given between parentheses in each statement.

11. Testosterone is secreted by (Leydig/hypothalamus) cells.
12. (Testosterone/FSH) governs the growth, form, and functions of the male reproductive tract.
13. Sexual behavior, aggressive behavior, and secondary sexual traits are associated with (LH/testosterone).
14. LH and FSH are secreted by the (anterior/posterior) lobe of the pituitary gland.
15. The (testes/hypothalamus) governs sperm production by controlling interactions among testosterone, LH, and FSH.
16. When blood levels of testosterone (increase/decrease), the hypothalamus stimulates the pituitary to release LH and FSH, which travel the bloodstream to the testes.
17. Within the testes, (LH/FSH) acts on Leydig cells; they secrete testosterone, which enters the sperm-forming tubes.
18. FSH enters the sperm-forming tubes and diffuses into (Sertoli/Leydig) cells to improve testosterone uptake.
19. When blood testosterone levels (increase/decrease) past a set point, negative feedback loops to the hypothalamus slow down testosterone secretion.

Fill-in-the-Blanks

The numbered items on the illustration below represent missing information; complete the numbered blanks in the narrative below to supply the missing information on the illustration. Some illustrated structures are numbered more than once to aid identification.

An immature egg (oocyte) is released from one (20) _____ of a pair. From each ovary, an

(21) _____ forms a channel for transport of the immature egg to the (22) _____, a hollow, pear-

shaped organ where the embryo grows and develops. The lower narrowed part of the uterus is the

(23) _____. The uterus has a thick layer of smooth muscle, the (24) _____, lined inside with con-

nective tissue, glands, and blood vessels; this lining is called the (25) _____. The (26) _____, a

muscular tube, extends from the cervix to the body surface; this tube receives sperm and functions as part of

the birth canal. At the body surface are external genitals (vulva) that include organs for sexual stimulation.

Outermost is a pair of fat-padded skin folds, the (27) _____ _____. Those folds enclose a smaller

pair of skin folds, the (28) _____ _____. The smaller folds partly enclose the (29) _____, an

organ sensitive to stimulation. The location of the (30) _____ is about midway between the clitoris and

the vaginal opening.

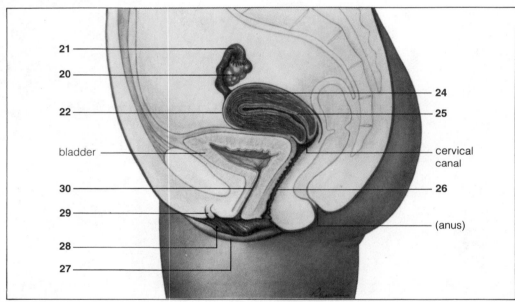

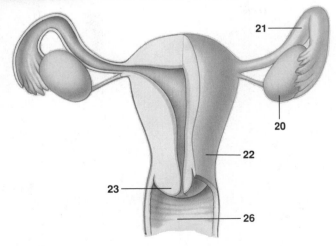

Fill-in-the-Blanks

At the beginning of a menstrual cycle, the (31) _____ signals the anterior pituitary to release (32) _____ and (33) _____. These hormones stimulate a (34) _____ to grow and its cells begin to secrete several estrogens; blood estrogens rise. The oocyte is stimulated to complete meiosis I, which produces a haploid secondary oocyte and a haploid (35) _____ body. Midway through the cycle, the (36) _____ gland detects the rising estrogen levels and briefly secretes relatively large amounts of (37) _____. This triggers (38) _____, the release of a secondary oocyte from the ovary. Follicle estrogens released during the early phase of the cycle also contribute to changes in endometrial growth that prepare the (39) _____ for pregnancy. Prior to the midcycle surge of LH, follicle cells also begin secreting (40) _____ that causes blood vessels to grow rapidly in the thickened endometrium. After ovulation, follicle cells form the (41) _____ _____, which secretes estrogen and progesterone. (42) _____ prepares the reproductive tract for embryo arrival and maintenance of the endometrium during pregnancy. During the twelve-day life of the corpus luteum, the (43) _____ calls for a decrease in FSH secretion to prevent development of other follicles. Without fertilization the corpus luteum breaks down near the end of the menstrual cycle. Following this, (44) _____ and (45) _____ levels decline and the (46) _____ breaks down. The (47) _____ flow continues for three to six days. The cycle begins again when rising (48) _____ levels initiate growth and repair of the endometrium.

FROM FERTILIZATION TO BIRTH (32-IV, pp. 520–526)

Terms

mature egg (521): _____

implantation (521): _____

HCG (521): _____

yolk sac (521): _____

allantois (521): _____

amnion (521): _____

chorion (521): _____

placenta (522): _____

fetus (522): _____

lactation (526): _____

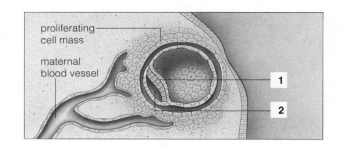

proliferating cell mass

maternal blood vessel

1

2

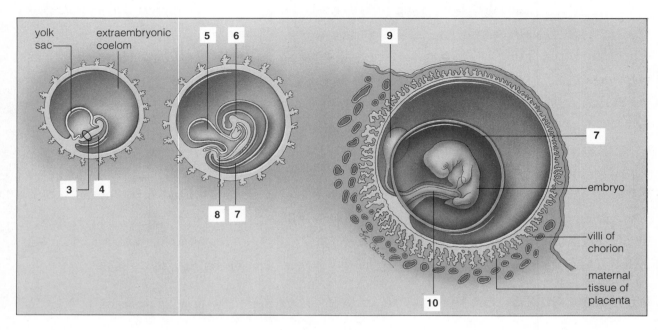

yolk sac

extraembryonic coelom

5   6

9

3   4

8   7

7

embryo

villi of chorion

maternal tissue of placenta

10

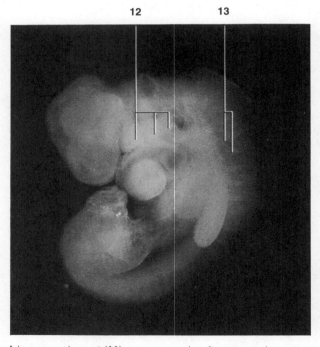

12   13

A human embryo at (11)_____ weeks after conception.

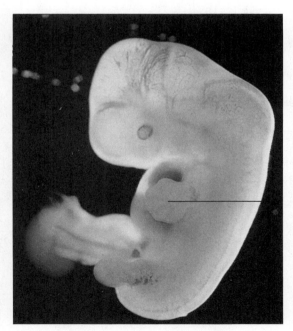

15

A human embryo at (14)_____ weeks after conception.

Labeling

Identify each indicated part of the illustrations on the facing page.

1. _____ _____
2. _____ _____
3. _____ _____
4. _____ _____
5. _____ _____
6. _____
7. _____
8. _____
9. _____ _____
10. _____ _____
11. _____
12. _____ _____
13. _____
14. _____
15. _____

Matching

Choose the one most appropriate answer for each.

16. _____ HCG
17. _____ second trimester
18. _____ umbilical cord
19. _____ first trimester
20. _____ mature egg
21. _____ fetus
22. _____ amnion
23. _____ placenta
24. _____ chorion
25. _____ lactation
26. _____ third trimester
27. _____ implantation
28. _____ FAS
29. _____ birth

A. The outermost extra-embryonic membrane that secretes HCG
B. Period of development extending from the start of the fourth month to the end of the sixth
C. Blastocyst adheres to uterine lining; cells invade maternal tissues
D. Fluid-filled sac immediately surrounding the embryo
E. Period of development extending from the seventh month until birth
F. Hormone secreted by the blastocyst; stimulates corpus luteum to continue secretion of estrogen and progesterone to maintain uterine lining
G. Found in 70% of the babies of alcoholic women in the U.S.
H. Sperm penetration of the oocyte's cytoplasm stimulates its formation in meiosis II
I. Period of development from fertilization to the end of the third month
J. Blood vessels of this structure are used to transport oxygen and nutrients for the embryo
K. Occurs about thirty-nine weeks after fertilization; involves uterine contractions, cervical dilation, amnion rupture, fetal expulsion, and severed umbilical cord
L. Period of development between recognizability as a developing human and birth
M. Spongy, blood-engorged tissue; composed partly of endometrium and extraembryonic membranes, especially chorion; diffusion link between mother and developing embryo
N. Period during which hormone-primed glands produce milk

POSTNATAL DEVELOPMENT AND AGING PROCESSES (32-V, p. 527)

Term

aging (527):_____

Dichotomous Choice

Circle one of two possible answers given between parentheses in each statement.

1. An individual from infancy to about twelve or thirteen years is in the (childhood/pubescent) period.
2. The individual from two weeks to about fifteen months after birth is a(an) (child/infant).
3. Human bone formation and growth is completed when an individual is an (adolescent/adult)
4. An individual is (pubescent/adolescent) when secondary sexual traits develop.
5. An individual from puberty until about three or four years later is (pubescent/adolescent).

CONTROL OF HUMAN FERTILITY (32-VI, pp. 528–532)

Fill-in-the-Blanks

By the time you go to bed tonight, there will be (1) (number) _____ more people on earth than there were last night at that hour. Each year in the United States, we still have about (2) (number) _____ unwed teenage mothers and (3) (number) _____ abortions. The most effective method of preventing conception is complete (4)_____. The idea of the (5) (number) _____ method is to avoid intercourse during the woman's fertile period, a few days before and a few days after ovulation. (6)_____ are about 85–93 percent reliable and help prevent venereal disease. A (7)_____ is a flexible, dome-shaped disk, used with a spermicidal foam or jelly, that is placed over the cervix. In the United States, the most widely used contraceptive is the Pill—an oral contraceptive of synthetic (8)_____ and (9)_____ that suppress the release of (10)_____ from the pituitary and thereby prevents the cyclic maturation and release of eggs. Two forms of surgical sterilization are male vasectomy and female (11)_____ _____. (12) _____ is the dislodging and removal of the embryo from the uterus. Fertilization outside of the body is possible if sperm and oocytes from the couple are normal; this procedure is called (13) _____ fertilization.

Choice

For questions 14–23, choose from the following.

a. AIDS   b. gonorrhea   c. syphilis   d. chlamydial infection
e. pelvic inflammatory disease   f. genital herpes   g. genital warts

___ 14. Benign, bumplike growths; most prevalent STD in the United States

___ 15. The virus slowly cripples the immune system and opens the door to "opportunistic" infections

___ 16. Caused by a motile, corkscrew-shaped bacterium, *Treponema pallidum*

___ 17. Serious complication of some STDs; severe abdominal pain, scarred oviducts leading to abnormal pregnancies and sterility

___ 18. A bacterium, *Neisseria gonorrhoeae*, enters mucous membranes and causes more noticeable symptoms in males; prompt treatment quickly cures this disease

___ 19. Requires direct contact with viruses or sores that contain them; small painful blisters occur; the virus is reactivated sporadically

___ 20. Caused by a parasitic bacterium that must spend parts of its life cycle inside cells of the genital and urinary tracts; causes several diseases, including NGU

___ 21. Persons who suspect they are at risk for this disease should know there may be a time lag from a few weeks to six months or more until detectable antibodies form in response to the infection

___ 22. Infection does not confer immunity to the bacterium causing this disease; oral contraceptives encourage infection

___ 23. Produces a chancre (localized ulcer) 1–8 weeks following infection

Self-Quiz

For questions 1–3, choose from the following answers:

    a. blastocyst
    b. allantois
    c. yolk sac
    d. oviduct
    e. cervix

___1. The _____ lies between the uterus and the vagina.

___2. The _____ is a pathway from the ovary to the uterus.

___3. The _____ results from the process known as cleavage.

For questions 4–7, choose from the following answers:

    a. Leydig cells
    b. seminiferous tubules
    c. vas deferens
    d. epididymis
    e. prostate

___ 4. The _____ connects a structure on the surface of the testis with the ejaculatory duct.

___ 5. Testosterone is produced by the _____.

___ 6. Meiosis occurs in the _____.

___ 7. Mature sperm are stored in the _____.

___ 8. Which of the following represents the most correct sequence of development?
a. cleavage → gamete formation → fertilization → blastula → gastrulation → organ formation
b. gamete formation→ fertilization → cleavage → blastula → gastrulation→ organ formation
c. gamete formation → cleavage → fertilization → blastula → gastrulation→ organ formation
d. fertilization→ cleavage → gamete formation → blastula → gastrulation→ organ formation

___ 9. In development, differentiated cells become organized into tissues and organs through _____.
a. gastrulation
b. cell differentiation
c. morphogenesis
d. cleavage

___ 10. Male reproductive functions are controlled by the hormones _____.
a. LH, FSH, and progesterone
b. estrogen, FSH, and LH
c. testosterone, LH, and FSH
d. testosterone and FSH

Chapter Objectives/Review Questions

This section lists general and detailed chapter objectives that can be used as review questions. You can make maximum use of these items by writing answers on a separate sheet of paper. Fill in answers where blanks are provided. To check for accuracy, compare your answers with information given in the chapter or glossary.

Page	Objectives/Questions
(510)	1. Most animals reproduce by _____ means.
(514–517)	2. Name and describe the functions of human primary reproductive organs.
(515)	3. In males, the hormones _____ and _____ act on different cells to stimulate testosterone secretion.
(514)	4. _____ stimulates sperm formation.
(515)	5. Describe how blood levels of FSH and LH are maintained by feedback loops in human males.
(516–518)	6. Describe how FSH and LH govern the menstrual cycle in human females.
(517)	7. A midcycle surge of _____ causes ovulation.
(517)	8. Ovulation is the release of a _____ oocyte from the ovary.
(518)	9. Describe the origin and functions of the corpus luteum.
(522–527)	10. Name the six stages of human embryonic development.
(522)	11. Tissues and organs of the embryo arise from three germ layers, the endoderm, ectoderm, and _____.
(521)	12. Name the four membranes that form around the early embryo and give the function of each.
(522)	13. The embryo and mother exchange substances through the _____.
(522, 524–525)	14. Describe limits of protection afforded a developing human embryo by the placental barrier.
(526)	15. Describe events occurring during the process of human birth.
(526)	16. Define lactation and relate the events that initiate this event.
(528–530)	17. Describe the most effective methods used in the control of human fertility.

Integrating and Applying Key Concepts

What rewards do you think a society should give a woman who has at most two children during her lifetime? In the absence of rewards or punishments, how can a society encourage women not to have abortions and yet ensure that the human birth rate does not continue to increase?

Critical Thinking Exercise

1. Spermatogenesis can fail because of loss or failure to form of any one of several cell types in the human testis. Each type of failure results in a characteristic pattern of hormonal changes. In each case below, state whether the levels of the following hormones would be elevated, normal, or depressed: FSH, LH, testosterone.
   a. Interstitial cells missing
   b. Sertoli cells missing
   c. Spermatogonia missing

Answers

Answers to Interactive Exercises

THE BEGINNING: REPRODUCTIVE MODES (32-I)
1. asexual; 2. environmental; 3. variation.

BASIC PATTERNS OF DEVELOPMENT (32-II)
1. F; 2. B; 3. C; 4. A; 5. E; 6. D; 7. a. mesoderm; b. ectoderm; c. endoderm; d. mesoderm; e. ectoderm; f. mesoderm; g. endoderm; h. mesoderm; i. mesoderm.

HUMAN REPRODUCTIVE SYSTEM (32-III)
1. mitotic (mitosis); 2. seminiferous; 3. meiosis (spermatogenesis); 4. sperm; 5. epididymis; 6. vas deferens; 7. urethra; 8. Seminal vesicles; 9. Prostate gland; 10. Bulbourethral; 11. Leydig; 12. Testosterone; 13. testosterone; 14. anterior; 15. hypothalamus; 16. decrease; 17. LH; 18. Sertoli; 19. increase; 20. ovary; 21. oviduct; 22. uterus; 23. cervix; 24. myometrium; 25. endometrium; 26. vagina; 27. labia major; 28. labia minor; 29. clitoris; 30. urethra; 31. hypothalamus; 32. FSH (LH); 33. LH (FSH); 34. follicle; 35. polar; 36. pituitary; 37. LH; 38. ovulation; 39. uterus; 40. progesterone; 41. corpus luteum; 42. Progesterone; 43. hypothalamus; 44. progesterone; 45. estrogen; 46. endometrium; 47. menstrual; 48. estrogen.

FROM FERTILIZATION TO BIRTH (32-IV)
1. embryonic disk; 2. amniotic cavity; 3. embryonic disk; 4. amniotic cavity; 5. yolk sac; 6. embryo; 7. amnion; 8. allantois; 9. yolk sac; 10. umbilical cord; 11. four; 12. gill arches; 13. somites; 14. five; 15. forelimb; 16. F; 17. B; 18. J; 19. I; 20. H; 21. L; 22. D; 23. M; 24. A; 25. N; 26. E; 27. C; 28. G; 29. K.

POSTNATAL DEVELOPMENT AND AGING PROCESSES (32-V)
1. childhood; 2. infant; 3. adult; 4. pubescent; 5. adolescent.

CONTROL OF HUMAN FERTILITY (32-VI)
1. 257,000; 2. 200,000; 3. 1,500,000; 4. abstinence; 5. rhythm; 6. Condoms; 7. diaphragm; 8. estrogens (progesterone); 9. progesterone (estrogens); 10. hormones; 11. tubal ligation; 12. Abortion; 13. in vitro; 14. g; 15. a; 16. c; 17. e; 18. b; 19. f; 20. d; 21. a; 22. b; 23. c.

Answers to Self-Quiz

1. e; 2. d; 3. a; 4. c; 5. a; 6. b; 7. d; 8. b; 9. c; 10. c.

Critical Thinking Analysis

a. Interstitial cells produce testosterone, so missing interstitial cells would cause very depressed levels of this hormone. Without the negative feedback from testosterone, the hypothalamus and pituitary would produce elevated levels of LH. The Sertoli cells, on the other hand, would continue to produce inhibin, and FSH levels would be expected to be normal.

b. Without Sertoli cells, inhibin would be absent, and FSH levels would rise. The testosterone-LH feedback loop would be intact, and their levels would be expected to be normal.

c. Spermatogonia are responsive to hormone levels but do not affect them by feedback mechanisms. All three hormone levels in this case would be normal.

33

POPULATION ECOLOGY

Interactive Exercises

FROM POPULATIONS TO THE BIOSPHERE (33-I, p. 537)

Terms

The page-referenced terms are important; they were in boldface type in the chapter. Refer to the instructions given in Chapter 1, p. 1 of this workbook.

ecology (537):_____

population (537):_____

community (537): _____

ecosystem (537): _____

biosphere (537):_____

Fill-in-the-Blanks

(1)_____ is the study of the ways in which organisms interact with their physical and chemical environment. A group of individuals of the same species occupying a given area is a(n) (2)_____. The place where a population lives is its (3)_____. Populations of all species occupying a habitat is defined as the (4) _____. A(n) (5) _____ is the ecological level that includes a biotic community plus its nonliving environment. The (6) _____ includes the entire earth realm in which organisms live.

POPULATION DYNAMICS (33-II, pp. 538–541)

Terms

population size (538): _____

population density (538): _____

distribution (538):_____

age structure (538):_____

zero population growth (538):_____

exponential growth (539):_____

doubling time (539):_____

net population growth rate (r) (538):_____

J-shaped curve (539):_____

biotic potential (539):_____

limiting factor (540): _____

carrying capacity (540):_____

logistic population growth (540):_____

S-shaped curve (540): _____

Dichotomous Choice

Circle one of two possible answers given between parentheses in each statement.

1. Population (size/density) is the number of individuals per unit of area or volume.
2. Population (size/density) refers to the number of members that make up the gene pool.
3. The general pattern of dispersal of members of a population through the habitat, e.g., clumped or random, is its (density/distribution).
4. Dividing a population into pre-reproductive, reproductive, and post-reproductive categories characterizes its (age/distribution) structure.
5. The "reproductive base" of a population refers to the number of individuals in the (pre-reproductive/reproductive) age structure category.
6. When the number of population members stabilizes due to a balance in births/immigrations and deaths/emigrations, a (biotic potential/zero population growth) is demonstrated.
7. Any population whose growth is not restricted in some way will show a pattern of (exponential/logistic) growth.
8. Any increase in population size enlarges the population (distribution/reproductive) base.
9. When the course of exponential growth is plotted on a graph, a (J-shaped/S-shaped) curve is obtained.
10. When limiting factors (essential resources in short supply) act on a population, population growth (increases/decreases).

Short Answer

For exercises 11–13, consider the equation $G = rN$, where G = the population growth rate, r = the net reproduction per individual, and N = the number of individuals in the population.

11. Assume that r remains constant at 0.2.

    a. As the value of G increases, what happens to the value of N? _____

    b. If the value of G decreases, what happens to the value of N? _____

    c. If the net reproduction per individual stays the same and the population grows faster, then what

    must happen to the number of individuals in the population? _____

12. If a society decides it is necessary to lower its value of N through reproductive means because supportive resources are dwindling, it must either lower its net reproduction per individual or its _____

    _____

13. The equation $G = rN$ expresses direct relationships with r and/or N. If G remains constant and N

    increases, what must the value of r do? (In this situation, r varies inversely with N.) _____

    _____

14. Look at line a in the graph below. After seven hours have elapsed, approximately how many individuals are in the population? _____

15. Look at line b in the graph below.

    a. After 24 hours have elapsed, approximately how many individuals are in the population?

    _____

    b. After 28 hours have elapsed, approximately how many individuals are in the population?

    _____

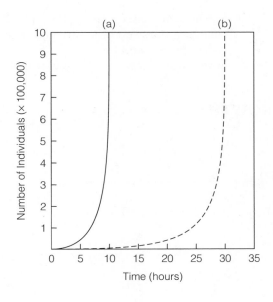

Fill-in-the-Blanks

(16)_____ _____ refers to the maximum number of individuals of a population that can be sustained indefinitely by the environment. S-shaped growth curves are characteristic of (17)_____ population growth. The plot of logistic growth levels once the (18) _____ _____ is reached. In the equation $G = r_{max}N (K- N/K)$, as the value of N approaches the value of K, and K and r_{max} remain constant, the value of G (19) [choose one] ( ) increases, ( ) decreases, ( ) cannot be determined by humans, even if they know algebra. As the value of r_{max} increases and G and N remain constant, the value of K (the carrying capacity) (20) [choose one] ( ) increases, ( ) decreases, ( ) cannot be determined by humans, even if they know algebra. In an overcrowded population, predators, parasites, and disease agents serve as (21) _____-_____ controls. When an event such as a freak summer snowstorm in the Colorado Rockies causes more deaths or fewer births in a butterfly population (with no regard to crowding or dispersion patterns), the controls are said to be (22) _____-_____.

LIFE HISTORY PATTERNS (33-III, pp. 542–543)

Terms

life table (542):_____

survivorship curves (542):_____

Matching

Choose the most appropriate answer for each.

____ 1. cohort

____ 2. type III survivorship curves

____ 3. life tables

____ 4. type I survivorship curves

____ 5. type II survivorship curves

A. Survivorship curves that reflect a fairly constant death rate at all ages; typical of some song birds, lizards, and small mammals
B. Survivorship curves that reflect high survivorship until fairly late in life; produce a few large offspring provided with extended parental care; examples are elephants and humans
C. A group tracked by researchers from birth until the last survivor dies
D. Survivorship curves that reflect a high death rate early in life; typical of sea stars and other invertebrate animals, insects, many fishes, plants, and fungi
E. Summaries of age-specific patterns of birth and death

HUMAN POPULATION GROWTH (33-IV, pp. 544–549)

Term

demographic transition model (547):_____

Graph Construction

1. Graph the following data in the space provided on page 414.

Year	Estimated World Population
1650	500,000,000
1850	1,000,000,000
1930	2,000,000,000
1975	4,000,000,000
1986	5,000,000,000
1993	5,500,000,000

From the graph constructed:

a. Estimate the year that the world contained 3 billion humans. _____

b. Estimate the year that Earth will house 8 billion humans. _____

c. Do you expect Earth to house 8 billion humans within your lifetime? _____

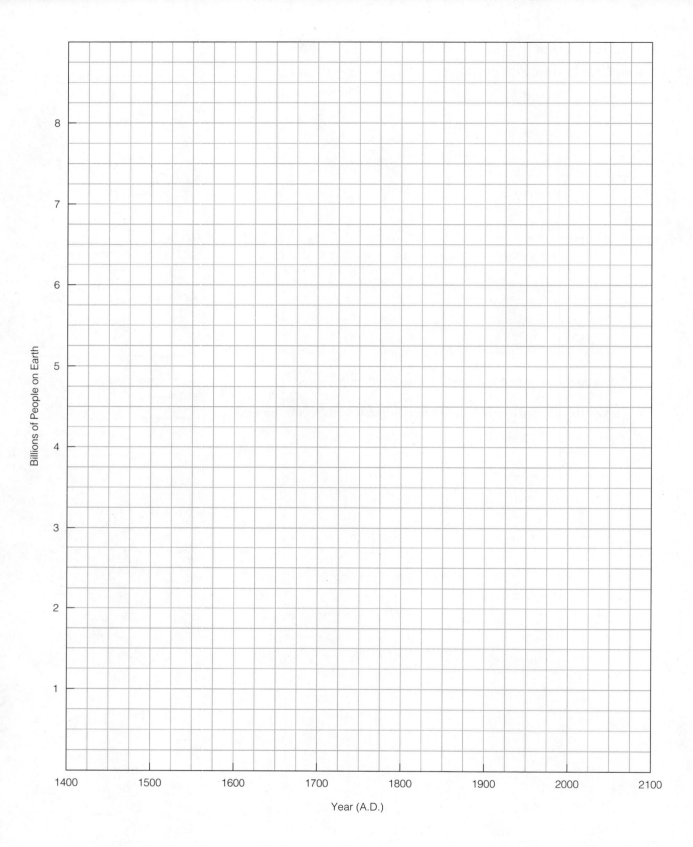

Year (A.D.)

True/False

If the statement is true, write a T in the blank. If the statement is false, make it correct by changing the underlined word(s) and writing the correct word(s) in the answer blank.

_____ 2. During 1993, one in <u>two</u> humans on the planet was malnourished or starving, without clean drinking water, and without adequate shelter.

_____ 3. Even if we could double our present food supply, death from starvation could still reach <u>20–40 million</u> people a year.

_____ 4. Compared with the geographic spread of other organisms, it has taken the human population a relatively <u>long</u> period of time to expand into new environments.

_____ 5. Managing food supplies through agriculture has had the effect of <u>increasing</u> the carrying capacity for human populations.

_____ 6. Humans have sidestepped the <u>biotic potential</u> of their environment by bringing many disease agents under control and by tapping into concentrated, existing stores of energy.

Fill-in-the-Blanks

If the present rate of human population growth continues, (7) _____ (number) billion people will live on the planet in just twenty years. These numbers will have adverse effects on resource supplies and such crowding will also invite the return of severe density- (8) _____ controls. An example is the largest (9) _____ epidemic that is now sweeping through South Asia and may claim up to 5 million lives. Many governments attempt to control population growth rates by restricting (10)_____, but others attempt to reduce population pressures by encouraging (11) _____. However, most countries focus on reducing (12)_____ rates through economic security and family planning. The key points of the (13)_____ _____ model is that changes in population growth can be correlated with changes that occur during four stages of economic development. The United States, Canada, Australia, Japan, the Soviet Union, and most countries of Western Europe are in the (14)_____ stage of the demographic transition model; Mexico and other less-developed countries are in the (15) _____ stage. To achieve global zero population growth, the average "replacement rate" is about (16) _____ (number) children per woman in less-developed countries, and (17) _____ (number) in more-developed countries. Age structure diagrams of actively growing human populations show that more than a third of the world population falls in the broad (18) _____ base category (high numbers of men and women of reproductive age). A simple method of slowing reproduction is to encourage (19) _____ reproduction. Either humans must make an effort to limit population growth to fit environmental (20) _____ _____ or the environment will eventually do it for us.

Arrange the following stages of the demographic transition model in correct chronological sequence. Write the letter of the first step next to 21, the letter of the second step next to 22, and so on.

21. ___ A. Industrial stage: population growth slows and industrialization is in full swing

22. ___ B. Preindustrial stage: harsh living conditions, high birth rates and low death rates, slow population growth

23. ___ C. Postindustrial stage: zero population growth is reached; birth rate falls below death rate, and population size slowly decreases

24. ___ D. Transitional stage: industrialization begins, food production rises, and health care improves; death rates drop, birth rates remain high, causing rapid population growth

---

Self-Quiz

___1. The total number of individuals of the same species that occupy a given area at a given time is _____.
   a. the population density
   b. the population growth
   c. the population birth rate
   d. the population size

___2. The average number of individuals of the same species per unit area at a given time is _____.
   a. the population density
   b. the population growth
   c. the population birth rate
   d. the population size

___3. A population that is growing exponentially in the absence of limiting factors can be illustrated accurately by a(n) _____.
   a. S-shaped curve
   b. J-shaped curve
   c. curve that terminates in a flattened phase
   d. survivorship curve

___4. If reproduction occurs early in the life cycle, _____.
   a. higher population levels tend to result
   b. it represents a density-independent factor that limits population size
   c. it represents a density-dependent factor that limits population size
   d. all of the above

___5. Which of the following is not characteristic of logistic growth?
   a. S-shaped curve
   b. leveling off of growth as carrying capacity is reached
   c. unrestricted growth
   d. slow growth of a low-density population followed by rapid growth

___6. The population growth rate (G) is equal to the _____ net reproduction per individual (r) and number of individuals (N).
   a. sum of
   b. product of
   c. doubling of
   d. difference between

___7. $G = r_{max} N (K - N/K)$ represents _____.
   a. exponential growth
   b. population density
   c. population size
   d. logistic growth

___8. The beginning of industrialization, a rise in food production, improvement of health care, rising birth rates and declining death rates describes the _____ stage of the demographic transition model.
   a. preindustrial
   b. transitional
   c. industrial
   d. postindustrial

___9. The maximum number of individuals of a population (or species) that can be sustained by a given environment defines _____.
   a. the carrying capacity
   b. exponential growth
   c. logistic growth
   d. density-independent factors

___10. The survivorship curve typical of human populations is type _____.
   a. I
   b. II
   c. III
   d. none of the above types

Chapter Objectives/Review Questions

This section lists general and detailed chapter objectives that can be used as review questions. You can make maximum use of these items by writing answers on a separate sheet of paper. Fill in answers where blanks are provided. To check for accuracy, compare your answers with information given in the chapter or glossary.

Page	Objectives/Questions
(538)	1. Distinguish between population size, population density, distribution, and age structure.
(538)	2. The letter r represents _____ _____ growth rate.
(538)	3. In the equation $G = rN$, as long as r holds constant, any population will show _____ growth.
(539–540)	4. Distinguish between exponential and logistic population growth; describe the growth curves associated with each type.
(539)	5. _____ potential is described as maximum population increase per individual under ideal conditions.
(540)	6. Define what is meant by a limiting factor on population growth and cite examples.
(540)	7. _____ capacity is the maximum number of individuals of a population (or species) that can be sustained indefinitely by a given environment.
(538, 549)	8. Define zero population growth and describe how achieving it would affect the human population of the United States.
(541)	9. Cite examples of density-dependent and density independent controls on population growth.
(543)	10. Briefly define Types I, II, and III survivorship curves.
(544)	11. List three reasons that humans have been able to sidestep natural controls on our own population growth.
(548)	12. Define age structure and explain why this is the principal reason it would be 70 to 100 years before the world population would stabilize even if the world average became 2.5 children per family.
(547)	13. Describe the four stages of the demographic transition model.
	14. After study of the conditions governing human population growth, and the description of present conditions, what is your personal prediction for conditions of human existence on our planet?

Integrating and Applying Key Concepts

Assume that the world has reached zero population growth. The year is 2110, and there are 10.5 billion individuals of *Homo pollutans* on Earth. You have seen stories on the community television screen about how people used to live 120 years ago. List the ways that life has changed and comment on the events that no longer happen because of the enormous human population.

Critical Thinking Exercise

1. The text says that the number of offspring per couple needed to achieve zero population growth is a little larger than 2.0, because some females die before they reach reproductive age. The text does not apply the same argument to pre-reproductive deaths of males. Which of the following assumptions are they most likely making?
   a. Mate choice is random.
   b. All matings are monogamous.
   c. Males may mate with more than one female.
   d. No females fail to mate.
   e. No males fail to mate.

Answers

Answers to Interactive Exercises

FROM POPULATIONS TO THE BIOSPHERE (33-I)
1. Ecology; 2. population; 3. habitat; 4. community; 5. ecosystem; 6. biosphere.

POPULATION DYNAMICS (33-II)
1. density; 2. size; 3. distribution; 4. age; 5. reproductive; 6. zero population growth; 7. exponential; 8. reproductive; 9. J-shaped; 10. decreases; 11.a. It increases; b. It decreases; c. It must increase; 12. population growth rate; 13. It must decrease; 14. 1000 or less; 15.a. 100,000; b. 300,000; 16. Carrying capacity; 17. logistic; 18. carrying capacity; 19. decreases; 20. decreases; 21. density-dependent; 22. density-independent.

LIFE HISTORY PATTERNS (33-III)
1. C; 2. D; 3. E; 4. B; 5. A.

HUMAN POPULATION GROWTH (33-IV)
1. a. 1962–63; b. 2025 or sooner; c. Depends on the age and optimism of the reader; 2. Five; 3. T; 4. short or brief; 5. T; 6. limiting factors; 7. eight (8); 8. dependent; 9. cholera; 10. immigration; 11. emigration; 12. birth; 13. demographic transition; 14. industrial; 15. transition; 16. 2.5; 17. 2.1; 18. reproductive; 19. delayed; 20. carrying capacity; 21. B; 22. D; 23. A; 24. C.

Answers to Self-Quiz

1. d; 2. a; 3. b; 4. a; 5. c; 6. b; 7. d; 8. b; 9. a; 10. a.

Critical Thinking Analysis

a. This assumption is not necessary. The conclusion concerns only the overall rate of reproduction, not the success of any particular segment of the population.

b. If all matings were monogamous, then loss of a male would lower the overall birthrate the same as loss of a female.

c. This means that if one male is lost before he reproduces, the female he would have impregnated will be mated by another male, and the total production of offspring will be unchanged. If the loss of unbred males is to have no effect on population dynamics, this must be assumed.

d. and e. These assumptions need not be made. The calculation requires only that a constant percentage of each sex fails to mate. However, if some do not reproduce, the number of offspring per couple must rise to compensate.

34

COMMUNITY INTERACTIONS

Interactive Exercises

CHARACTERISTICS OF COMMUNITIES (34-I, pp. 552–553)

Terms

The page-referenced terms are important; they were in boldface type in the chapter. Refer to the instructions given in Chapter 1, p. 1 of this workbook.

habitat (552): _____

community (552): _____

niche (552):_____

commensalism (553): _____

mutualism (553):_____

interspecific competition (553):_____

predation (553):_____

parasitism (553): _____

Matching

Choose the most appropriate answer to match with each term.

1. ___ habitat
2. ___ community
3. ___ commensalism
4. ___ mutualism
5. ___ interspecific competition
6. ___ predation
7. ___ parasitism

A. An interaction that directly benefits one species but does not harm or help the other
B. Has adverse effects on both of the interacting species
C. The parasite benefits, the host is harmed
D. The type of place one finds a particular organism; characterized by physical and chemical features as well as other species
E. The predator benefits, the prey is harmed
F. An interaction from which both species benefit
G. The associations of all populations of species in any given habitat

MUTUALLY BENEFICIAL INTERACTIONS (34-II, pp. 553–554)

Terms

symbiosis (553): _____

Complete the Table

1. Complete the following table to describe how each of the organisms listed is intimately dependent on the other for survival and reproduction in a mutualistic symbiotic interaction.

Organism	Dependency
a. Yucca moth	
b. Yucca plant	

COMPETITIVE INTERACTIONS (34-III, p. 555)

Term

competitive exclusion (555): _____

Dichotomous Choice

Circle one of two possible answers given between parentheses in each statement.

1. Intraspecific competition is (more/less) fierce than interspecific competition.
2. When all individuals have equal access to a required resource and some are better at exploiting it, the interaction tends to (reduce/increase) the common supply of the shared resource unless it is abundant.
3. Two species are (less/more) likely to coexist in the same habitat when they are very similar in their use of scarce resources.
4. Gause used two species of *Paramecium* competing for (the same/different) bacterial cells to illustrate competitive exclusion.
5. Gause also found that two species of *Paramecium* that did not overlap as much in their requirements were (more/less) likely to coexist.

CONSUMER-VICTIM INTERACTIONS (34-IV, pp. 556–560)

Terms

predator (556):_____

parasite (556): _____

coevolution (558):_____

mimicry (558):_____

camouflage (558): _____

parasitoids (560):_____

Matching

Match each of the following with the most appropriate answer. The same letter may be used more than once. Use only one letter per blank.

1. ___ polar bears against snow

2. ___ cornered earwigs, skunks, and stink beetles producing awful odors

3. ___ tapeworms and humans

4. ___ yucca moth and yucca plant

5. ___ Canadian lynx and snowshoe hare

6. ___ tigers against tall-stalked and golden grasses

7. ___ striped skunk, yellow-banded wasp, and bright-orange monarch butterfly

8. ___ fungal mycelia and plant root hairs

9. ___ insect larvae that always kill larvae or pupae of other insect species and eat them

10. ___ baboon on the run turns to give canine tooth display to a pursuing leopard

11. ___ an aggressive yellowjacket is the probable model for similar-looking but edible flies

12. ___ least bittern with coloration similar to surrounding withered reeds

A. Parasitism
B. Mutualism
C. Predator-prey relationship
D. Camouflage
E. Warning coloration
F. Mimicry
G. Moment-of-truth defense
H. Parasitoid

COMMUNITY ORGANIZATION, DEVELOPMENT, AND DIVERSITY (34-V, pp. 560–566)

Terms

resource partitioning (560):_____

succession (562):_____

climax community (562): _____

pioneer species (562):_____

True/False

If the statement is true, write a T in the blank. If the statement is false, make it correct by changing the underlined word(s) and writing the correct word(s) in the answer blank.

_____ 1. Nine species of fruit-eating pigeons living in the same forest are a good example of resource <u>disturbance</u>.

_____ 2. Water hyacinths and killer bees are examples of <u>predator</u> introductions.

_____ 3. In resource partitioning, <u>similar</u> species generally share the same kind of resource in different ways, in different areas, or at different times.

_____ 4. The number of algal species is greatest in tidepools with <u>high</u> densities of algae-eating periwinkles.

_____ 5. Predation reduces the density of prey populations but can also <u>increase</u> competition between prey species and promote their coexistence.

_____ 6. In the rocky intertidal zone, mussels are the main prey of sea stars; the reduction in numbers of mussels maintains a <u>greater</u> diversity of other invertebrate prey species.

Choice

For questions 7–16, choose from the following.

   a. primary succession   b. secondary succession

___ 7. Following a disturbance, a patch of habitat or a community moves once again toward the climax state.

___ 8. Successional changes begin when a pioneer population colonizes a barren habitat.

___ 9. Involves populations that are adapted to growing in habitats that cannot support most other populations.

___ 10. Many plants in this succession arise from seeds or seedlings that are already present when the process begins.

___ 11. Early successional populations inhibit the growth of later ones, which become dominant only when some disturbance removes the established competitors.

___ 12. The first plants are typically small with short life cycles; each year they produce an abundance of quickly dispersed small seeds.

___ 13. A successional pattern that occurs in ponds, shallow lakes, abandoned fields, and in parts of established forests.

___ 14. On land, early and late species often are able to grow together under prevailing conditions.

___ 15. Might occur on a new volcanic island or on land exposed by the retreat of a glacier.

___ 16. Includes populations adapted to growing in areas exposed to intense sunlight, wide temperature swings, and nutrient-deficient soil.

Short Answer

17. Explain how modest fire disturbances among groves of dominant giant sequoia trees of the Sierra Nevada in California benefit that climax community.

_____

_____

_____

_____

18. After consideration of the distance effect, the area effect, and species diversity patterns as related to the equator, answer the following question. There are two islands (B and C) of the same size and topography that are equidistant from the African coast (A), as shown in the illustration below. Which will have the higher species diversity values?

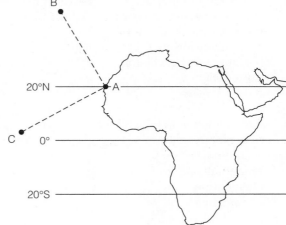

_____

19. Name two factors responsible for creating the higher species diversity values as related to the distance of land and sea from the equator.

a. _____

b. _____

Self-Quiz

___1. All the populations of different species that occupy and are adapted to a given habitat are referred to as a(n) _____.
  a. biosphere
  b. community
  c. ecosystem
  d. niche

___2. The range of all factors that influence whether a species can obtain resources essential for survival and reproduction is called the _____ of a species.
  a. habitat
  b. niche
  c. carrying capacity
  d. ecosystem

___3. A one-way relationship in which one species benefits and the other is directly harmed is called _____.
  a. commensalism
  b. competitive exclusion
  c. parasitism
  d. mutualism

___4. A lopsided interaction that directly benefits one species but does not harm or help the other much, if at all, is _____.
  a. commensalism
  b. competitive exclusion
  c. predation
  d. mutualism

___5. An interaction in which both species benefit is best described as _____.
  a. commensalism
  b. mutualism
  c. predation
  d. parasitism

___6. The brown-headed cowbird removes an egg from the nest of another kind of bird and lays one as a "replacement." This is an example of _____.
  a. commensalism
  b. competitive exclusion
  c. mutualism
  d. social parasitism

___7. When an inexperienced predator attacks a yellow-banded wasp, the predator receives the pain of a stinger and will not attack again. This is an example of _____.
  a. mimicry
  b. camouflage
  c. a prey defense
  d. warning coloration
  e. both c and d

___8. _____ is represented by foxtail grass, mallow plants, and smartweed because their root systems exploit different areas of the soil in a field.
  a. Succession
  b. Resource partitioning
  c. A climax community
  d. A disturbance

___9. During the process of community succession, _____.
  a. pioneer populations adapt to growing in habitats that cannot support most species
  b. pioneers set the stage for their own replacement
  c. later successional populations crowd out the pioneers
  d. species composition eventually is stable in the form of the climax community
  e. all of the above

___10. G. Gause utilized two species of *Paramecium* in a study that described _____.
  a. interspecific competition and competitive exclusion
  b. resource partitioning
  c. the establishment of territories
  d. coevolved mutualism

___11. The most striking patterns of species diversity on land and in the seas relate to _____.
  a. distance effect
  b. area effect
  c. immigration rate for new species
  d. distance from the equator

Chapter Objectives/Review Questions

This section lists general and detailed chapter objectives that can be used as review questions. You can make maximum use of these items by writing answers on a separate sheet of paper. Fill in answers where blanks are provided. To check for accuracy, compare your answers with information given in the chapter or glossary.

Page	Objectives/Questions
(552)	1. The type of place where you normally find a maple is its _____.
(552)	2. List five factors that shape the structure of a biological community.
(552)	3. The full range of environmental and biological conditions under which its members can live, grow, and reproduce is called the _____ of that species.
(553)	4. Define symbiosis.
(553)	5. The interdependence of the yucca plant and yucca moth is an example of _____.
(553)	6. The interaction of a bird's nest and a tree is known as _____.
(555)	7. Describe a study that demonstrates laboratory evidence in support of the competitive exclusion concept.
(556)	8. A predator gets food from other living organisms, its _____.
(556–557)	9. List three factors that influence the outcome of predator-prey interactions.
(558)	10. Suggest why coevolving might serve the interests of the populations concerned.
(558–559)	11. Be able to completely define and give examples of the following prey defenses: warning coloration and mimicry, moment-of-truth defenses, and camouflage.
(560)	12. Parasites tend to _____ with their hosts in ways that produce less-than-fatal effects.
(560)	13. _____ are insect larvae that always kill larvae or pupae of other insect species for food.
(560)	14. Cite one example of resource partitioning.
(565)	15. Explain how the introduction of nonnative species can be disastrous. List five specific examples of species introductions into the United States that have had adverse results (see Table 34–2).
(562)	16. Distinguish between primary and secondary succession.
(563)	17. Describe how fire disturbances positively affect sequoia communities.
(566)	18. Estimate qualitatively the differences in species diversity and abundance of organisms likely to exist on two islands with the following characteristics: Island A has an area of 6,000 square miles, and Island B has an area of 60 square miles; both islands lie at 10° N latitude and are equidistant from the same source area of colonizers.

Integrating and Applying Key Concepts

If you were Ruler of All People on Earth, how would you organize industry and human populations in an effort to solve our most pressing pollution problems?

Is there a fundamental niche that is occupied by humans? If you think so, describe the minimal abiotic and biotic conditions required by populations of humans in order to live and reproduce. (Note that "thrive and be happy" are not criteria.) If you do not think so, state why.

These minimal niche conditions can be viewed as resource categories that must be protected by populations if they are to survive. Do you believe that the cold war between the United States and the Soviet Union primarily involved protection of minimal niche conditions, or do you believe that the cold war was based on other, more (or less) important factors?

a. If the former, how do you think minimal niche conditions might have been guaranteed for all humans willing and able to accept certain responsibilities as their contribution toward enabling this guarantee to be met?

b. If the latter, identify what you think those factors are and explain why you consider them more (or less) important than minimal niche conditions.

Critical Thinking Exercise

1. Not long after feral dogs were introduced into Australia, the native doglike marsupial disappeared. One hypothesis proposed to explain this observation was that the dogs were stronger competitors for food than the marsupials. Which of the following assumptions was most likely made in proposing this hypothesis?
   a. Dogs produce larger litters of pups than marsupials.
   b. Marsupials could not switch to alternative food sources.
   c. Dogs introduced diseases that are lethal to marsupials.
   d. The marsupials prey on puppies of the feral dogs.
   e. Feral dogs and coyotes compete for food in America.

Answers

Answers to Interactive Exercises

CHARACTERISTICS OF COMMUNITIES (34-I)
1. D; 2. G; 3. A; 4. F; 5. B; 6. E; 7. C.

MUTUALLY BENEFICIAL INTERACTIONS (34-II)
1. a. obtains pollen only from the yucca plant and its larvae eat only yucca seeds; b. the yucca moth is the plant's only pollinator.

COMPETITIVE INTERACTIONS (34-III)
1. more; 2. reduce; 3. less; 4. the same; 5. more.

CONSUMER-VICTIM INTERACTIONS (34-IV)
1. D; 2. G; 3. A; 4. B; 5. C; 6. D; 7. E; 8. B; 9. H; 10. G (also predator-prey); 11. F; 12. D.

COMMUNITY ORGANIZATION, DEVELOPMENT, AND DIVERSITY (34-V)
1. partitioning; 2. species; 3. T; 4. intermediate or moderate; 5. reduce or decrease; 6. T; 7. b; 8. a; 9. a; 10. b; 11. b; 12. a; 13. b; 14. b; 15. a; 16. a; 17. Sequoia seeds germinate only in the absence of smaller, shade-tolerant plant species. Modest fires eliminate trees and shrubs that compete with young sequoias but do not damage mature sequoias. 18. Island C; 19. a. Greater annual amount of sunlight promotes greater resource availability. b. Species diversity is self-reinforcing. When more plant species coexist, more herbivore species emerge. More predators and parasites evolve in response to the diversity of prey and hosts.

Answers to Self-Quiz

1. b; 2. b; 3. c; 4. a; 5. b; 6. d; 7. e; 8. b; 9. e; 10. a; 11. d.

Critical Thinking Analysis

a. This could make competition more intense, but it is not a necessary assumption. Even producing smaller litters, but with higher survival rates, dogs could be more effective competitors for food species.
b. If the marsupials switched to preying on alternative food species, they would escape the competition, and both dogs and marsupials could coexist. The assumption of inability to partition resources is necessary in order to conclude that competition operates between two species.
c. This is an alternative hypothesis that would also explain the disappearance without competition.
d. This assumption would tend to lead to the opposite outcome, extinction of the dogs, and therefore does not contribute to an explanation of the observations.
e. This would show only that dogs can compete for food with some species under some circumstances. It is not a necessary assumption in order to conclude that they do compete in this situation.

35

ECOSYSTEMS

Interactive Exercises

CHARACTERISTICS OF ECOSYSTEMS (35-I, pp. 569–570)

Terms

The page-referenced terms are important; they were in boldface type in the chapter. Refer to the instructions given in Chapter 1, p. 1 of this workbook.

primary producers (569):_____

consumers (569): _____

detritivores (569): _____

decomposers (569):_____

ecosystem (570): _____

Choice

For questions 1–15, choose from the following:

a. primary producer   b. consumer   c. detritivore   d. decomposer

_____ 1. mule deer

_____ 2. earthworm

_____ 3. parasites

_____ 4. the only category lacking heterotrophs

_____ 5. omnivores

_____ 6. tapeworm

_____ 7. herbivores

_____ 8. wolf

_____ 9. fungi and bacteria

_____ 10. carnivores

_____ 11. green plants

_____ 12. *Homo sapiens*

_____ 13. crabs

_____ 14. autotrophs

_____ 15. grasshopper

Dichotomous Choice

Circle one of two possible answers given between parentheses in each statement.

16. Ecosystems are (open/closed) systems, and so are not self-sustaining.
17. Minerals carried by erosion into a lake represent nutrient (input/output).
18. Energy (can/cannot) be recycled.
19. Nutrients typically (can/cannot) be cycled.
20. Ecosystems have energy and nutrient inputs as well as nutrient output and (have/lack) energy output.

STRUCTURE OF ECOSYSTEMS (35-II, pp. 570–571)

Terms

trophic levels (570):_____

food chains (570):_____

food webs (570):_____

Matching

Match each organism in the Antarctic food web given below with the principal trophic level it occupies. (See Figure 35.3.)

1. _____ emperor penguin
2. _____ krill
3. _____ blue whale
4. _____ diatoms
5. _____ leopard seal
6. _____ fishes, small squid
7. _____ killer whale
8. _____ petrel
9. _____ skua
10. _____ Adélie penguin

A. Primary producers
B. Primary consumer, herbivore
C. Consumer, primary carnivore
D. Tertiary consumers
E. Consumer, secondary carnivore
F. Consumer, secondary or tertiary carnivore
G. Consumer, top carnivore

ENERGY FLOW THROUGH ECOSYSTEMS (35-III, pp. 572–574)

Terms

primary productivity (572):_____

grazing food webs (572):_____

detrital food webs (572):_____

energy pyramid (573):_____

Complete the Table

1. Complete the following table of productivity definitions.

Productivity Type	Definition
a. Primary productivity	
b. Gross primary productivity	
c. Net primary productivity	

Dichotomous Choice

Circle one of two possible answers given between parentheses in each statement.

2. The amount of energy actually stored in an ecosystem depends on how many (plants/animals) are present and the balance between photosynthesis and aerobic respiration in the plants.
3. In a harsh ecosystem environment, productivity would be expected to be (lower/higher).
4. Heat losses represent a one-way flow of (energy/materials) out of the ecosystem.

5. In (grazing/detrital) food webs, energy flows from plants to herbivores, then through some array of carnivores.
6. In (grazing/detrital) food webs, energy flows mainly from plants through decomposers and detritivores.

Fill-in-the-Blanks

Energy flows into ecosystems from an outside source, which in most cases is the (7) _____. Energy flows through the food (8) _____ of ecosystems. Living tissues of photosynthesizers are the basis of (9) _____ food webs. The remains of photosynthesizers and consumers are the basis of (10) _____ food webs. Energy leaves ecosystems mainly by loss of metabolic (11) _____, which each organism generates. The trophic structure of an ecosystem is often represented by an "ecological (12) _____." In such schemes, (13) _____ form the base for successive tiers of consumers above them. Some pyramids are based on (14) "_____" as determined by the weight of all the members at each trophic level. Some pyramids of (15) _____ are "upside down" with the smallest tier at the bottom; they represent an ecosystem in which reproductive rates and body sizes differ at each trophic level. An (16) _____ pyramid reflects the energy losses at each transfer to a different trophic level and are always "right-side up" with the largest energy tier at the base.

BIOGEOCHEMICAL CYCLES (35-IV, pp. 575–584)

Terms

biogeochemical cycles (575):_____

water cycle (576):_____

watersheds (576): _____

carbon cycle (579): _____

greenhouse effect (580):_____

nitrogen cycle (582):_____

nitrogen fixation (582): _____

ammonification (582): _____

nitrification (582):_____

leaching (582):_____

denitrification (582):_____

ecosystem modeling (583) _____

Short Answer

1. In what form are elements used as nutrients usually available to producers?

   _____

2. How is the ecosystem's reserve of nutrients maintained? _____

   _____

3. How does the amount of a nutrient being cycled through most major ecosystems compare with the amount entering or leaving in a given year? _____

   _____

4. What are the input sources for an ecosystem's nutrient reserves? _____

   _____

5. What are the output sources of nutrient loss for land ecosystems? _____

   _____

Complete the Table

6. Complete the following table to summarize the functions of the three types of biogeochemical cycles.

Biogeochemical Cycle Type	General Function(s)
a. Hydrologic cycle	
b. Atmospheric cycles	
c. Sedimentary cycles	

Matching

Choose the one most appropriate answer for questions 7–10; 11–13 may have more than one answer.

7. _____ solar energy

8. _____ source of most water

9. _____ watershed

10. _____ water and plants taking up water

11. _____ forms of precipitation falling to land

12. _____ deforestation

13. _____ have important roles in the global hydrologic cycle

14. _____ forms of atmospheric water

A. Mostly rain and snow
B. Mechanisms by which nutrients move into and out of ecosystems
C. Water vapor, clouds, and ice crystals
D. Where precipitation of a specified region becomes funneled into a single stream or river
E. May have long-term disruptive effects on nutrient availability for an entire ecosystem
F. Slowly drives water through the atmosphere, on or through land mass surface layers, to oceans, and back again
G. Ocean currents and wind patterns
H. Evaporation from the oceans

Matching

Choose the one most appropriate answer for each.

15. _____ greenhouse gases

16. _____ carbon dioxide fixation

17. _____ carbon cycle

18. _____ ways carbon enters the atmosphere

19. _____ greenhouse effect

20. _____ carbon dioxide (CO_2)

21. _____ oceans and plant biomass

        accumulation

A. Form of most of the atmospheric carbon

B. Aerobic respiration, fossil fuel burning, and volcanic eruptions

C. CO_2, CFCs, CH_4, and N_2O

D. Photosynthesizers incorporate carbon atoms into organic compounds

E. Annual "holding stations" for about half of all atmospheric carbon

F. Carbon reservoirs → atmosphere and oceans→ through organisms→ carbon reservoirs

G. Warming of the earth's lower atmosphere due to accumulation of greenhouse gases

Choice

For questions 22–26, choose from the following.

    a. nitrogen cycle   b. nitrogen fixation   c. ammonification   d. nitrification   e. denitrification

_____ 22. Ammonia or ammonium in soil are stripped of electrons, and nitrite (NO_2) is the result; other bacteria convert nitrite to nitrate (NO_3^-).

_____ 23. Bacteria convert nitrate or nitrite to N_2 and a bit of nitrous oxide (N_2O).

_____ 24. Bacteria and fungi break down nitrogen-containing wastes and plant and animal remains; released amino acids and proteins are used for growth with the excess given up as ammonia or ammonium that plants can use.

_____ 25. Occurs in the atmosphere (largest reservoir); only certain bacteria, volcanic action, and lightning can convert N_2 into forms that can enter food webs.

_____ 26. A few kinds of bacteria convert N_2 to ammonia (NH_3), which dissolves quickly in water to form ammonium (NH_4^+).

Short Answer

27. List reasons that a sufficient soil nitrogen supply is a problem for land plants.

_____

_____

_____

_____

_____

TRANSFER OF HARMFUL COMPOUNDS THROUGH ECOSYSTEMS (35-V, p 584)

Terms

biological magnification (584):_____

ecosystem modeling (583):_____

Fill-in-the-Blanks

During World War II, DDT was sprayed in the tropical Pacific to control (1)_____ responsible for trans-mitting the organisms that cause a dangerous disease, (2)_____. Fill in the blank for this food web:

DDT sprayings

detrital particles → bugs, worms → songbirds → peregrine falcons

dead insects → salmon → osprey → eggs with (3) _____ _____

Because of its stability, DDT is a prime candidate for (4)_____ _____—the increasing concentra-tion of a nondegradable substance as it moves up through trophic levels. (5)_____ _____ is a method of combining crucial bits of information about an ecosystem through computer programs and mod-els in order to predict the outcome of the next disturbance.

Self-Quiz

___1. A network of interactions that involve the cycling of materials and the flow of energy between a community and its physical environment is a(n) _____.
a. population
b. community
c. ecosystem
d. biosphere

___2. _____ consume dead or decomposing particles of organic matter.
a. Herbivores
b. Parasites
c. Detritivores
d. Carnivores

___3. In the Antarctic, blue whales feed mainly on _____.
a. petrels
b. krill

c. seals
d. fish and small squids

___4. _____ is a process in which nitroge-nous waste products or organic remains of organisms are decomposed by soil bacteria and fungi that use the amino acids being released for their own growth and release the excess as ammonia or ammonium, which plants take up.
a. Nitrification
b. Ammonification
c. Denitrification
d. Nitrogen fixation

___5. In a natural community, the primary con-sumers are _____.
a. herbivores
b. carnivores
c. scavengers
d. decomposers

___6. A straight-line sequence of who eats whom in an ecosystem is sometimes called a(n) _____.
   a. trophic level
   b. food chain
   c. ecological pyramid
   d. food web

___7. Which of the following is a primary consumer?
   a. cow
   b. dog
   c. hawk
   d. all of the above

___8. Of the 1,700,000 kilocalories of solar energy that entered an aquatic ecosystem in Silver Springs, Florida, investigators determined that about _____ percent of incoming solar energy was trapped by photosynthetic autotrophs.
   a. 1
   b. 10

   c. 25
   d. 74

___9. In the carbon cycle, carbon enters the atmosphere through _____.
   a. carbon dioxide fixation
   b. respiration, burning, and volcanic eruptions
   c. oceans and accumulation of plant biomass
   d. release of greenhouse gases

___10. _____ refers to an increase in concentration of a nondegradable (or slowly degradable) substance in organisms as it is passed along food chains.
   a. Ecosystem modeling
   b. Nutrient input
   c. Biogeochemical cycle
   d. Biological magnification

Chapter Objectives/Review Questions

This section lists general and detailed chapter objectives that can be used as review questions. You can make maximum use of these items by writing answers on a separate sheet of paper. Fill in answers where blanks are provided. To check for accuracy, compare your answers with information given in the chapter or glossary.

Page	Objectives/Questions
(570–571)	1. List the principal trophic levels in an ecosystem of your choice; state the source of energy for each trophic level and give one or two examples of organisms associated with each trophic level.
(570)	2. Explain why nutrients can be completely recycled but energy cannot.
(570)	3. An _____ is a complex of organisms and their physical environment, all interacting through a flow of energy and a cycling of materials.
(570)	4. Members of an ecosystem fit somewhere in a hierarchy of energy transfers (feeding relationships) called _____ levels.
(570–571)	5. Distinguish between food chains and food webs.
(572)	6. Understand how materials and energy enter, pass through, and exit an ecosystem.
(572)	7. Distinguish between net and gross primary productivity.
(572)	8. Compare grazing food webs with detrital food webs. Present an example of each.
(572–573)	9. Ecological pyramids are based on _____ as determined by the weight of all the members of each trophic level; _____ pyramids reflect the energy losses at each transfer to a different trophic level.
(575)	10. In _____ cycles, the nutrient is transferred from the environment to organisms, then back to the environment—which serves as a large reservoir for it.
(576)	11. Be able to discuss water movements through the hydrologic cycle.
(576–577)	12. Explain what studies in the Hubbard Brook watershed have taught us about the movement of substances (water, for example) through a forest ecosystem.

(578–579) 13. The carbon cycle traces carbon movement from reservoirs in the _____ and oceans, through organisms, then back to reservoirs.

(580) 14. Certain gases cause heat to build up in the lower atmosphere, a warming action known as the _____ effect.

(582) 15. A major element found in all proteins and nucleic acids moves in an atmospheric cycle called the _____ cycle.

(582) 16. Define the chemical events that occur during nitrogen fixation, nitrification, ammonification, and denitrification.

(582–583) 17. Explain why agricultural methods in the United States tend to put more energy into mechanized agriculture in the form of fertilizers, pesticides, food processing, storage, and transport than is obtained from the soil in the form of energy stored in foods.

(584) 18. Describe how DDT damages ecosystems; discuss biological magnification.

(583) 19. Through _____ modeling, crucial bits of information about different ecosystem components are identified and used to build computer models for predicting outcomes of ecosystem disturbances.

Integrating and Applying Key Concepts

In 1971, *Diet for a Small Planet* was published. Frances Moore Lappé, the author, felt that people in the United States of America wasted protein and ate too much meat. She said, "We have created a national consumption pattern in which the majority, who can pay, overconsume the most inefficient livestock products [cattle] well beyond their biological needs (even to the point of jeopardizing their health), while the minority, who can not pay, are inadequately fed, even to the point of malnutrition." Cases of marasmus (a nutritional disease caused by prolonged lack of food calories) and kwashiorkor (caused by severe, long-term protein deficiency) have been found in Nashville, Tennessee, and on an Indian reservation in Arizona, respectively. Lappé's partial solution to the problem was to encourage people to get as much of their protein as possible directly from plants and to supplement that with less meat from the more efficient converters of grain to protein (chickens, turkeys, and hogs) and with seafood and dairy products. Most of us realize that feeding the hungry people of the world is not just a matter of distributing the abundance that exists—that it is being prevented in part by political, economic, and cultural factors. Devise two full days of breakfasts, lunches, and dinners that would enable you to exploit the lowest acceptable trophic levels to sustain yourself healthfully.

Critical Thinking Exercise

1. A sealed chamber was prepared containing five different organisms under optimum conditions for extended survival. One organism was a plant containing starch labeled with radioactive carbon. Another was a carnivore. If you predict that radioactive carbon would eventually be found in all the organisms, which of the following are you most likely assuming?

   a. The other three organisms were carnivores.
   b. At least one of the other organisms was an herbivore.
   c. The plant would oxidize some of its own starch to carbon dioxide.
   d. At least two of the other organisms were also plants.
   e. The carnivore respired more rapidly than the plant.

Answers

Answers to Interactive Exercises

CHARACTERISTICS OF ECOSYSTEMS (35-I)

1. b; 2. c; 3. b; 4. a; 5. b; 6. b; 7. b; 8. b; 9. d; 10. b; 11. a; 12. b; 13. b, c; 14. a; 15. b; 16. open; 17. input; 18. cannot; 19. can; 20. have.

STRUCTURE OF ECOSYSTEMS (35-II)

1. E; 2. B; 3. C; 4. A; 5. F; 6. C; 7. G; 8. C; 9. E; 10. C.

ENERGY FLOW THROUGH ECOSYSTEMS (35-III)

1. a. the rate at which the ecosystem's producers capture and store a given amount of energy in a given length of time; b. the total rate of photosynthesis for an ecosystem during a specified period; c. the rate of energy storage in plant tissues in excess of the rate of aerobic respiration by the plants themselves; 2. plants; 3. lower; 4. energy; 5. grazing; 6. detrital; 7. Sun; 8. webs; 9. grazing; 10. detrital; 11. heat; 12. pyramid; 13. producers; 14. biomass; 15. biomass; 16. energy.

BIOGEOCHEMICAL CYCLES (35-IV)

1. As mineral ions such as ammonium (NH_4^+); 2. Inputs from the physical environment and the cycling activities of decomposers and detritivores; 3. The amount of a nutrient being cycled through the ecosystem is greater; 4. Rainfall or snowfall, metabolism (such as nitrogen fixation), and weathering of rocks; 5. Losses by runoff and evaporation; 6. a. oxygen and hydrogen move in the form of water molecules; b. a large portion of the nutrient is in the form of atmospheric gas such as carbon and nitrogen; c. the nutrient is not in gaseous forms; it moves from land to the seafloor and only "returns" to land through geological uplifting of long duration; phosphorus is an example; 7. F; 8. H; 9. D; 10. B; 11. A (C); 12. E (B); 13. G (H); 14. C; 15. C; 16. D; 17. F; 18. B; 19. G; 20. A; 21. E; 22. d; 23. e; 24. c; 25. a; 26. b; 27. Soil nitrogen compounds are vulnerable to being leached and lost from the soil; some fixed nitrogen is lost to air by denitrification;

nitrogen fixation comes at high metabolic cost to plants that are symbionts of nitrogen-fixing bacteria; losses of nitrogen are enormous in agricultural regions through the tissues of harvested plants, soil erosion, and leaching processes.

TRANSFER OF HARMFUL COMPOUNDS THROUGH ECOSYSTEMS (35-V)

1. mosquitoes; 2. malaria; 3. thin shells (brittle shells); 4. biological magnification; 5. Ecosystem modeling.

Answers to Self-Quiz

1. c; 2. c; 3. b; 4. b; 5. a; 6. b; 7. a; 8. a; 9. b; 10. d.

Critical Thinking Analysis

Your answer depends on your assumption about the radioactive carbon to be found in the organisms. If you assume that this radioactive carbon could simply be in some carbon dioxide molecules in the animals' body fluids, then you are most likely assuming (c). Oxidation of its own starch would be one way to release radiocarbon from the plant into the atmosphere, from where it could diffuse into the blood of any animal present. However, if one of the other organisms was an herbivore, it could eat the plant and release some radioactive carbon dioxide, or it could incorporate radiocarbon into its own molecules and then be eaten by the carnivore. The latter point illustrates the other interpretation of the question. If you assume that the radioactive carbon must be found in the organisms' own molecules, then you must be assuming that at least one of the animals is an herbivore. Only via an herbivore can radioactive carbon move from plant starch into a carnivore's molecules. This also assumes that the carnivore is a strict carnivore and never eats a little plant material.

36

THE BIOSPHERE

CHARACTERISTICS OF THE BIOSPHERE
    Biosphere Defined
    Global Patterns of Climate

THE WORLD'S BIOMES
    Deserts
    Dry Shrublands and Woodlands
    Grasslands and Savannas

    Forests
    Tundra

THE WATER PROVINCES
    Lake Ecosystems
    Stream Ecosystems
    Marine Ecosystems

Interactive Exercises

CHARACTERISTICS OF THE BIOSPHERE (36-I, pp. 587–589)

Terms

The page-referenced terms are important; they were in boldface type in the chapter. Refer to the instructions given in Chapter 1, p. 1 of this workbook.

biosphere (587):_____

hydrosphere (587): _____

atmosphere (587): _____

climate (587): _____

ozone layer (587): _____

rain shadow (589): _____

Matching

Choose the one most appropriate answer for each.

1. _____ biosphere
2. _____ global patterns of air circulation
3. _____ surface currents and drifts in the ocean
4. _____ ozone layer
5. _____ influence the distribution and dominant features of ecosystems
6. _____ rain shadow
7. _____ greenhouse effect
8. _____ Earth's seasonal climate variations
9. _____ Earth's regional climates
10. _____ climate

A. Due to tilt of Earth's axis, which yields annual incoming solar radiation variation; provides daylength and temperature differences
B. Caused by molecules of the lower atmosphere absorbing some heat and then reradiating it back to Earth; this heat drives earth's weather systems
C. Influenced by mountains, valleys, and other land formations
D. The entire Earth realm where organisms live
E. Due to warm equatorial air rising, then spreading southward and northward; Earth's rotation creates worldwide belts of prevailing east and west winds
F. Reduction in rainfall on the leeward side of high mountains
G. Created by Earth's rotation, prevailing surface winds, and water temperature variations
H. Refers to prevailing weather conditions such as temperature, humidity, wind speed, cloud cover, and rainfall
I. Contains atmospheric molecules that absorb potentially lethal ultraviolet wavelengths
J. Influenced by interaction of atmospheric circulation patterns, ocean currents, and landforms

THE WORLD'S BIOMES (36-II, pp. 590–599)

Terms

biogeographic realms (590): _____

biome (590): _____

soil (592): _____

deserts (593): _____

desertification (593): _____

dry shrublands (594): _____

dry woodlands (594): _____

grasslands (594): _____

forests (596): _____

arctic tundra (598): _____

permafrost (598): _____

alpine tundra (598): _____

Choice

For questions 1–15, choose from the following.

    a. deserts   b. dry shrublands   c. dry woodlands   d. grasslands   e. evergreen broadleaf forests
    f. deciduous broadleaf forests   g. evergreen coniferous forests   h. tundra

_____ 1. Nearly continuous sunlight in summer.

_____ 2. Near the equator, highly productive, rainfall regular and heavy.

_____ 3. Decomposition and mineral cycling are rapid in the hot, humid climate.

_____ 4. Biome where the potential for evaporation greatly exceeds rainfall.

_____ 5. Home to deep-rooted evergreen shrubs, fleshy-stemmed, shallow-rooted cacti, saguaro, short prickly pear, and ocotillo.

_____ 6. Steinbeck's *Grapes of Wrath* speaks eloquently of the disruption of this biome.

_____ 7. Maple and beech forests.

_____ 8. Boreal forests or taiga.

_____ 9. The plants are woody, multibranched and a few meters tall; fynbos and chaparral.

_____ 10. Spruce and balsam fir dominate this North American biome.

_____ 11. Shortgrass prairie, tallgrass prairie, tropical savannas, and monsoon grasslands.

_____ 12. Tall trees are here; eucalyptus woodlands of southwestern Australia and oak woodlands of the Pacific states.

_____ 13. Biome of the temperate zone, cold winter temperatures, many trees drop all their leaves in winter.

_____ 14. Lies between the polar ice cap and belts of boreal forests in North America, Europe, and Asia.

_____ 15. Soils are highly weathered, humus-poor, and not good nutrient reservoirs.

THE WATER PROVINCES (36-III, pp. 600–609)

Terms

lake (600):_____

plankton (600): _____

spring overturn (601): _____

fall overturn (601): _____

eutrophication (601): _____

streams (602):_____

estuary (602):_____

intertidal zone (603):_____

benthic province (605):_____

pelagic province (605):_____

hydrothermal vents (605):_____

upwelling (608): _____

El Niño (608): _____

Complete the Table

1. Complete the following table, which describes the three zones of a lake.

Lake Zone	Description
a. Littoral	
b. Limnetic	
c. Profundal	

Dichotomous Choice

Circle one of two possible answers given between parentheses in each statement.

2. Water near the freezing point is the (least/most) dense and accumulates beneath the ice.
3. Water at 4°C is densest; it accumulates in deeper layers, which are a bit (warmer/cooler) than the surface layer in midwinter.
4. Daylength increases, air warms, lake ice melts, and surface water slowly warms to 4°C so that temperatures become uniform throughout; winds blowing across a lake cause a (spring/fall) overturn.
5. In (spring/fall) overturns, strong vertical movements carry dissolved oxygen from the lake's warming surface layer to its depths, and nutrients released by decomposition are brought from the bottom sediments to the surface layer.
6. By midsummer, the surface layer is well above 4°C and the lake may have a "thermocline," a(n) (upper/middle) layer that changes abruptly in temperature and prevents vertical mixing.
7. At midsummer, (warmer/cooler) and less dense surface water floats on the thermocline.
8. In summer, decomposers may deplete the (upper/lower) water of dissolved oxygen.
9. The lake's upper layer cools, becomes denser, and sinks, and the thermocline vanishes. This describes the (spring/fall) overturn when the dissolved oxygen again moves down and nutrients move up.
10. Primary productivity of a lake increases following spring overturn, but as the season progresses, the thermocline prevents mixing of nutrients accumulating in (surface/deeper) waters and primary productivity declines.
11. By late summer, nutrient shortages limit photosynthesis. Following fall overturn, cycled nutrients drive a (short/long) burst of primary productivity.

Complete the Table

12. Complete the following table that summarizes the trophic nature of lakes.

Trophic Nature	Description
a. Oligotrophic lake	
b. Eutrophic lake	

Labeling

Label each numbered part of the accompanying illustration (see text, Figure 36.29).

13. _____ zone

14. _____ zone

15. _____ zone

16. _____ province

17. _____ _____

18. _____ zone

19. _____ zone

20. _____ zone

21. _____ province

22. _____ water

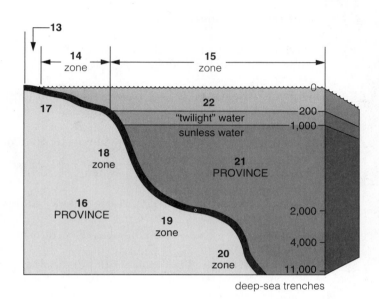

deep-sea trenches

Choice

For questions 23–47, choose from the following.

    a. stream ecosystem   b. estuary   c. intertidal zone   d. benthic province   e. pelagic province

_____ 23. Includes all sediments and rocky formations of the ocean bottom.

_____ 24. Riffles, pools, and runs.

_____ 25. A partly enclosed coastal ecosystem where seawater mixes with nutrient-rich freshwater from rivers, streams, and runoff from land.

_____ 26. Along rocky and sandy coastlines where resident organisms are battered by waves.

_____ 27. The entire volume of ocean water.

_____ 28. Begin as freshwater springs or seeps.

_____ 29. Chesapeake Bay and San Francisco Bay are broad shallow examples; narrow and deep examples are found in Alaska, British Columbia, and Norway.

_____ 30. Rocky shores, sandy and muddy shores.

_____ 31. Photosynthetic activity is restricted to its upper waters.

_____ 32. Includes the neritic and oceanic zones.

_____ 33. Remarkable communities thrive at hydrothermal vents of its abyssal zone.

_____ 34. Average flow volume and temperature depend on rainfall, snowmelt, geography, altitude, and even shade cast by plants.

_____ 35. Primary producers here include salt-tolerant grasses, algae that grow on surfaces of plants and mud, and phytoplankton.

_____ 36. Tide pools with algae, small invertebrates, and small fishes.

_____ 37. Phytoplankton drifting with currents are the basis of food webs that include zooplankton, copepods, shrimplike krill, squids, and fishes.

_____ 38. Organic remains and wastes from pelagic communities sink down to become the basis of detrital food webs for most communities here.

_____ 39. They grow and merge as they flow downslope, then often combine to form a river.

_____ 40. The most productive aquatic ecosystems on earth.

_____ 41. Upper, mid-, and lower littoral zones of rocky shores.

_____ 42. Chemosynthetic bacteria are the primary producers of hydrothermal vents in the abyssal zone.

_____ 43. Tropical reefs form in both the neritic and oceanic zones here.

_____ 44. Home to a great diversity of invertebrates and a rest stop for migratory birds; subject to much human pollution.

_____ 45. Since cities formed, these have been sewers for industrial and municipal wastes.

_____ 46. Alternately submerged and exposed by tides; land animals feed here at low tides, and high tides bring predatory fishes.

_____ 47. Begins with the continental shelf and extends down to deep-sea trenches.

Fill-in-the-Blanks

(48)_____ is an upward movement of deep, nutrient-rich ocean water along the margins of continents. This occurs when (49)_____ force surface waters to move away from a coastline and deep water moves in vertically to replace it. Every two to seven years, warm surface waters of the western equatorial Pacific move eastward. This is a massive displacement of (50)_____ water that influences the prevailing winds to accelerate eastward flow. This flow displaces the cooler waters of the Humboldt Current and prevents upwelling. Meteorologists call the phenomenon ENSO, or the (51)_____ _____ Southern Oscillation. ENSO has a catastrophic effect on (52)_____, particularly on anchoveta-eating birds, and the anchoveta industry.

Self-Quiz

___1. The distribution of different types of ecosystems is influenced by _____.
a. air currents
b. variation in the amount of solar radiation reaching the earth through the year
c. ocean currents
d. all of the above

___2. In a(n)_____, water draining from the land mixes with seawater carried in on tides.
a. pelagic province
b. rift zone
c. upwelling
d. estuary

___3. A biome with grasses as primary producers and scattered trees adapted to prolonged dry spells is known as a _____.
a. warm desert
b. savanna
c. tundra
d. taiga

___4. Located at latitudes of about 30° north and south, limited vegetation, and rapid surface cooling at night describes a _____ biome.
a. shrubland
b. savanna
c. taiga
d. desert

___5. In tropical rain forests, _____.
a. competition for available sunlight is intense
b. diversity is limited because the tall forest canopy shuts out most of the incoming light
c. conditions are extremely favorable for growing luxuriant crops
d. all of the above

___6. In a lake, the open sunlit water with its suspended phytoplankton is referred to as its _____ zone.
a. epipelagic
b. limnetic
c. littoral
d. profundal

___7. The lake's upper layer cools, the thermocline vanishes, lake water mixes vertically, and once again dissolved oxygen moves down and nutrients move up. This describes the _____.
a. spring overturn
b. summer overturn
c. fall overturn
d. winter overturn

___8. The _____ is a permanently frozen, water-impermeable layer just beneath the surface of the _____ biome.
a. permafrost; alpine tundra
b. hydrosphere; alpine tundra
c. permafrost; arctic tundra
d. taiga; arctic tundra

___9. _____ refers to a season of heavy rain that corresponds to a shift in prevailing winds over the Indian Ocean.
a. Geothermal ecosystem
b. Upwelling
c. Taiga
d. Monsoon

___10. All of the water above the continental shelves is in the _____.
a. neritic zone of the benthic province
b. oceanic zone of the pelagic province
c. neritic zone of the pelagic province
d. oceanic zone of the benthic province

Chapter Objectives/Review Questions

This section lists general and detailed chapter objectives that can be used as review questions. You can make maximum use of these items by writing answers on a separate sheet of paper. Fill in answers where blanks are provided. To check for accuracy, compare your answers with information given in the chapter or glossary.

Page   *Objectives/Questions*

(587)   1. The _____ is the entire realm in which organisms live; the portion containing oceans, polar ice caps, and all other forms of water constitute the _____; the envelope of gases and airborne particles is the _____.

(587)   2. _____ refers to prevailing weather conditions, such as temperature, humidity, wind speed, cloud cover, and rainfall.

(587)   3. State the reason that most forms of life depend on the ozone layer.

(587)   4. _____ energy drives the earth's weather systems.

(587–588)   5. Be able to describe the causes of global air circulation patterns.

(589)   6. Describe how the tilt of the earth's axis affects annual variation in the amount of incoming solar radiation.

(589)   7. Atmospheric and oceanic _____ patterns influence the distribution of different types of ecosystems.

(589)   8. Mountains, valleys, and other land formations influence _____ climates.

(589)   9. Describe the cause of the rain shadow effect.

(590)   10. Broadly, there are six distinct land realms, the _____ realms that were named by W. Sclater and Alfred Wallace.

(590)   11. Each major type of ecosystem is a _____.

(592)   12. _____ is a mixture of rock, mineral ions, and organic matter in some state of physical and chemical breakdown.

(593–599)   13. Be able to list the major biomes and briefly characterize them in terms of climate, topography, and organisms.

(600)   14. A _____ is a body of fresh water with littoral, limnetic, and profundal zones.

(600)   15. Define plankton, phytoplankton, and zooplankton.

(601)   16. Describe the spring and fall overturn in a lake in terms of causal conditions and physical outcomes.

(601)   17. _____ refers to nutrient enrichment of a lake or some other body of water.

(601)   18. _____ lakes are often deep, poor in nutrients, and low in primary productivity; _____ lakes are often shallow, rich in nutrients, and high in primary productivity.

(602)   19. Describe a stream ecosystem.

(602–605)   20. Be able to fully describe estuary, intertidal zone, benthic, and pelagic provinces of the ocean.

(605)   21. In the abyssal zone, remarkable communities thrive at _____ vents.

(608)   22. State the significance of ocean upwelling.

(608)   23. Describe conditions of ENSO occurrence and its significance to the ocean.

Integrating and Applying Key Concepts

One species, *Homo sapiens*, uses about 40 percent of all of Earth's productivity, and its representatives have invaded every biome, either by living there or by dumping waste products there. Many of Earth's residents are being denied the minimal resources they need to survive, while human populations continue to increase exponentially. Can you suggest a better way of keeping Earth's biomes healthy while providing at least the minimal needs of all Earth's residents (not just humans)? If so, outline the requirements of such a system and devise a way in which it could be established.

Critical Thinking Exercise

1. Eutrophic lakes are characterized by enormous growth of photosynthetic algae that release oxygen. Why, then, do these lakes undergo severe oxygen depletion?

    a. The algae exhaust the available supply of phosphorus and stop photosynthesis.
    b. Fish in the lakes die as a result of the low oxygen concentration.
    c. The rate of decomposition of dead algae becomes greater than the rate of photosynthesis by living algae.
    d. Oxygen diffuses into the lake water from the atmosphere.
    e. Oxygen diffuses out of the lake water into the atmosphere.

Answers

Answers to Interactive Exercises

CHARACTERISTICS OF THE BIOSPHERE (36-I)
1. D; 2. E; 3. G; 4. I; 5. J (C); 6. F; 7. B; 8. A; 9. C (J); 10. H.

THE WORLD'S BIOMES (36-II)
1. h; 2. e; 3. e; 4. a; 5. a; 6. d; 7. f; 8. g; 9. b; 10. g; 11. d; 12. c; 13. f; 14. h; 15. e (h).

THE WATER PROVINCES (36-III)
1. a. shallow, usually well-lit zone extending all around the shore to the depth at which rooted aquatic plants stop growing; diversity of organisms is greatest in the littoral; b. Open, sunlit water beyond the littoral, down to a depth where photosynthesis is insignificant; has plankton; c. all the open water below the depth at which wavelengths suitable for photosynthesis can penetrate; detritus sinks through the profundal to the bacterial decomposers of bottom sediments; decomposition releases nutrients into the water; 2. least; 3. warmer; 4. spring; 5. spring; 6. middle; 7. warmer; 8. lower; 9. fall; 10. deeper; 11. short; 12. a. often deep, poor in nutrients, and low in primary productivity; b. often shallow, rich in nutrients, and high in primary productivity; 13. intertidal; 14. neritic; 15. oceanic; 16. benthic; 17. continental shelf; 18. bathyal; 19. abyssal; 20. hadal; 21. pelagic; 22. sunlit; 23. d; 24. a; 25. b; 26. c; 27. e; 28. a; 29. b; 30. c; 31. e; 32. e; 33. d; 34. a; 35. b; 36. c; 37. e; 38. d; 39. a; 40. b; 41. c; 42. d; 43. e; 44. b; 45. a; 46. c; 47. d; 48. Upwelling; 49. winds; 50. warm; 51. El Niño; 52. productivity.

Answers to Self-Quiz

1. d; 2. d; 3. b; 4. d; 5. a; 6. b; 7. c; 8. c; 9. d; 10. c.

Critical Thinking Analysis

a. Even if the production of new algal cells stops because of exhaustion of phosphorus, photosynthesis will continue in the cells already there. Phosphorus is a required nutrient for ATP production, nucleic acid synthesis, and cell reproduction, but not for photosynthesis.
b. This is a common consequence of eutrophic oxygen depletion, but it does not explain the depletion.
c. Decomposition is respiration of decomposers, which consumes oxygen. The rate of change of oxygen concentration is the difference between processes that add oxygen and processes that remove oxygen. As the amount of dead algae becomes larger, the population of decomposers also grows, the rate of oxygen consumption by decomposers can exceed the rate of oxygen production by the living algae, and the oxygen concentration will decrease.
d. This would tend to reverse the depletion, not account for it.
e. Diffusion can only equalize concentrations of a substance between two compartments. This case is development of an inequality. Diffusion cannot accomplish that.

37

HUMAN IMPACT ON THE BIOSPHERE

Interactive Exercises

ENVIRONMENTAL EFFECTS OF HUMAN POPULATION GROWTH (37-I, p. 612)

Short Answer

1. Briefly list the major environmental effects of human population growth.

_____

_____

_____

_____

CHANGES IN THE ATMOSPHERE (37-II, pp. 612–615)

Terms

The page-referenced terms are important; they were in boldface type in the chapter. Refer to the instructions given in Chapter 1, p. 1 of this workbook.

pollutants (612): _____

thermal inversion (613): _____

industrial smog (613): _____

photochemical smog (613): _____

acid rain (614): _____

ozone hole (615): _____

chlorofluorocarbons (615): _____

Choice

For questions 1–22, choose from the following aspects of atmospheric pollution.

    a. thermal inversion   b. industrial smog   c. photochemical smog   d. acid deposition
    e. chlorofluorocarbons   f. ozone layer

_____ 1. Develops as a brown, smelly haze over large cities.

_____ 2. Includes the "dry" and "wet" types.

_____ 3. Contributes to ozone reduction more than any other factor.

_____ 4. Where winters are cold and wet, this develops as a gray haze over industrialized cities that burn coal and other fossil fuels.

_____ 5. Weather conditions trap a layer of cool, dense air under a layer of warm air.

_____ 6. The cause of London's 1952 air pollution disaster, in which 4,000 died.

_____ 7. Each year, from September through mid-October, it thins down by as much as *half* above Antarctica.

_____ 8. By some estimates, nearly all of this released between 1955 and 1990 is still making its way up to the stratosphere.

_____ 9. Intensifies a phenomenon called smog.

_____ 10. Today most of this forms in cities of China, India, and other developing countries, as well as in Hungary, Poland, and other countries of eastern Europe.

_____ 11. Contains airborne pollutants, including dust, smoke, soot, ashes, asbestos, oil, bits of lead and other heavy metals, and sulfur oxides.

____ 12. Depending on soils and vegetation, some regions are more sensitive than others to this.

____ 13. Have contributed to some of the worst local air pollution disasters.

____ 14. Chemically attack marble buildings, metals, mortar, rubber, plastic, and even nylon stockings.

____ 15. Its reduction allows more ultraviolet radiation to reach the earth's surface.

____ 16. Reaches harmful levels where the surrounding land forms a natural basin, as it does around Los Angeles and Mexico City.

____ 17. Tall smokestacks were added to power plants and smelters in an unsuccessful attempt to solve this problem.

____ 18. Those already in the air will be there for over a century, before they are neutralized by natural processes.

____ 19. A dramatic rise in skin cancers, eye cataracts, immune system weakening, and harm to photosynthesizers is related to its reduction.

____ 20. Widely used as propellants in aerosol spray cans, refrigerator coolants, air conditioners, industrial solvents, and plastic foams; enter the atmosphere slowly and resist breakdown.

____ 21. The main culprit is nitric oxide, produced mainly by cars and other vehicles with internal combustion engines.

____ 22. Oxides of sulfur and nitrogen dissolve in water to form weak solutions of sulfuric acid and nitric acid that may fall with rain or snow.

CHANGES IN THE HYDROSPHERE (37-III, pp. 616–617)

Terms

salination (616): _____

water table (616): _____

Fill-in-the-Blanks

There is a tremendous amount of water in the world, but most is too (1)_____ for human consumption. Today, we produce about one-third of our food supply on (2)_____ land; this often changes the land's suitability for (3)_____. Irrigation water is often loaded with (4)_____ salts. In regions with poorly draining soils, evaporation may cause (5)_____; such soils can stunt growth, decrease yields, and in time kill crop plants. Improperly drained irrigated lands can become (6)_____. Water accumulating underground can gradually raise the water (7)_____; when this is close to the ground's surface, soil becomes (8)_____ with saline water that can damage plant roots. Salinity and waterlogging can be corrected at great cost. (9)_____ is often the principal use of groundwater. Removing the salt from seawater is known as (10)_____ but is not cost-effective for large-scale agriculture.

Complete the Table

11. Complete the following table, which summarizes three levels of treatment methods for maintaining the water quality of polluted wastewater.

Treatment	Description
a. Primary treatment	
b. Secondary treatment	
c. Tertiary treatment	

CHANGES IN THE LAND (37-IV, pp. 618–621)

Terms

green revolution (618): _____

deforestation (620): _____

shifting cultivation (620): _____

desertification (619): _____

Matching

Choose the most appropriate answer to match with each term.

1. _____ throwaway mentality

2. _____ green revolution

3. _____ shifting cultivation

4. _____ animal-assisted agriculture

5. _____ deforestation

6. _____ recycling

7. _____ forested watersheds

8. _____ subsistence agriculture

9. _____ desertification

10. _____ mechanized agriculture

A. Runs on energy inputs from sunlight and human labor
B. An affordable, technologically feasible alternative to "throwaway technology"
C. Act like giant sponges that absorb, hold, and gradually release water
D. Conversion of large tracts of grasslands, rain-fed cropland, or irrigated cropland to a more desertlike state with a 10 percent or more productivity decrease
E. Runs on energy inputs from oxen and other draft animals
F. An attitude prevailing in the United States and other developed countries that greatly adds to solid waste accumulation
G. Requires massive inputs of fertilizers, pesticides, fossil fuel energy, and ample irrigation to sustain high-yield crops
H. Research directed toward improving crop plants for higher yields and exporting modern agricultural practices and equipment to developing countries
I. Trees are cut and burned, then ashes tilled into the soil; crops are grown for one to several seasons on quickly leached soils
J. Removal of all trees from large land tracts; leads to loss of fragile soils and disrupts watersheds

A QUESTION OF ENERGY INPUTS (37-V, pp. 622–625)

Terms

net energy (622): _____

fossil fuels (622): _____

meltdown (623): _____

photovoltaic cells (625): _____

Dichotomous Choice

Circle one of two possible answers given between parentheses in each statement.

1. Paralleling the (S-shaped/J-shaped) curve of human population growth is a steep rise in total and per capita energy consumption.
2. The increase in per capita energy consumption is due to (increased numbers of energy users and to extravagant consumption and waste/energy used to locate, extract, transport, store, and deliver energy to consumers).
3. (Total energy/Net energy) is that left over after subtracting the energy used to locate, extract, transport, store, and deliver energy to consumers.
4. Fossil fuels are the carbon-containing remains of (plants/plants and animals) that lived hundreds of millions of years ago.
5. Even with strict conservation efforts, known petroleum and natural gas reserves may be used up during the (current/next) century.
6. The net energy (decreases/increases) as costs of extraction and transportation to and from remote areas increase.
7. Collecting, concentrating, heating, and converting kerogen into shale oil in (the Middle East/Colorado, Utah, and Wyoming) may cost so much that the net energy yield would be low.
8. World coal reserves can meet human energy needs for several centuries, but burning releases sulfur dioxides into the atmosphere and adds to the global problem of (photochemical smog/acid deposition).
9. By 1990 in the United States, it cost slightly (less/more) to generate electricity by nuclear energy than by using coal.
10. The danger in the use of radioactivity as an energy supply during normal operation is with potential (radioactivity escape/meltdown).

Sequence-Classify

Arrange the consumption of world resources in correct hierarchical order. Enter the letter of the energy source of highest consumption next to 11, the letter of the next highest next to 12, and so on. Enter an (n) in the parentheses following the letter of the resource if it is nonrenewable and an (r) if the resource is renewable.

11. ____ ( )   A. Hydropower

12. ____ ( )   B. Natural gas

13. ____ ( )   C. Oil

14. ____ ( )   D. Nuclear power

15. ____ ( )   E. Biomass

16. ____ ( )   F. Coal

Self-Quiz

____ 1.   Which of the following processes is not generally considered a component of secondary wastewater treatment?
a. screens and settling tanks remove sludge
b. microbial populations are used to break down organic matter
c. removal of all nitrogen, phosphorus, and toxic substances
d. chlorine is often used to kill pathogens in the water

____ 2.   When fossil-fuel burning gives dust, smoke, soot, ashes, asbestos, oil, bits of lead, other heavy metals, and sulfur oxides, we have _____.
a. photochemical smog
b. industrial smog
c. a thermal inversion
d. both (a) and (c)

____ 3.   _____ result(s) when nitrogen dioxide and hydrocarbons react in the presence of sunlight.
a. Photochemical smog
b. Industrial smog
c. A thermal inversion
d. Both (a) and (c)

____ 4.   When weather conditions trap a layer of cool, dense air under a layer of warm air, _____ occurs.
a. photochemical smog
b. a thermal inversion
c. industrial smog
d. acid deposition

____ 5.   About _____ of the waste water in the United States is not even receiving primary treatment.
a. 20 percent
b. 50 percent
c. 30 percent
d. 70 percent

____ 6.   Sulfur and nitrogen dioxides dissolve in atmospheric water to form a weak solution of sulfuric acid and nitric acid; this describes _____.
a. photochemical smog
b. industrial smog
c. ozone and PANs
d. acid rain

____ 7.   Which of the following statements is false?
a. Ozone reduction allows more ultraviolet radiation to reach the earth's surface.
b. CFCs enter the atmosphere and resist breakdown.
c. Salination of soils aids plant growth and increases yields.
d. CFCs already in the air will be there for over a century.

____ 8.   For every million liters of water in the world, only about _____ liters are in a form that can be used for human consumption or agriculture.
a. 6
b. 60
c. 600
d. 6,000

___ 9. Each day in the United States, more than _____ metric tons of pollutants are discharged into the atmosphere.
a. 1,000
b. 100,000
c. 700,000
d. 5,000,000

___ 10. The most abundant fossil fuel in the United States is _____.
a. carbon monoxide
b. oil
c. natural gas
d. coal

Chapter Objectives/Review Questions

This section lists general and detailed chapter objectives that can be used as review questions. You can make maximum use of these items by writing answers on a separate sheet of paper. Fill in answers where blanks are provided. To check for accuracy, compare your answers with information given in the chapter or glossary.

Page	Objectives/Questions
(612)	1. Briefly list the major environmental effects of human population growth.
(612)	2. _____ are substances with which ecosystems have had no prior evolutionary experience, in terms of kinds or amounts, and so have no mechanisms for dealing with them.
(614)	3. Distinguish between dry and wet acid deposition; list the major acids present in acid rain.
(614)	4. List the major sources of acid pollutants
(615)	5. Pronounced seasonal thinning of the ozone layer is called an ozone _____.
(615)	6. More than any other factor, _____ are bringing about ozone reduction
(615)	7. Ozone reduction allows more _____ radiation to reach the earth's surface; a dramatic rise in skin _____ is one consequence.
(615)	8. Chlorine reacts with ozone, forming an oxygen molecule and a _____ _____ molecule.
(613)	9. Describe the conditions of a thermal inversion.
(613)	10. Distinguish between industrial smog and photochemical smog.
(615–616)	11. Suggest what you believe would be the very best solution for saving the ozone layer on a global basis.
(616)	12. _____ (salt buildup) of soils can stunt growth, decrease yields, and in time kill crop plants.
(616)	13. The water _____ is the upper limit at which the ground is fully saturated with water.
(617)	14. What is the major reason that desalination may never solve the problem of freshwater shortage?
(617)	15. Describe the three levels of wastewater treatment.
(618)	16. The major problem arising from our throwaway mentality is what to do with _____ _____.
(619)	17. _____ leads to loss of fragile soils and disrupts watersheds, especially in steeply sloped regions.
(620)	18. Shifting _____ disrupts the forest ecosystem through the cutting of trees and then using ashes to till the soil.
(619)	19. Describe the process of desertification and its effects.
(622)	20. _____ energy is that left over after subtracting the energy used to locate, extract, transport, store, and deliver energy to consumers.
(622)	21. List the disadvantages of the use of fossil fuels as energy sources.
(622–623)	22. The danger in the use of nuclear reactors lies with the potential for _____.
(624)	23. Hope for the future use of nuclear fuel lies in the development of _____ reactors, which convert uranium isotopes into fissionable _____ isotopes; _____ power might also provide energy by fusing hydrogen atoms to form _____ atoms.

Integrating and Applying Key Concepts

If you were Ruler of All People on Earth, how would you encourage people to depopulate the cities and adopt a way of life by which they could supply their own resources from the land and dispose of their own waste products safely on their own land?

Explain why some biologists believe that the endangered species list now includes all species.

Critical Thinking Exercise

1. The two curves below represent two models for the long-term future of the human population. From the following list, select the assumptions inherent in each of the two models.

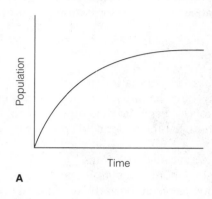

A

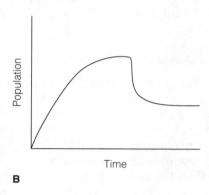

B

a. The environment imposes a carrying capacity.
b. All essential resources are renewable or inexhaustible.
c. The environment is stable with only minor fluctuations around a mean condition.
d. The carrying capacity can be enlarged indefinitely by technology.
e. The carrying capacity can be enlarged temporarily by technology.

Answers

Answers to Interactive Exercises

ENVIRONMENTAL EFFECTS OF HUMAN POPULATION GROWTH (37-I)
1. Human population growth became exponential and may be demanding more than the biosphere can sustain; as we take energy and resources, we give back monumental amounts of waste; we are destroying the stability of ecosystems, tainting the hydrosphere, and changing the composition of the atmosphere; millions in developing countries are already starving to death and hundreds of millions more suffer from malnutrition and inadequate health care.

CHANGES IN THE ATMOSPHERE (37-II)
1. c; 2. d; 3. e; 4. b; 5. a; 6. b; 7. f; 8. e; 9. a; 10. b; 11. b; 12. d; 13. a; 14. d; 15. f; 16. c; 17. d; 18. e; 19. f; 20. e; 21. c; 22. d.

CHANGES IN THE HYDROSPHERE (37-III)
1. salty; 2. irrigated; 3. agriculture; 4. mineral; 5. salination; 6. waterlogged; 7. table; 8. saturated; 9. Irrigation; 10. desalination; 11. a. screens and settling tanks remove sludge, which is dried, burned, dumped in landfills, or treated further; chlorine is often used to kill pathogens in water, but does not kill them all; b. microbial populations are used to break down organic matter after primary treatment but before chlorination; c. remove nitrogen, phosphorus, and toxic substances, including heavy metals, pesticides, and industrial chemicals; it is largely experimental and expensive.

CHANGES IN THE LAND (37-IV)
1. F; 2. H (G); 3. I; 4. E; 5. J; 6. B; 7. C; 8. A; 9. D; 10. G.

A QUESTION OF ENERGY INPUTS (37-V)

1. J-shaped; 2. increased number of energy users and to extravagant consumption and waste; 3. Net energy; 4. plants; 5. next; 6. decreases; 7. Colorado, Utah, and Wyoming; 8. acid deposition; 9. more; 10. meltdown; 11. C(n); 12. F(n); 13. B(n); 14. E(r); 15. A(r); 16. D(n).

Answers to Self-Quiz

1. a; 2. b; 3. a; 4. b; 5. c; 6. d; 7. c; 8. a; 9. c; 10. d.

Critical Thinking Analysis

a. Both models assume a carrying capacity. In A, the population rises to the capacity and remains stable thereafter. In B, the population rises beyond the carrying capacity, and then crashes and stabilizes.
b. Both models assume that no essential resources become exhausted. If any required resources did disappear, the population would become extinct.
c. Model A makes this assumption. Major environmental changes would reset the carrying capacity at a new level or cause temporary swings in population. Model B assumes environmental changes that trigger the crash. Some essential resource becomes temporarily reduced to a level insufficient to sustain the inflated population.
d. Model A may or may not make this assumption. The graph does not indicate whether the stable population is at a natural or a technologically inflated carrying capacity.
e. Model B makes this assumption. The final stable population is at the natural carrying capacity. The precrash population was artificially inflated by technology until an unmanageable environmental change occurred.

38

ANIMAL BEHAVIOR

MECHANISMS UNDERLYING BEHAVIOR
    Genetic Effects on Behavior
    Hormonal Effects on Behavior

INSTINCTIVE AND LEARNED BEHAVIOR
    Instinct and Learning Defined
    Genes and the Environment

THE ADAPTIVE VALUE OF BEHAVIOR
    Selection Theory and Mating Behavior
    Selection Theory and Feeding Behavior

MECHANISMS OF SOCIAL LIFE
    Functions of Communication Signals

    Types of Communication Signals

COSTS AND BENEFITS OF SOCIAL LIFE
    Advantages to Sociality
    Disadvantages to Sociality
    Self-Sacrifices in Dominance Hierarchies
    Costs and Benefits of Parenting

THE EVOLUTION OF ALTRUISM
    Focus on Science: About Those Self-Sacrificing
      Naked Mole-Rats

HUMAN SOCIAL BEHAVIOR

Interactive Exercises

MECHANISMS UNDERLYING BEHAVIOR (38-I, pp. 629–630)

Terms

The page-referenced terms are important; they were in boldface type in the chapter. Refer to the instructions given in Chapter 1, p. 1 of this workbook.

animal behavior (629):_____

sound system (630):_____

Dichotomous Choice

Circle one of two possible answers given between parentheses in each statement.

1. For snake populations living along the California coast, the food of choice is (the banana slug/tadpoles and small fishes).
2. In Stevan Arnold's experiments, newborn garter snakes that were offspring of coastal parents usually (ate/ignored) a chunk of slug as the first meal.
3. Newborn garter snake offspring of (coastal/inland) parents ignored cotton swabs drenched in essence of slug and only rarely ate the slug meat.
4. The differences in the behavioral eating responses of coastal and inland snakes (were/were not) learned.
5. Hybrid garter snakes with coastal and inland parents exhibited a feeding response that indicated a(n) (environmental/genetic) basis for this behavior.

6. In white-throated sparrows and some other songbirds, singing behavior begins with melatonin, a hormone secreted by the (gonads/pineal gland).
7. In spring, melatonin secretion is suppressed and gonads grow; their secretion of estrogen and testosterone is increased to (indirectly/directly) influence singing behavior.
8. In very young male songbirds, a high (estrogen/testosterone) level triggers development of a masculinized brain.
9. Later, at the start of the breeding season, a male's enlarged gonads secrete even more (estrogen/testosterone) that acts on cells in the sound system to prepare the bird to sing when properly stimulated.
10. (Hormones/Genes) influence the organization and activation of mechanisms required for particular forms of behavior.

INSTINCTIVE AND LEARNED BEHAVIOR (38-II, pp. 630–632)

Terms

instinctive behavior (630):_____

fixed action pattern (630):_____

learned behavior (630):_____

Complete the Table

1. Complete the following table to examine examples of instinctive and learned behavior.

Category	Example(s)
a. Instinctive behavior	
b. Fixed action pattern	
c. Learned behavior	

Matching

Choose the one most appropriate answer for each.

2. ___ imprinting

3. ___ classical conditioning

4. ___ operant conditioning

5. ___ habituation

6. ___ spatial or latent learning

7. ___ insight learning

A. Birds living in cities learn not to flee from humans or cars, which pose no threat to them
B. Chimpanzees abruptly stack several boxes and use a stick to reach suspended bananas out of reach
C. In response to a bell, dogs salivate even in the absence of food
D. Bluejays storing information about dozens or hundreds of places where they have stashed food
E. Adult roosters courting ducks instead of hens of their own species
F. A toad learns to avoid stinging or bad-tasting insects after voluntary attempts to eat them

THE ADAPTIVE VALUE OF BEHAVIOR (38-III, pp. 632–634)

Terms

reproductive success (632):_____

adaptive behavior (632):_____

selfish behavior (632):_____

altruistic behavior (632):_____

natural selection (632):_____

territory (634):_____

sexual selection (632):_____

Matching

Match the following organisms with the appropriate description relating to feeding or mating selection theory.

1. ___ bighorn sheep

2. ___ wandering ravens

3. ___ hangingflies

4. ___ sage grouse

5. ___ elk and bison

A. Females widely dispersed; during mating season, males congregate in a lek where a display ground is staked out; females select and mate with one male only and nest by themselves

B. Males compete intensely for access to clusters of sexually receptive females; competition for ready-made harems favors combative males

C. Males fight only to control areas where receptive females gather during winter rutting season; winners mate often with many females; rather than challenging stronger, larger males, gangs of losers invade guarded areas and overwhelm male winners; losers attempt to mate on the run

D. Selfishly advertise carcasses; this gives them a chance at otherwise off-limits food (within a territory defended by powerful adults)

E. Male captures and kills an insect, then releases pheromones to attract females to the nuptial gift; females choose choicest offering and permit mating following eating

MECHANISMS OF SOCIAL LIFE (38-IV, pp. 634–636)

Terms

social behavior (634):_____

communication signals (634):_____

pheromones (634):_____

visual signals (635):_____

courtship displays (635):_____

acoustical signals (636):_____

tactile signal (636):_____

Choice

For questions 1–10, choose from the following.

a. chemical signal   b. visual signal   c. acoustical signal   d. tactile signal

___ 1. Worker termites bang their heads to alert brown soldier termites.

___ 2. A handshake, hug, caress, or shove.

___ 3. The bioluminescent signals of male and female fireflies of the same species.

___ 4. Sex pheromones of hangingfly males.

___ 5. Illegitimate signals of some predatory female fireflies seeking a meal of males belonging to other firefly species.

___ 6. Information-seeking bees stay in physical contact with a dancing bee who has returned to the beehive after foraging.

___ 7. Fringe-lipped bat as an illegitimate receiver of the tungara frog's call.

___ 8. Soldier termite shooting thin jets of silvery goo from its nose; the goo emits volatile odors that attract more soldiers to battle danger.

___ 9. Distinctive song of the male white-throated sparrow.

___ 10. Exposing formidable canine teeth, male baboons "yawn" at each other when they compete for a receptive female; this is a threat display that may precede an attack.

COSTS AND BENEFITS OF SOCIAL LIFE (38-V, pp. 637–640)

Terms

selfish herd (639):_____

dominance hierarchy (639): _____

Matching

Choose the one most appropriate answer for each concept.

1. ___ disadvantages to sociality
2. ___ cooperative predator avoidance
3. ___ the selfish herd
4. ___ dominance hierarchy
5. ___ costs and benefits of parenting

A. Disturbed clumps of Australian sawfly caterpillars collectively rear up, writhe about, and regurgitate partially digested eucalyptus leaves; the leaves contain chemical compounds toxic to most animals
B. Male baboons relinquish safe sleeping sites, choice bits of food, even receptive females to others upon receiving a threat signal from another male
C. Building of adjacent male bluegill nests on lake bottoms; the largest, most powerful males claim the central locations
D. Breeding pairs of herring gulls in a nesting colony, if given opportunity, will cannibalize the eggs or young chicks of their neighbors in an instant
E. A pair of Caspian terns incubate their eggs, then shelter and feed the nestlings, defend them from predators, and accompany them when they begin to fly and forage; in this case, the benefit of immediate reproductive success may outweigh the cost of reduced reproductive success at some later time

THE EVOLUTION OF ALTRUISM (38-VI, p. 641)

Term

indirect selection (641):_____

Dichotomous Choice

Circle one of two possible answers given between parentheses in each statement.

1. By one theory, individuals (as with sterile workers in insect societies) can (directly/indirectly) pass on their genes by helping relatives survive and reproduce.
2. When a sexually reproducing parent helps offspring, it (is/is not) helping exact genetic copies of itself.
3. According to Hamilton's theory of (direct/indirect) selection, caring for nondescendant relatives favors the genes associated with helpful behavior.
4. An uncle helps a niece or nephew survive long enough to reproduce; if the cost of his altruistic action is (more/less) than the benefit, genes the uncle has in common with his relatives will be propagated.
5. DNA fingerprints made for naked mole-rat colonies indicate that individuals from a given colony are very close relatives due to high inbreeding; such lineages have extremely (increased/reduced) genetic variability.
6. An altruistic naked mole-rat helps to perpetuate a very (low/high) proportion of the genes it carries.

HUMAN SOCIAL BEHAVIOR (38-VII, pp. 642–643)

Dichotomous Choice

Circle one of two possible answers given between parentheses in each statement.

1. Many people seem to believe that attempts to identify the adaptive value of a particular (animal/human) trait is an attempt to define its moral or social advantage.
2. "Adaptive" (does/does not) mean "moral" in terms of the value of the transmission of an individual's genes.

3. The human adoption process (can/cannot) be considered adaptive when indirect selection favors adults who direct parenting assistance to relatives.
4. Joan Silk showed that in some traditional societies, children (are not/are) adopted overwhelmingly by relatives.
5. In modern societies in which agencies and other means of adoption exist, adoption of relatives (is/is not) predominant.

Self-Quiz

___ 1. The observable responses that animals make to stimuli are what we call

_____.
a. imprinting
b. instinct
c. behavior
d. learning

___ 2. Newly hatched goslings follow any large moving objects to which they are exposed shortly after hatching; this is an example of

_____.
a. homing behavior
b. imprinting
c. piloting
d. migration

___ 3. In _____, components of the nervous system allow an animal to carry out complex, stereotyped responses to certain environmental cues, which are often simple.
a. natural selection
b. altruistic behavior
c. sexual selection
d. instinctive behavior

___ 4. Newly hatched blind cuckoos contact an egg, maneuver it onto their back, and then push it out of the nest. This instinctive response is called _____.
a. resource-defense
b. fixed action pattern
c. learned behavior
d. altruism

___ 5. A young toad flips its sticky-tipped tongue and captures a bumblebee that stings its tongue; in the future, the toad leaves bumblebees alone; this is _____.
a. instinctive behavior
b. a fixed action pattern
c. altruistic
d. learned behavior

___ 6. Female insects often attract mates by releasing sex pheromones. This is an example of a(n) _____ signal.
a. chemical
b. visual
c. acoustical
d. tactile

___ 7. Worker termites bang their heads specifically to attract soldiers; male birds sing to stake out territories, attract females, and discourage males. These are examples of _____ signals.
a. chemical
b. visual
c. acoustical
d. tactile

___ 8. The "selfish herd" is illustrated by

_____.
a. the largest, most powerful male sunfish claiming nests in the central location of many adjacent male nests
b. members of wolf packs helping others by sharing food or fending off predators
c. clumps of regurgitating Australian sawfly caterpillars
d. nesting colonies of herring gulls who cannibalize neighbors' eggs or young chicks

___ 9. Usually only one male and one female of a wolf pack produce pups; this is an indication of _____.
a. a selfish herd
b. cooperative predator avoidance
c. group defense
d. dominance hierarchy

___ 10. Caring for nondescendant relatives favors the genes associated with helpful behavior and is classified as _____.
a. dominance hierarchy
b. indirect selection
c. altruism
d. both b and c

Chapter Objectives/Review Questions

This section lists general and detailed chapter objectives that can be used as review questions. You can make maximum use of these items by writing answers on a separate sheet of paper. Fill in answers where blanks are provided. To check for accuracy, compare your answers with information given in the chapter or glossary.

Page	Objectives/Questions
(629)	1. Animal _____ refers to the observable responses an animal makes to stimuli.
(629)	2. By influencing the development of the nervous system, _____ contribute in an indirect yet major way to behavior.
(630)	3. _____ influence the organization and activation mechanisms required for particular forms of behavior.
(630)	4. Define instinctive behavior.
(630)	5. Give an example of a fixed action pattern.
(630)	6. When animals incorporate and process information gained from specific experiences and then use the information to vary or change responses to stimuli, it is _____ behavior.
(632)	7. Genes and the _____ contribute to mechanisms underlying both instinctive and learned behavior.
(631)	8. A rooster was exposed to a mallard duck during a critical period in the rooster's life and is attracted to the mallard; this time-dependent form of learning is called _____.
(629)	9. What explains the fact that coastal and inland garter snakes of the same species have different food preferences?
(632)	10. _____ behavior is self-sacrificing behavior.
(632)	11. _____ selection is the outcome of individuals of one sex competing for matings with individuals of the opposite sex.
(632)	12. _____ behavior promotes the propagation of an individual's genes and so tends to occur at increased frequency in successive generations.
(632)	13. _____ behavior is a behavior by which an individual protects or increases its own chance of producing offspring, regardless of the consequences for the group to which it belongs.
(632)	14. Describe altruistic behavior.
(634)	15. At the heart of social behavior is the ability of animals to _____ with one another, using a complex array of signals to exchange information.
(634)	16. Examples of _____ signals are chemical, visual, acoustical, and tactile.
(634–635)	17. Lekking sage grouse and sexual contortions of albatrosses are examples of _____ displays.
(634)	18. _____ behavior refers to cooperative, interdependent relationships among animals of the same species.
(637)	19. Studies of Australian sawfly caterpillars indicate _____ predator avoidance.
(637)	20. Define the selfish herd and cite an example.
(639)	21. List two examples of self-sacrifice in a dominance hierarchy.
(641)	22. Hamilton's theory of _____ selection relates to caring for nondescendant relatives and how this favors genes associated with helpful behavior.
(642–643)	23. Describe how hypotheses about "selfish" and "altruistic" behavior of humans can be tested without attempts to justify the behavior.

Integrating and Applying Key Concepts

Explain whether or not you think humans have any critical periods for establishing the ability to learn certain kinds of knowledge. State whether or not you think humans undergo imprinting. Do you think humans employ resource-defense behavior? Female-defense behavior? If so, can you cite an example?

Critical Thinking Exercise

1. Ants commonly move between the nest and a food source along an indirect path and follow the same track on repeated trips. To test the hypothesis that ants communicate the location of food by laying a trail of chemical signals as they walk, you set up a good food source some distance from an ant nest on a table in the laboratory. When the ants have found the food and many ants are going to and from it along a single path, you rub your finger on the table across the path of ants. The ants begin to walk aimlessly in loops and can't find the food or the nest. You assume that your finger removed any chemicals from the table surface and conclude that your hypothesis is supported. Which of the following is the best criticism of this experiment?

   a. The experiment should have been performed with more than one species of ant.
   b. The effect of varying the direction of the light source should have been investigated.
   c. Chemicals from human skin might have disrupted the ants' behavior.
   d. The trail should have been disrupted by rubbing a dead ant across it.
   e. The experiment should have been performed on a natural surface of loose dirt.

Answers

Answers to Interactive Exercises

MECHANISMS UNDERLYING BEHAVIOR (38-I)
1. the banana slug; 2. ate; 3. inland; 4. were not; 5. genetic; 6. pineal gland; 7. directly; 8. estrogen; 9. testosterone; 10. Hormones.

INSTINCTIVE AND LEARNED BEHAVIOR (38-II)
1. a. cuckoo birds are social parasites in that adult females lay eggs in the nests of other bird species; young cuckoos eliminate the natural-born offspring and then receive the undivided attention of their unsuspecting foster parents; b. the reaction pattern whereby newly hatched cuckoos contact an egg, maneuver it onto their back, then push it out of the nest; this is an instinctive response, triggered by a well-defined, simple stimulus—once set in motion, it is performed in its entirety; this is a fixed reaction pattern; c. young toads instinctively use their sticky-tipped tongues to capture edible insects; if a bumblebee is captured and then stings the tongue, the toad learns to leave bumblebees alone; 2. E; 3. C; 4. F; 5. A; 6. D; 7. B.

THE ADAPTIVE VALUE OF BEHAVIOR (38-III)
1. C; 2. D; 3. E; 4. A; 5. B.

MECHANISMS OF SOCIAL LIFE (38-IV)
1. c; 2. d; 3. b; 4. a; 5. b; 6. d; 7. c; 8. a; 9. c; 10. b.

COSTS AND BENEFITS OF SOCIAL LIFE (38-V)
1. D; 2. A; 3. C; 4. B; 5. E.

THE EVOLUTION OF ALTRUISM (38-VI)
1. indirectly; 2. is not; 3. indirect; 4. less; 5. reduced; 6. high.

HUMAN SOCIAL BEHAVIOR (38-VII)
1. human; 2. does not; 3. can; 4. are; 5. is not.

Answers to Self-Quiz

1. c; 2. b; 3. d; 4. b; 5. d; 6. a; 7. c; 8. a; 9. d; 10. d.

Critical Thinking Analysis

a. Results obtained with only one species cannot be generalized to all ants, but they can still lead to valid conclusions about that one species. Because an experiment cannot be repeated with all the almost 9,000 species of ants, how many species must be used before the results can be generalized?

b. This investigation would test an alternative hypothesis that ants navigate by the direction of the light source. However, this is a separate investigation, not a criticism of the experiment.

c. This is a strong criticism of the experiment. The procedure not only may disrupt any chemical trails the ants have laid but also would deposit on the surface any chemicals present on the investigator's finger as a result of secretion or contamination. If these chemicals caused a response in the ants, the response would be an experimental artifact, not an interpretable result.

d. This procedure would create an even stronger possibility of artifactual result. A dead ant would deposit all chemical signaling substances the individual carried, and the response to the mixed signals would be uninterpretable.

e. In this design, the disorientation of the ants might be due to disruption of a chemical trail, but it might also be due to changes in topography. The experiment must be designed so that only one variable is changed at a time. The best design would be to allow the ants to establish their pathway across a movable surface such as a piece of paper, then change the orientation of the movable segment. If the ants' direction does not change, they are not navigating by a chemical trail. If they follow their old path in its new direction, they probably are using chemical communication.